PROBABILITY
WITH
APPLICATIONS

**McGRAW-HILL
BOOK COMPANY**

New York
St. Louis
San Francisco
Düsseldorf
Johannesburg
Kuala Lumpur
London
Mexico
Montreal
New Delhi
Panama
Paris
São Paulo
Singapore
Sydney
Tokyo
Toronto

MICHAEL WOODROOFE

*Professor of Mathematics and Statistics
University of Michigan*

Probability
with Applications

This book was set in Times New Roman.
The editors were Robert H. Summersgill and Shelly Levine Langman;
the cover was designed by Nicholas Krenitsky;
the production supervisor was Dennis J. Conroy.
The drawings were done by Textart Service, Inc.
Kingsport Press, Inc., was printer and binder.

Library of Congress Cataloging in Publication Data

Woodroofe, Michael.
    Probability with applications.

    Bibliography: p.
    1. Probabilities.   I. Title.
QA273.W72   519.2   74-7202
ISBN 0-07-071718-4

**PROBABILITY
WITH
APPLICATIONS**

1 2 3 4 5 6 7 8 9 0 KP KP 7 9 8 7 6 54

CONTENTS

[1] Material may be omitted without loss of continuity.

[1] Material may be omitted without loss of continuity.

[1] Material may be omitted without loss of continuity.

---

[1] Material may be omitted without loss of continuity.

PREFACE

This book has grown out of several different courses I have given at the University of Michigan during the past few years. The students in these courses were primarily juniors, seniors, and first-year graduate students in mathematics and such related fields as engineering, statistics, mathematical psychology, and econometrics, and this book is designed for such an audience. Preliminary versions of the book used in these courses have benefited from the students' comments.

The book has several aims. First, as a textbook, it is intended to introduce its readers to the basic concepts of probability and to acquaint them with the mathematical theory of probability. A working knowledge of the unstarred sections of Chapters 1 to 10 should enable the reader to understand applications of probability theory to many scientific and social phenomena.

Another, equally important objective is to develop the reader's intuition about probability. The book contains numerous examples and several applications to scientific and statistical problems. Moreover, many topics have been approached from more than one point of view.

The book is also intended to serve as a reference for those whose formal training in probability does not continue beyond the introductory level. Thus, the book contains much material that finds manifold application but would not usually be presented in an introductory course. This additional material has been placed in footnoted sections and may be omitted without loss of continuity. It may also provide an instructor with a wide selection of special topics from which he may choose one or two for in-depth studies.

The book divides naturally into three parts. Chapters 1 to 4 treat combinatorial probability and introduce the notions of sample space, statistical and subjective probability, conditional probability, and independence. The notions of random variable, probability distribution, and expectation are then introduced and developed in Chapters 5 to 10. Finally, Chapters 11 and 12 introduce the reader to stochastic processes and develop random walks and martingales.

The unstarred sections of Chapters 1 to 10 form the basis for a strong introductory course in probability theory. The prerequisite for an intelligent reading of this book is 2 years of calculus. Additional background is desirable for Chapters 11 and 12, but not essential.

The book has benefited from discussion which I have had with Bill Ericson, Richard Olshen, Herb Robbins, Norman Starr, and Jim Wendel. Patricia Holly did an efficient job with the typing, and Charles Keller and Francis Smock helped with the proofreading. To all my sincere thanks.

MICHAEL WOODROOFE

1

THE CLASSICAL MODEL

1.1 INTRODUCTION

We shall begin our study of probability theory with games of chance. In this chapter we study games of chance which must result in one of a finite number of possible outcomes, the totality of which can be specified before the game is played. For example, most card games are of this nature. Our goal in this chapter is to construct a mathematical model for such games and to develop some of the simpler properties of the model. The model we choose is called the *classical model* because it was the first probability model to be studied.[1]

Given a particular game of chance, as described above, we shall denote the set of possible outcomes of the game by S, and we shall call S the *sample space*.[2] Subsets of S will be called *events*, and an event $A \subset S$ will be said to *occur* if and only if the actual outcome of the game is an element of A. For example, if our game consists of rolling a balanced die once, we could take S to be the set $\{1,2,3,4,5,6\}$ with the convention that $k \in S$ represents the outcome that a face showing exactly k spots

---

[1] References to works on the history of probability are given at the end of the chapter.

[2] *Outcome space* would be better, but we shall use the conventional terminology *sample space*.

appears. The event that an odd number of spots appears is then $A = \{1,3,5\}$, and the event that only one spot appears is $\{1\}$.

If S is the sample space for a particular game and $A \subset S$ is an event, we define the *probability of A* to be

$$P(A) = \frac{|A|}{|S|} \qquad (1.1)$$

where for any subset $B \subset S$, $|B|$ denotes the number of distinct elements of B. Thus, within the classical model, the probability of an event is the ratio of the number of outcomes which imply the occurrence of the event to the total number of possible outcomes. For example, in the dice game mentioned above, the probability of the event $A = \{1,3,5\}$ that an odd number of spots appear is $P(A) = 3/6 = \frac{1}{2}$, while the probability that only one spot appears is $P(\{1\}) = \frac{1}{6}$.

Equation (1.1) defines a function whose domain is the class (or set) of all subsets of S. Thus, probability is a property of sets (events) $A \subset S$, not of points $s \in S$. In particular, if $s \in S$, we shall refer to $\{s\}$, the set whose only element is s, as the event that the outcome of our game will be s. Equation (1.1) then requires $P(\{s\}) = 1/|S|$. The symbol $P(s)$ has not been defined.

For later reference, we observe that the function P of Equation (1.1) has the following properties:

$$0 \le P(A) \le P(S) = 1 \qquad (1.2)$$

$$P(A \cup B) = P(A) + P(B) \qquad \text{if } AB = \varnothing \qquad (1.3)$$

$$P(A') = 1 - P(A) \qquad (1.4)$$

where $A \cup B$ denotes the union of A and B, AB denotes the intersection of A and B, A' denotes the complement of A, and $\varnothing$ denotes the empty set.[1] For example, to establish (1.3) simply observe that if $AB = \varnothing$, then $|A \cup B| = |A| + |B|$, so that $P(A \cup B) = P(A) + P(B)$ by (1.1). Equation (1.4) then follows from $P(A) + P(A') = P(S) = 1$, and (1.2) is obvious. These properties are sometimes useful in reducing a complicated calculation to a series of simpler ones.

EXAMPLE 1.1.1 If two balanced, distinguishable dice are tossed, we can describe the outcome of the game by the ordered pair (x,y), where x denotes the number of spots on the first die and y the number on the second. Thus, we may take S to be the set of ordered pairs (x,y), where x and y are integers between 1 and 6. Inspection

---

[1] Some elements of set theory are reviewed in Appendix A.

shows that there are $|S| = 36$ elements in S. Let us compute the probability of the event A that the sum of spots on the two dice is 7. Clearly,

$$A = \{(1,6),(2,5),(3,4),(4,3),(5,2),(6,1)\}$$

so that $|A| = 6$. Therefore, $P(A) = \frac{1}{6}$. ////

Although the example is quite simple, it will repay careful study, for the technique employed in Example 1.1.1 will be employed throughout this chapter. Observe that we gave a careful description of the sample space (set of possible outcomes) and of the event A whose probability we wished to compute. After this was done, the calculation of $P(A)$ merely involved counting the number of elements in A, counting the number of elements in S, and dividing. Conceptually, all the problems we shall encounter in this chapter are just as simple as Example 1.1.1, although the actual counting may become a bit more complicated. Many students have difficulty with elementary probability theory because they do not conceptualize the problems properly. That is, they do not take the time and effort to define their sample space and event carefully. As a result, they do not know what to count. Therefore, we repeat: *The first step in computing any probability in this chapter should be a careful definition of the sample space and of the event whose probability is to be computed.*

Equation (1.1) reflects an assumption about the game under consideration. Namely, it assumes that the various outcomes of the experiment are *equally likely* in the sense that $P(\{s\}) = 1/|S|$ for all $s \in S$. We are therefore confronted with the following question: To what games does the classical model apply? We discuss this question in Section 2.1. For the present, we assume that the reader has had enough experience with such terms as "chance," "likely," and "probability" to recognize games to which the classical model applies.

1.2 COMBINATORIAL ANALYSIS

In principle, all computations which are derived from the classical model are completely straightforward. Indeed, to compute $P(A)$ from (1.1) one has merely to count the number of distinct elements in A, count the number of distinct elements in S, and divide the former by the latter. In practice, however, it is often impossible to count the number of elements in A or S by simple inspection. For example, if we wished to compute the probability of being dealt a full house in a poker game, we could not realistically hope to list all possible poker hands and count the number which contained three cards of one denomination and two of another. Why? Because, as we shall see in Example 1.2.4a, there are 2,598,960 distinct poker hands. What we need is an efficient method of counting, one which will allow us to count the total number of distinct poker hands without actually listing them, for example. The body of techniques which comprise this efficient method of counting is known as *combinatorial analysis* and is the topic of the present section.

If Z is a nonempty set and k is a positive integer, then we define an *ordered k-tuple of elements of Z* to be an array $(z_1, z_2, \ldots, z_k)$ with $z_i \in Z$ for $i = 1, 2, \ldots, k$. z_i is called the ith component of $(z_1, \ldots, z_k)$ for $i = 1, \ldots, k$. When there is no danger of confusion, we shall omit the phrase "of elements of Z," and when $k = 2$ or 3, we shall refer to ordered k-tuples as *ordered pairs* and *ordered triples*, respectively. Two ordered k-tuples are equal if and only if they list the same components in the same order. That is, $(z_1, \ldots, z_k) = (w_1, \ldots, w_k)$ if and only if $z_i = w_i$ for $i = 1, \ldots, k$.

The notion of an ordered k-tuple of elements of Z should be contrasted with the notion of a subset $\{z_1, \ldots, z_k\} \subset Z$ of Z. Two subsets $\{z_1, \ldots, z_k\}$ and $\{w_1, \ldots, w_j\}$ are equal if they list the same elements, even if they list them in different orders or with repetition. For example, $\{1,2\} = \{2,1\} = \{2,1,2\}$, but $(1,2) \neq (2,1)$. The distinction is simple but important.

An ordered k-tuple of elements of Z, say $(z_1, \ldots, z_k)$, with distinct components (that is, $z_i \neq z_j$ for $i \neq j$) is called a *permutation* of k elements of Z. A subset $\{z_1, \ldots, z_k\}$ with k distinct elements is called a *combination* of k elements of Z. Many of the problems in this chapter will be phrased in terms of permutations and combinations.

The combinatorial analysis which we shall need can be derived from the following basic principle, which we adopt as an axiom.

The basic principle of combinatorial analysis   *Suppose that we may select two objects x and y in that order. If we have m distinct choices for x and n distinct choices for y, where m and n are positive integers, then we can select the ordered pair (x, y) in mn distinct ways.*

   More generally, suppose that $k \geq 2$ is an integer and that objects $x_1, \ldots, x_k$ are selected sequentially; that is, first x_1, then $x_2, \ldots$. If x_i can be selected in n_i distinct ways, $i = 1, \ldots, k$, then the ordered k-tuple $(x_1, \ldots, x_k)$ can be selected in

$$n = n_1 n_2 \cdots n_k$$

   distinct ways.

The second assertion of the basic principle can, in fact, be derived from the first by mathematical induction. We leave the derivation as an exercise for the interested reader and turn directly to some examples.

EXAMPLE 1.2.1   From a menu containing 3 soups, 2 salads, 6 entrées, and 3 desserts, $3 \times 2 \times 6 \times 3 = 108$ different dinners can be ordered. Simply take x_1

to be the soup, x_2 to be the salad, x_3 to be the entrée, and x_4 to be the dessert and apply the basic principle with $k = 4$. ////

It should be emphasized that the basic principle allows the set of objects from which x_i is chosen to depend on the choice of $x_1, \ldots, x_{i-1}$. Only the number of possible choices n_i must be fixed in advance.

EXAMPLE 1.2.2 If a man has red, green, and gold shirts and red, green, and gold neckties, how many ways can he choose different colors for his shirt and necktie? $3 \times 2 = 6$, for he has 3 choices for the color of his shirt and, thereafter, only 2 for the color of his necktie. Here, of course, the two colors from which he selects the color of his necktie will depend on the color he selected for his shirt. ////

Theorem 1.2.1 *Let Z be a set containing $n \geq 1$ distinct elements, and let $k \geq 1$ be an integer. Then, there are n^k distinct ordered k-tuples $(z_1, \ldots, z_k)$ with $z_i \in Z$, $i = 1, \ldots, k$. If $k \leq n$, then there are*

$$(n)_k = n(n - 1) \cdots (n - k + 1) \qquad (2.1)$$

distinct ordered k-tuples with distinct components, that is, $z_i \neq z_j$ for $i \neq j$.

PROOF   In choosing an ordered k-tuple $(z_1, \ldots, z_k)$ with $z_i \in Z$, $i = 1, \ldots, k$, we have n choices for z_1, n choices for z_2, and in general, n choices for z_i, $i = 1, \ldots, k$. Therefore, by the basic principle, we have $nn \cdots n = n^k$ choices for $(z_1, \ldots, z_k)$. If $k \leq n$ and we require the z_i to be distinct, then we still have n choices for z_1 but only $n - 1$ for z_2, which must be different from z_1, and only $n - 2$ for z_3, which must differ from both z_1 and z_2. In general, we shall have $n - i + 1$ choices for z_i, $i = 1, \ldots, k$, and therefore $n(n - 1) \cdots (n - k + 1) = (n)_k$ choices for $(z_1, \ldots, z_k)$. ////

EXAMPLE 1.2.3 If four distinguishable dice are cast, there are $6^4 = 1296$ distinguishable outcomes. Of these there are $(6)_4 = 360$ distinguishable outcomes for which no two dice show the same number of spots. Indeed, we can apply the theorem with $Z = \{1, \ldots, 6\}$ by letting z_i denote the number of spots which appear on the ith die, $i = 1, \ldots, 4$. ////

The notation $(n)_k$ has been defined by (2.1) when n and k are positive integers for which $k \leq n$. We now extend this notation by defining

$$(n)_0 = 1 = n^0 \qquad (2.2a)$$

$$(n)_k = 0 \qquad \text{if either } k < 0 \text{ or } k > n \qquad (2.2b)$$

for $n = 0, 1, 2, \ldots$. We shall also find it convenient to write $n!$ (read "n factorial") for $(n)_n$. Thus, $0! = 1$, and

$$n! = n(n - 1) \cdots 2 \times 1 \qquad (2.3)$$

for $n = 1, 2, \ldots$. Theorem 1.2.1 then asserts that if Z contains $n \geq 1$ distinct elements, there are $n!$ permutations of the n elements of Z.

For later reference, we observe that

$$(n)_k = \frac{n!}{(n - k)!} \qquad (2.4a)$$

$$(n)_{i+j} = (n)_i(n - i)_j \qquad (2.4b)$$

for nonnegative integers n, i, j, and k with $k \leq n$.

Our next result gives the number of combinations of k elements which can be selected from a set which contains n elements.

Theorem 1.2.2 *Let Z be a set containing $n \geq 0$ distinct elements, and let k be an integer for which $0 \leq k \leq n$. Then there are*

$$\binom{n}{k} = \frac{n!}{k! \, (n - k)!} \qquad (2.5)$$

distinct subsets of size k contained in Z. Here (2.5) defines the notation $\binom{n}{k}$.

PROOF   If either $k = 0$ or $n = 0$, the result is obvious, for the only subset of size zero is the empty set, and, by definition, $\binom{n}{0} = 1, n = 0, 1, \ldots$. Therefore, we may restrict our attention to positive n and k. An ordered k-tuple with k distinct components may be selected in two steps: first, select a subset of size k; then arrange the subset into a definite order. Choosing a permutation of k elements of Z is therefore equivalent to choosing an ordered pair $(Z_0, \mathscr{P})$, where Z_0 is a subset of size k and $\mathscr{P}$ is a permutation of the k elements of Z_0. Let A denote the number of subsets of size k. Then, since there are $(n)_k$ distinct ordered k-tuples with distinct components and $k!$ ways in which to arrange a subset of size k into a definite order (both by Theorem 1.2.1), we have $(n)_k = Ak!$ by the basic principle. Solving for A, we find

$$A = \frac{(n)_k}{k!} = \frac{n!}{k!\,(n-k)!} = \binom{n}{k}$$

as asserted.   ////

Theorem 1.2.2 is especially useful in problems which involve card games. To make this precise, we define a poker hand to be a combination of five cards (subset of size 5) drawn from a standard deck of 52 cards.[1] Similarly, we define a bridge hand to be a combination of 13 cards drawn from a standard deck. Thus, two hands which contain the same cards arranged in different orders are regarded as identical.

EXAMPLE 1.2.4

a There are $\binom{52}{5} = 2{,}598{,}960$ distinct poker hands.

b There are $\binom{52}{13}$ distinct bridge hands.

c m indistinguishable red balls and n indistinguishable white balls can be arranged in a row to form $\binom{n+m}{m} = \binom{n+m}{n}$ distinguishable configurations. Indeed, a distinguishable configuration is determined by the m places occupied by the red balls.   ////

The numbers $\binom{n}{k}$ are known as *binomial coefficients* because they appear in the binomial theorem, which states that for real numbers a and b and for nonnegative integers n,

$$(a + b)^n = \sum_{k=0}^{n} \binom{n}{k} a^k b^{n-k} \tag{2.6}$$

In fact, the binomial theorem follows easily from Theorem 1.2.2, for if $(a + b)^n = (a + b)(a + b) \cdots (a + b)$ is expanded into a sum of powers of a times powers of b, then $a^k b^{n-k}$ will appear as many times as we can select a from k of the factors and b from the remaining $n - k$. By Example 1.2.4c this can be done in $\binom{n}{k}$ ways.

---

[1] That is, a deck which consists of 4 suits, spades, hearts, diamonds, and clubs, and the 13 denominations aces, twos, threes,..., queens, kings, with exactly one of each denomination in each suit.

In the sequel, it will often be convenient to use the notation $\binom{n}{k}$ when k is either a negative integer or a positive integer which exceeds n. We define $\binom{n}{k} = 0$ in both cases. Observe that with the extended definition it is still true that there are $\binom{n}{k}$ subsets of size k contained in a set of n elements.

We conclude this section with an extension of Theorem 1.2.2. Let Z be a finite, nonempty set. We define a *partition* of Z to be an ordered k-tuple $(Z_1, \ldots, Z_k)$, where $Z_1, \ldots, Z_k$ are disjoint subsets of Z for which

$$\bigcup_{i=1}^{k} Z_i = Z$$

We expressly allow some of the Z_i to be empty. If $(Z_1, \ldots, Z_k)$ is a partition of the set Z, then the numbers $r_i = |Z_i|$, $i = 1, \ldots, k$, will be called *partition numbers*. Clearly, $r_1, \ldots, r_k$ satisfy

$$r_i \geq 0 \quad i = 1, \ldots, k \quad \text{and} \quad \sum_{i=1}^{k} r_i = |Z| \qquad (2.7)$$

For example, if $Z = \{1,2,3,4\}$, then taking $Z_1 = \{1\}$, $Z_2 = \{2,3\}$, and $Z_3 = \{4\}$ defines a partition for which $r_1 = 1$, $r_2 = 2$, and $r_3 = 1$. In our next theorem we propose to answer the following question: Given integers $r_1, \ldots, r_k$ which satisfy (2.7), how many partitions $(Z_1, \ldots, Z_k)$ for which $|Z_i| = r_i$, $i = 1, \ldots, k$, exist?

Theorem 1.2.3 *Let Z be a set which contains n distinct elements, and let $r_1, \ldots, r_k$ be integers which satisfy (2.7). Then there are*

$$\frac{n!}{r_1! \cdots r_k!} \qquad (2.8)$$

distinct partitions $(Z_1, \ldots, Z_k)$ of Z with $|Z_i| = r_i$, $i = 1, \ldots, k$.

PROOF   We shall apply the basic principle. In choosing Z_1, we are simply selecting a subset of size r_1 from Z, a set containing n elements. By Theorem 1.2.2, this can be done in $\binom{n}{r_1}$ distinct ways. Thereafter, we must select Z_2 from the remaining $n - r_1$ elements in $Z - Z_1$. This can be done in $\binom{n - r_1}{r_2}$ distinct ways. In general, we must select Z_i from the $n - (r_1 + \cdots + r_{i-1})$ elements of $Z - (Z_1 \cup \cdots \cup Z_{i-1})$, and this can be done in

$$n_i = \binom{n - r_1 - \cdots - r_{i-1}}{r_i}$$

distinct ways, $i = 2, \ldots, k$. Therefore, by the basic principle, $(Z_1, \ldots, Z_k)$ can be selected in

$$\binom{n}{r_1}\binom{n - r_1}{r_2} \cdots \binom{n - r_1 - \cdots - r_{k-1}}{r_k} \qquad (2.9)$$

distinct ways. Finally, writing the binomial coefficients in terms of factorials we now find that (2.9) is

$$\frac{n!}{r_1(n - r_1)!}\frac{(n - r_1)!}{r_2!(n - r_1 - r_2)!} \cdots \frac{(n - r_1 - \cdots - r_{k-1})!}{r_k!(n - r_1 - \cdots - r_k)!}$$

$$= \frac{n!}{r_1! \cdots r_k!}$$

as asserted. ////

EXAMPLE 1.2.5

a If $Z = \{1,2,3,4\}$, then there are $4!/2! = 12$ partitions of Z for which $r_1 = 1, r_2 = 2$, and $r_3 = 1$.

b A deck of cards can be partitioned into four bridge hands in $52!/(13!)^4$ different ways. ////

The numbers

$$\binom{n}{r_1, \ldots, r_k} = \frac{n!}{r_1! \cdots r_k!} \qquad (2.10)$$

are called *multinomial coefficients*. There is also a multinomial theorem which states that for real numbers $a_1, \ldots, a_k$ and nonnegative integers n

$$(a_1 + \cdots + a_k)^n = \sum \binom{n}{r_1, \ldots, r_k} a_1^{r_1} \cdots a_k^{r_k}$$

where the summation extends over all nonnegative integers $r_1, \ldots, r_k$ for which $r_1 + \cdots + r_k = n$. The proof of the multinomial theorem is similar to that of the binomial theorem and will be omitted.

Let us briefly review. In this section, we have presented four rules for counting— *the basic principle, formulas for the number of ordered k-tuples, a formula for the number of combinations, and a formula for the number of partitions.* When used with a slight amount of ingenuity, these four rules will allow us to compute a wide variety of interesting probabilities. *Since they comprise the minimal amount of combinatorial analysis with which probability theory can be mastered, they should be understood and memorized.*

More combinatorial analysis will be found in Section 1.6 and in the problems at the end of this chapter.

1.3 URN MODELS

In this section and the next two, we shall study models for the following game: from an urn which contains balls of various colors, a sample is drawn and examined. That is, some of the balls are taken out of the urn and examined. We are interested in the probability that the sample has some particular property, such as containing three balls of a specified color. Here the terms "balls," "colors," and "urn" are not to be taken literally but as substitutes for the more prosaic terms "objects," "types of objects," and "group of objects." Thus, our model has a wider applicability than might at first appear. Indeed, with a proper interpretation of the terms "balls," "colors," and "urn," each of the following examples can be phrased as an urn problem.

EXAMPLE 1.3.1

a Opinion polls A group of people (the sample) is selected from a larger group of people (the urn) and asked their opinion on some political issue or candidate. Here we may regard the people as balls and different opinions as different colors.

b Acceptance sampling From a lot of manufactured items (the urn) a sublot (the sample) is selected and examined for defective items. Here we may regard the defective items as balls of one color and the nondefective items as balls of another.

c Gambling We may regard a poker hand as a sample of five cards from a deck of cards (the urn) and cards of different denominations (or of different suits) as balls of different colors. Similarly, if a die is tossed repeatedly, the numbers of spots which appear on the successive tosses may be regarded as a sample from the integers $1, \ldots, 6$, which, in turn, may be regarded as balls of six different colors.

d Coupon collecting If a manufacturer gives away various types of coupons with his product, we may regard the coupons as balls, the types as colors, and the coupons collected by a particular individual as the sample.   /////

There are several types of samples which may be drawn from an urn, and it will be convenient to distinguish them. First, the balls may be drawn either sequentially (that is, one at a time) or simultaneously (all at once). Let Z denote the set of balls in the urn. If the balls are drawn sequentially, then we may describe the outcome of our game by the ordered k-tuple $(z_1, \ldots, z_k)$ of elements of Z, where z_1 denotes the first ball drawn from the urn, z_2 the second,..., and k denotes the total number of balls drawn. Thus, we shall refer to $(z_1, \ldots, z_k)$ as an *ordered sample of size k*. If the balls are drawn simultaneously, it no longer makes sense to speak of a first ball or second ball and we may describe the outcome of our sampling only by the subset (combination) $\{z_1, \ldots, z_k\}$ of distinct elements of Z which were selected. We shall refer to $\{z_1, \ldots, z_k\}$ as an *unordered sample of size k*. We must, of course, have $k \leq |Z|$ in the case of unordered samples.

There is a further distinction to be drawn in the case of ordered samples. We may either replace each ball after it has been drawn and examined, or not. In the first instance, we shall say that the sampling was done *with replacement*, and in the second, we shall say that the sampling was done *without replacement*. We shall not consider here the more complicated scheme in which some of the balls are replaced and others are not.

We shall now state models for each of the three types of sampling.

Unordered samples   If an unordered sample of size k is drawn from an urn containing n balls, then we take the sample space S to be the set of all subsets of size k which may be drawn from the urn. By Theorem 1.2.2, there are then $|S| = \binom{n}{k}$ possible outcomes.

Ordered samples with replacement   If an ordered sample of size k is drawn with replacement from an urn with n balls, then we may take the sample space S to be the set of all ordered k-tuples $(z_1, \ldots, z_k)$ with $z_i \in Z$, the set of balls, $i = 1, \ldots, k$. In this case, there are $|S| = n^k$ possible outcomes by Theorem 1.2.1.

Ordered samples without replacement   If an ordered sample of size k is drawn without replacement from an urn containing n balls, then we may take the sample space S to be the set of all ordered k-tuples $(z_1, \ldots, z_k)$ with $z_i \neq z_j$ for $i \neq j$ and $z_i \in Z$, the set of balls in the urn, $i = 1, \ldots, k$. In this case, there are $|S| = (n)_k$ possible outcomes by Theorem 1.2.1.

We shall say that a sample has been drawn *at random* when we are assuming all samples of the size and type in question to be equally likely. In this case we can compute many interesting probabilities from (1.1) and the results of Section 1.2. For these computations, it is imperative that the reader not confuse the sample space S with the set of balls in the urn. The appropriate sample space depends on the type of sampling and has been defined above.

EXAMPLE 1.3.2   All parts of the example refer to an urn which contains 4 red balls and 4 white balls. Thus, there are $n = 8$ balls in the urn.

    a   If an ordered sample of size 2 is drawn at random with replacement, what is the probability that the sample will contain 2 red balls? The sample space S consists of all ordered pairs (z_1, z_2) which may be drawn from the urn. Therefore,

$|S| = 8^2$ by Theorem 1.2.1. We require the probability of the event A, which consists of all ordered pairs (z_1, z_2) for which z_1 and z_2 are both red. Thus, in selecting an element of A, we have 4 choices for z_1 and 4 choices for z_2 (since the sampling is with replacement). Therefore, there are $|A| = 4^2 = 16$ elements in A, so that $P(A) = 16/64 = \frac{1}{4}$.

b If the sampling is without replacement, we would find $|S| = 8 \times 7 = 56$, $|A| = 4 \times 3 = 12$, and $P(A) = 12/56 = \frac{3}{14}$.

c Let us compute the probability of drawing 2 red balls when an unordered, random sample of size 2 is drawn from the urn. In this case the sample space S consists of all subsets of size 2 which may be drawn from the 8 balls, so that $|S| = \binom{8}{2} = 28$. The event A now consists of all subsets of size 2 which may be drawn from the 4 red balls, so that $|A| = \binom{4}{2} = 6$. Therefore, $P(A) = 6/28 = \frac{3}{14}$. As we shall see in Section 1.5, it is no accident that the answers in parts *b* and *c* are the same.   ////

Examples 1.3.2*a* to *c* may be generalized considerably, and we shall consider these generalizations in the next two sections. We conclude this section with two simple but interesting results.

If an ordered sample of size k is drawn (either with or without replacement) from an urn which contains m red and $n - m$ white balls, it is intuitively clear that the probability of drawing a red ball on the first draw is m/n. This is also the probability of drawing a red ball on the second, or third, or jth draw, $j = 1, \ldots, k$, as we shall now show.

Theorem 1.3.1   *Let an ordered random sample of size $k \geq 1$ be drawn either with or without replacement from an urn containing m red and n − m white balls, and let A_i be the event that the ith ball drawn is red for $i = 1, \ldots, k$. Then, $P(A_i) = m/n, i = 1, \ldots, k$.*

PROOF   If the sampling is with replacement, then there are $|S| = n^k$ possible outcomes and A_i consists of all ordered k-tuples $(z_1, \ldots, z_k)$ for which z_i is red. Thus, there are m possible choices for z_i and n choices for z_j for $j \neq i$ since z_j is not restricted by A_i for $j \neq i$. By the basic principle, there are $|A_i| = n \cdots nmn \cdots n = mn^{k-1}$ outcomes in A_i, and therefore $P(A_i) = m/n$, as asserted.

If the sampling is without replacement, the situation is slightly more complicated, and we shall give the proof only for the special case that $i = 2$. Clearly, $A_2 = A_1 A_2 \cup A_1' A_2$ with $A_1 A_2 \cap A_1' A_2 \subset A_1 A_1' = \varnothing$, so that $P(A_2) = P(A_1 A_2) + P(A_1' A_2)$. Thus, we need only compute $P(A_1 A_2)$ and $P(A_1' A_2)$.

In sampling without replacement there are $|S| = (n)_k$ possible outcomes. Now $A_1 A_2$ consists of all ordered k-tuples $(z_1, \ldots, z_k)$ for which z_1 is red and z_2 is red and z_j is unrestricted for $j = 3, \ldots, k$, so that there are m choices for z_1, $m - 1$ choices for z_2, and $(n - 2)_{k-2}$ choices for $(z_3, \ldots, z_k)$. Thus, $|A_1 A_2| = m(m - 1)(n - 2)_{k-2}$ by the basic principle. Therefore, $P(A_1 A_2) = m(m - 1) \times (n - 2)_{k-2}/(n)_k = m(m - 1)/n(n - 1)$. Similarly, $P(A_1' A_2) = m(n - m)/n \times (n - 1)$, so that

$$P(A_2) = \frac{m(m - 1) + m(n - m)}{n(n - 1)} = \frac{m}{n}$$

as asserted.   ////

EXAMPLE 1.3.3   In the national draft lottery, balls labeled with the days of the year are drawn sequentially and without replacement from an urn. What is the probability that the last ball drawn will be labeled with a day in January? We may regard the balls labeled with days in January as red balls and the others as white balls. We then have an ordered random sample without replacement of size $k = 365$ from an urn containing $m = 31$ red balls and $n - m = 334$ white balls. The desired probability is therefore $m/n = 31/365 = 0.085.$[†]   ////

Let us now consider an urn which contains n balls of different colors. If an ordered random sample of size k is drawn with replacement, what is the probability that the k balls drawn will be of different colors? That is, if repetition is allowed in the sample, what is the probability that no repetition occurs?

Theorem 1.3.2   *If an ordered random sample of size k is drawn with replacement from an urn containing n balls of different colors, then the probability that all balls in the sample are of different colors is*

$$p_{n,k} = \prod_{i=1}^{k} \left(1 - \frac{i - 1}{n}\right)$$

PROOF   The sample space S consists of all ordered k-tuples $(z_1, \ldots, z_k)$ which may be selected from the n balls, and so $|S| = n^k$ by Theorem 1.2.1. The event A that all balls in the sample are of different colors consists of all ordered k-tuples $(z_1, \ldots, z_k)$ with distinct components, so that $|A| = (n)_k$, again by Theorem 1.2.1. Thus,

$$P(A) = \frac{(n)_k}{n^k} = 1 \left(1 - \frac{1}{n}\right)\left(1 - \frac{2}{n}\right) \cdots \left(1 - \frac{k - 1}{n}\right)$$

as asserted.   ////

[†] Numerical answers will often be rounded. They are accurate to the number of decimals given.

EXAMPLE 1.3.4

a If a balanced die is tossed six times, what is the probability that no face appears more than once? By (3.1) this probability is simply $(6)_6/6^6 = 6!/6^6 = 0.0154$, since the six rolls select a sample of size $k = 6$ from the integers $\{1, \ldots, 6\}$. Thus, although the faces are equally likely to appear on any toss, the probability that they all appear during six tosses is less than 1 in 50.

b If 25 people gather at a party, what is the probability that they all have different birthdays? Let us regard the 365 days of the year as balls of different colors and the birthdays of the people at the party as a random sample with replacement from the 365 balls. Let A be the event that no two people have the same birthday. Then, $P(A) = p_{365, 25} = 0.44$. That is, if 25 people gather at a party, the probability that no two have the same birthday is less than 0.5.   ////

A simple approximation to $p_{n,k}$ will be given in Example 1.7.2.

1.4   UNORDERED SAMPLES

In this section we consider problems which arise when an unordered, random sample of size $k \geq 1$ is drawn from an urn containing m red balls and $n - m$ white balls. Here m and n are nonnegative integers with $n \geq k$. What is the probability of obtaining exactly r red balls in the sample, where r is a nonnegative integer with $r \leq k$? The answer is provided by the following theorem, which generalizes Example 1.3.2c.

Theorem 1.4.1   *If an unordered random sample of size k is drawn from an urn which contains m red balls and $n - m$ white balls with $k \leq n$, then the probability that the sample will contain exactly r red balls is*

$$p_r = \frac{\binom{m}{r}\binom{n-m}{k-r}}{\binom{n}{k}} \qquad (4.1)$$

for $r = 0, 1, \ldots, k$.

PROOF   The sample space S for this problem is the set of all unordered samples which may be drawn from the urn. Therefore, there are $|S| = \binom{n}{k}$ possible outcomes. Let $A \subset S$ be the event consisting of all unordered samples which contain exactly r red balls. We need to find $|A|$. An unordered sample which contains exactly r red balls may be selected in two steps. First, select a subset of size r from the m red balls in the urn; then select a subset of size $k - r$

from the $n - m$ white balls in the urn. That is, an element of A corresponds uniquely to an ordered pair (Z_0, Z_1), where Z_0 is a combination of r red balls and Z_1 is a combination of $k - r$ white balls. The first step requires the selection of a subset of size r from a set of m elements and can therefore be performed in $\binom{m}{r}$ ways by Theorem 1.2.2. Similarly, the second step can be performed in $\binom{n - m}{k - r}$ ways by the same theorem. Therefore,

$$|A| = \binom{m}{r}\binom{n - m}{k - r}$$

by the basic principle. Therefore, $P(A) = |A|/|S| = \binom{m}{r}\binom{n - m}{k - r}\bigg/\binom{n}{k}$, as asserted.   ////

The probability of getting exactly r red balls is, of course, zero if either $r > m$ or $k - r > n - m$. The reader should check that our conventions about binomial coefficients give $p_r = 0$ in these cases.

The numbers p_r are known as the *hypergeometric probabilities.* For tables of the hypergeometric probabilities for $0 \leq r \leq k$, $0 \leq m \leq n$, $1 \leq k \leq n$, and $1 \leq n \leq 20$, see Beyer (1966).

EXAMPLE 1.4.1   In these examples, we regard a poker hand as an unordered, random sample of size 5 drawn from a standard deck of 52 cards.

   a   The probability that a poker hand contains exactly 3 aces is

$$\frac{\binom{4}{3}\binom{48}{2}}{\binom{52}{5}} = 0.001736 \qquad (4.2)$$

for we may regard the 4 aces as red balls and the 48 nonaces as white balls. Theorem 1.4.1 then applies with $m = 4$, $n = 52$, $k = 5$, and $r = 3$. More generally, Equation (4.2) gives the probability of getting exactly three cards of any specified denomination, such as kings, queens, etc.....

   b   What is the probability that a poker hand contains exactly 3 cards of an unspecified denomination (3 of a kind)? Let A be the event that the hand contains 3 cards of some denomination. Then we may select an element of A in three steps. First, select a denomination; then select 3 cards from the 4 cards of that denomination; then select 2 cards from the remaining 48 cards. The first step can be performed in 13 ways since there are 13 denominations, and the last two can be performed in $\binom{4}{3}\binom{48}{2}$ ways by part a. Therefore, the desired probability is

$$13 \frac{\binom{4}{3}\binom{48}{2}}{\binom{52}{5}} = 0.0226$$

c The probability of getting exactly 4 aces is

$$\frac{\binom{4}{4}\binom{48}{1}}{\binom{52}{5}} = 0.0000184$$

again by Theorem 1.4.1. Therefore, the probability of getting at least 3 aces is .001736 + .0000184 = 0.00175 by Equation (1.3). The probability of getting at least 3 of an unspecified denomination can now be computed as in part b.

d The probability of getting exactly 2 aces is

$$\frac{\binom{4}{2}\binom{48}{3}}{\binom{52}{5}} = 0.03993$$

which also give the probability of getting exactly 2 cards of any specified denomination. However, the probability of getting exactly 2 cards of an unspecified denomination is not $13\binom{4}{2}\binom{48}{3}/\binom{52}{5}$, since it is possible to get more than one pair in a single hand.

e The probability that a poker hand contains exactly 3 hearts is $\binom{13}{3}\binom{39}{2}/\binom{52}{2}$. Here we may regard the hearts as red balls. ////

EXAMPLE 1.4.2 Acceptance sampling Consider a company which markets its goods in lots of size $n = 100$. Suppose that each lot contains an unknown number m of defective items and that it is unprofitable for the company to release a lot which contains more than 5 defective items. Suppose also that the process of inspecting the items in a lot is expensive. Then the company might wish to inspect only a randomly selected sample from each lot, to release immediately those lots from which the samples contain no defectives, and to inspect all items in those lots from which the samples contain at least one defective. The probability that a particular lot will be released (i.e., that the sample will contain no defectives) is then

$$q(k,m) = \frac{\binom{100 - m}{k}}{\binom{100}{k}}$$

for we may regard the defective items as red balls and nondefective items as white balls. Of course, if $m > 5$, then $q(k,m)$ is the probability of releasing a bad lot, one which contains too many defectives. How large is this probability? The answer depends on the parameters m and k. A few typical values are given in Table 1.

The company might wish to control the probability of releasing a bad lot by choice of the sample size k. That is, the company might wish to choose k in such a manner that the probability of releasing a bad lot is at most a specified number α. How large should k be in order that the probability of releasing a bad lot will be at most $\alpha = 0.05$? Since $q(k,m)$ is a decreasing function of m, it will suffice to choose k in such a manner that $q(k,6) \leq 0.05$. The table then indicates that 40 is a sufficiently large sample size. In fact, 39 is the smallest value of k for which $q(k,6) \leq 0.05$. ////

Theorem 1.4.1 extends from the case of two colors to the case of several. Thus, consider an urn which contains balls of c different colors. Let n_1 be the number of balls of the first color, n_2 the number of balls of the second color, and, in general, let n_i be the number of balls of the ith color, $i = 1, \ldots, c$. Then there are $n = n_1 + \cdots + n_c$ balls in the urn. Suppose now that an unordered sample of size k is drawn at random from the urn, and let $k_1, \ldots, k_c$ be nonnegative integers for which $k_1 + \cdots + k_c = k$. Then we can compute the probability that the sample contains exactly k_1 balls of the first color, exactly k_2 balls of the second color, etc.

Table 1

		k		
m	10	25	40	50
3	0.727	0.418	0.212	0.121
6	0.522	0.169	0.042	0.013
9	0.371	0.066	0.007	0.001

Theorem 1.4.2 *With the notation of the previous paragraph, the probability that the sample contains exactly k_i balls of color i, $i = 1, \ldots, c$ is*

$$\frac{\binom{n_1}{k_1} \cdots \binom{n_c}{k_c}}{\binom{n}{k}}$$

Since the notation is somewhat complicated, we shall illustrate Theorem 1.4.2 with some examples before proving it.

EXAMPLE 1.4.3

a What is the probability that a poker hand contains 3 aces and 2 kings? Let us regard the aces as red balls, the kings as black balls, and the remaining cards as white balls. Then, we have $n_1 = 4$ red balls, $n_2 = 4$ black balls, and $n_3 = 44$ white balls, and we require the probability of obtaining a sample which contains $k_1 = 3$ red balls, $k_2 = 2$ black balls, and $k_3 = 0$ white balls. By Theorem 1.4.2, this is

$$\frac{\binom{4}{3}\binom{4}{2}\binom{44}{0}}{\binom{52}{5}} = \frac{\binom{4}{3}\binom{4}{2}}{\binom{52}{5}} = 0.00000923 \qquad (4.3)$$

More generally, (4.3) gives the probability that a poker hand will contain 3 cards of one specified denomination and 2 of another.

b What is the probability of getting 3 cards of one unspecified denomination and 2 of another (a full house)? We can select an ordered pair of distinct denominations in $(13)_2$ ways by Theorem 1.2.1; then we can select 3 cards of the first denomination and 2 of the second in $\binom{4}{3}\binom{4}{2}\binom{44}{0} = \binom{4}{3}\binom{4}{2}$ ways by part *a*. Therefore, the desired probability is $(13)_2 \binom{4}{3}\binom{4}{2} \Big/ \binom{52}{5} = 0.00144$.

c The probability of getting 2 aces, 2 kings, and 1 card which is neither an ace nor a king is $\binom{4}{2}\binom{4}{2}\binom{44}{1} \Big/ \binom{52}{5} = 0.00061$, again by Theorem 1.4.2. This is also the probability that a poker hand will contain exactly 2 cards of one specified denomination, exactly 2 cards of another specified denomination, and 1 which belongs to neither of the specified denominations.

d The probability of getting exactly 2 cards of each of two unspecified denominations is $\binom{13}{2}\binom{4}{2}^2\binom{44}{1} \Big/ \binom{52}{5} = 0.0475$. In fact, we may select a set of two distinct denominations in $\binom{13}{2}$ ways; then we may select a hand with exactly 2 cards of each of these two denominations in $\binom{4}{2}^2\binom{44}{1}$ ways by part c. Observe that we multiplied by $(13)_2$ in a similar situation in part b.   ////

EXAMPLE 1.4.4 Opinion polls Suppose that an electorate consists of n individuals of which n_a favor candidate A, n_b favor candidate B, and n_u are undecided. In order to learn about the collective opinion of the electorate, an unordered random sample of size k is selected from the electorate, and the members of the sample are asked their opinions. If k_a, k_b, and k_u are nonnegative integers for which $k_a + k_b + k_u = k$, what is the probability that k_a members of the sample will favor A, k_b will favor B, and k_u will be undecided? The answer can be obtained by a direct application of Theorem 1.4.2 as $\binom{n_a}{k_a}\binom{n_b}{k_b}\binom{n_u}{k_u}\Big/\binom{n}{k}$.   ////

PROOF of Theorem 1.4.2   As in the proof of Theorem 1.4.1, the sample space for our game is the set of all unordered samples which may be drawn from the urn. Therefore, $|S| = \binom{n}{k}$. We now require the probability of the event A, which consists of all unordered samples containing exactly k_i balls of color i, $i = 1, \ldots, c$. An element of A may be chosen in c steps. First, choose a subset of size k_1 from the n_1 balls of color 1. Then, select a subset of size k_2 from the n_2 balls of color 2. In general, we must select a subset of size k_i from the n_i balls of color i, $i = 1, \ldots, k$. The ith step can therefore be performed in $\binom{n_i}{k_i}$ distinct ways by Theorem 1.2.2. Therefore, by the basic principle,

$$|A| = \binom{n_1}{k_1}\binom{n_2}{k_2}\cdots\binom{n_c}{k_c}$$

The theorem follows from (1.1).   ////

1.5 ORDERED SAMPLES[1]

Let us now consider ordered samples. As in the previous section, we shall consider an urn which contains m red balls and $n - m$ white balls from which a sample of size k is to be drawn, and we shall find the probability that the sample contains exactly r red balls. This time, however, we shall consider ordered samples.

[1] The main results of this section will be derived again in a more general context in Sections 4.1 and 4.2.

In the case of ordered samples, there is an important distinction to be made between drawing r red balls in the sample and drawing red balls on r specified draws. For example, if an ordered, random sample of size $k = 3$ is drawn with replacement from an urn which contains $m = 1$ red ball and $n - m = 1$ white ball, then the probability that the first two balls drawn are red and the third is white is simply $1/2^3 = \frac{1}{8}$. For the sample space S (which consists of all ordered triples which can be drawn from the 2 balls) contain $n^k = 2^3 = 8$ elements, only one of which results in 2 red balls followed by 1 white one. Similarly, the probability that the first and third balls drawn are red while the second is white is also $\frac{1}{8}$, as is the probability that the first ball drawn is white while the second and third are red. Thus, the probability that red balls are drawn on any two specified draws is $\frac{1}{8}$. The event that the sample contains exactly 2 red balls may occur in three ways, however, namely, (red, red, white), (red, white, red), and (white, red, red). Therefore, the probability that the sample contains exactly 2 red balls is $\frac{3}{8}$. Having, we hope, made the distinction clear, we shall now develop some general formulas. We begin with the case of r specified draws.

Lemma 1.5.1   *Let an ordered, random sample of size $k \geq 1$ be drawn from an urn which contains m red balls and $n - m$ white balls. Then the probability that red balls are drawn on r specified draws and white balls are drawn on the remaining draws is*

$$\frac{m^r(n - m)^{k-r}}{n^k} \tag{5.1}$$

if the sampling is with replacement and is

$$\frac{(m)_r(n - m)_{k-r}}{(n)_k} \tag{5.2}$$

if the sampling is without replacement and $k \leq n$.

PROOF   We shall prove the lemma for sampling with replacement only, since the proof for sampling without replacement is similar. The sample space S is then the set of all ordered k-tuples $(z_1, \ldots, z_k)$ which can be drawn from the urn, so $|S| = n^k$. Let $J \subset \{1, \ldots, k\}$ denote the set consisting of the r specified draws, and let A be the event that red balls are drawn on draws $i \in J$ and that white balls are drawn on draws $i \notin J$. In selecting an element of A, we then have n_i choices for the ith ball, where $n_i = m$ (the number of red balls in the urn), if $i \in J$ and $n_i = n - m$ if $i \notin J$. Thus, there are $n_1 n_2 \cdots n_k = m^r(n - m)^{k-r}$ distinct elements in A by the basic principle. Expression (5.1) now follows easily.   ////

As a corollary to Lemma 1.5.1, we now compute the probability that the first red ball to be drawn is drawn on the kth (last) draw.

Theorem 1.5.1   *If an ordered, random sample of size k is drawn from an urn which contains m red balls and n − m white balls, then the probability that the first red ball to be drawn is drawn on the kth draw is*

$$\frac{m(n - m)^{k-1}}{n^k} \qquad (5.3a)$$

if the sampling is with replacement and is

$$\frac{m(n - m)_{k-1}}{(n)_k} \qquad (5.3b)$$

if the sampling is without replacement and k ≤ n.

PROOF   The event that the first red ball is drawn on the kth draw requires a red ball to be drawn on one specified draw, the last one. Thus, (5.3a) and (5.3b) are special cases of (5.1) and (5.2), respectively.   ////

Expression (5.3a) defines a special case of the *geometric probabilities*, which we shall encounter again in Section 4.2.

EXAMPLE 1.5.1

a   If a fair coin is tossed k times, the probability that the first head will appear on the kth toss is simply 2^{-k}, for we may regard the first k tosses as an ordered sample with replacement from the set {heads, tails}.

b   If a man has n keys, only one of which will unlock his door, and if he tries them in a random order (without replacement), what is the probability that he will try exactly $k - 1$ wrong keys before finding the correct one? If we regard the correct key as a red ball and the incorrect ones as white balls, the answer is given by (5.3b) as

$$\frac{(1)_1(n - 1)_{k-1}}{(n)_k} = \frac{1}{n}$$

for $k = 1, 2, \ldots$. Thus, the man is as likely to try one key, as two keys, as three keys, etc.   ////

We shall now compute the probability that the sample will contain exactly r red balls.

Theorem 1.5.2   *Let an ordered, random sample of size k be drawn from an urn which contains m red balls and n − m white balls. If the sampling is with replacement, then the probability that the sample will contain exactly r red balls is*

$$\frac{\binom{k}{r} m^r(n - m)^{k-r}}{n^k} \qquad (5.4)$$

for r = 0, ..., k. If the sampling is without replacement, and k ≤ n, then the probability that the sample will contain exactly r red balls is

$$\frac{\binom{k}{r}(m)_r(n-m)_{k-r}}{(n)_k} \tag{5.5}$$

for r = 0, ..., k.

PROOF   Again, we shall prove the theorem only for sampling with replacement, since the proof for sampling without replacement is similar. Thus, the sample space contains $|S| = n^k$ elements. Let B denote the event that the sample contains exactly r red balls. Then, an element of B may be selected in two steps. First, select a subset J of size $|J| = r$ from the integers $1, \ldots, k$. Then, draw red balls on those draws $i \in J$ and draw white balls on those draws $i \notin J$. The first step can be performed in $\binom{k}{r}$ distinct ways by Theorem 1.2.2, and the second in $m^r(n-m)^{k-r}$ by Lemma 1.5.1. Therefore,

$$|B| = \binom{k}{r} m^r(n-m)^{k-r}$$

by the basic principle. The theorem follows.   ////

EXAMPLE 1.5.2

a   If a balanced die is rolled 5 times, the probability of getting exactly 1 spot on the first and last rolls and more than 1 spot on the other three rolls is $(\frac{1}{6})^2(\frac{5}{6})^3 = 0.0161$ by Lemma 1.5.1. The probability of getting exactly 1 spot on exactly two tosses is $\binom{5}{2}\left(\frac{1}{6}\right)^2\left(\frac{5}{6}\right)^3 = 0.161$ by Theorem 1.5.2.

b   If a balanced coin is tossed k times, what is the probability of obtaining exactly r heads? We may regard heads as a red ball and tails as a white ball. Thus, the k tosses constitute an ordered, random sample from an urn containing $m = 1$ red ball and $n - m = 1$ white ball, and the required probability is therefore $\binom{k}{r} 2^{-k}$.   ////

In Equation (5.4), let $p = m/n$ and $q = 1 - p = (n - m)/n$. Then, the first conclusion in Theorem 1.5.1 may be stated: the probability of obtaining exactly r red balls when sampling with replacement is

$$\binom{k}{r} p^r q^{k-r} \qquad r = 0, \ldots, k \tag{5.6}$$

These numbers are known as the *binomial probabilities*. We shall meet them again in Chapters 4 and 5. Tables of the binomial probabilities for $0 \le r \le k$, $1 \le k \le 10$, and selected values of p will be found in Appendix C. For more extensive tables see, for example, Beyer (1966) or Selby (1965).

It is interesting that the probability of obtaining exactly r red balls in an ordered, random sample which is drawn without replacement is the same as the probability of drawing exactly r red balls in an unordered sample. To see this observe that, by (5.5), the probability that an ordered, random sample contains exactly r red balls is

$$\frac{\binom{k}{r}(m)_r(n-m)_{k-r}}{(n)_k} = \frac{k!}{r!\,(k-r)!}\frac{(m)_r(n-m)_{k-r}}{(n)_k}$$

$$= \frac{(m)_r}{r!}\frac{(n-m)_{k-r}/(k-r)!}{(n)_k/k!}$$

$$= \frac{\binom{m}{r}\binom{n-m}{k-r}}{\binom{n}{k}} \tag{5.7}$$

which is also the probability that an unordered, random sample contains exactly r red balls.

It is also interesting that if m, n, and $n - m$ are all large, then the difference between the binomial probabilities (5.4) and the hypergeometric probabilities (5.5) is small. To see this observe that

$$\frac{(n)_k}{n^k} = \prod_{i=1}^{k}\frac{n-i+1}{n} \to 1$$

as $n \to \infty$ for each fixed $k = 1, 2, \ldots$. Thus, if $n \to \infty$ and $m \to \infty$ in such a manner that $m/n \to p$, $0 < p < 1$, then

$$\lim\frac{\binom{k}{r}(m)_r(n-m)_{k-r}}{(n)_k}$$

$$= \lim\binom{k}{r}\left(\frac{m}{n}\right)^r\left(\frac{n-m}{n}\right)^{k-r}\frac{(m)_r}{m^r}\frac{(n-m)_{k-r}}{(n-m)^{k-r}}\frac{n^k}{(n)_k} = \binom{k}{r}p^r q^{k-r} \tag{5.8}$$

where $q = 1 - p$, for $r = 0, \ldots, k$ for each fixed k. The practical value of (5.8) is that the left side of (5.8) may be approximated by the right side if m and n are sufficiently large. In fact, the approximation (5.8) will be good provided only that k^2/n, r^2/m, and $(k-r)^2/(n-m)$ are all small (see Problems 1.62 and 1.63).

EXAMPLE 1.5.3   Opinion polls   From an electorate of $n = 70{,}000{,}000$ a random sample of size k is drawn, and members of the sample are asked whether they prefer candidate A or candidate B. What is the probability that exactly r members of the sample will prefer candidate A? Let m denote the number of people in the electorate who prefer candidate A, and suppose, for simplicity, that the remaining $n - m$ prefer candidate B. Then, the exact probability is given by Theorem 1.4.1 as $\binom{m}{r}\binom{n-m}{k-r} / \binom{n}{k}$. By Equations (5.7) and (5.8), this is approximately $\binom{k}{r} p^r q^{k-r}$, where $p = m/n$ and $q = 1 - p$, provided that k^2/n, r^2/m, and $(k - r)^2/(n - m)$ are small. In particular, if $20{,}000{,}000 \leq m \leq 50{,}000{,}000$, the approximation is excellent for $k \leq 500$.   ////

1.6   OCCUPANCY PROBLEMS[1]

In the previous three sections, we have dealt extensively with problems which arise when balls are drawn from an urn. We now turn our attention to problems which arise when balls are placed in urns, or cells, as we shall call them in this section. Suppose, then, that we have k balls which we wish to place in n cells, and let us inquire as to how many distinguishable configurations of balls in cells can be so formed.

As in Section 1.3, we must consider several cases. We may either have *distinguishable* balls or *indistinguishable* balls, and we may either allow *repetition* (that is, more than 1 ball in a cell) or we may not. There is a definite relation with the sampling theory of Section 1.3 here, for *we may think of the k balls as selecting a sample from the n cells.* In this analogy, we see that the distinction between distinguishable and indistinguishable balls made here corresponds to the distinction between ordered and unordered samples made in Section 1.3. Moreover, the concept of repetition introduced above corresponds to the concept of replacement in Section 1.3.

Therefore, we have the following theorem.

Theorem 1.6.1   *Let n and k be positive integers. If k distinguishable balls are placed in n cells, then there are n^k distinguishable arrangements of balls in cells if repetition is allowed and there are $(n)_k$ distinguishable arrangements of balls in cells if repetition is not allowed and $k \leq n$. Moreover, if k indistinguishable balls are placed in n cells, where $k \leq n$ and repetition is not allowed, then there are $\binom{n}{k}$ distinguishable arrangements of balls in cells.*

---

[1] This section treats a special topic and may be omitted without loss of continuity.

The novel feature which we encounter when placing balls in cells is that we may place indistinguishable balls in cells with repetition, whereas we did not define an unordered sample with replacement. The number of distinguishable arrangements in this case is given by the following theorem.

Theorem 1.6.2 *Let n and k be positive integers. If k indistinguishable balls are placed in n cells with repetition allowed, then there are*

$$\binom{n + k - 1}{k} = \binom{n + k - 1}{n - 1}$$

distinguishable arrangements of balls in cells; and if $k \geq n$, then there are $\binom{k - 1}{n - 1}$ *such arrangements in which no cell remains empty.*

PROOF Let us divide the cells by the lines and represent the balls by circles. Thus, if $n = 5$ and $k = 4$, we represent the five cells as $1 \mid 2 \mid 3 \mid 4 \mid 5$. The array

$$|\circ\circ||\circ|\circ$$

represents the arrangement with no balls in the first cell, 2 in the second, none in the third, and 1 each in the fourth and fifth cells. Observe that we need only $n - 1 = 4$ lines to represent the $n = 5$ cells since the outer walls of the first and last cells are not explicitly drawn. In general, we can represent any distinguishable arrangements of balls in cells by such an array, where the number of circles to the left of the first line gives the number of balls in the first cell, the number of circles between the first and second lines gives the number of balls in the second cell, etc. The number of distinguishable arrangements of balls in cells is therefore equal to the number of distinguishable arrays which can be formed from k circles and $n - 1$ lines. Since we can choose k of the $n + k - 1$ places to be occupied by circles in exactly

$$\binom{n + k - 1}{k} = \binom{n + k - 1}{n - 1}$$

ways by Theorem 1.2.2 (compare Example 1.2.4c), the first assertion of the theorem has been proved. The second now follows easily. Indeed, if $k \geq n$ and we require that every cell contain at least 1 ball, then we are free to place only $k' = k - n$ of the balls as we please, and we can do so in

$$\binom{n + k' - 1}{k'} = \binom{k - 1}{n - 1}$$

distinct ways by the first assertion of Theorem 1.6.2. ////

EXAMPLE 1.6.1

a If five indistinguishable dice are tossed, then there are $\binom{10}{5} = 252$ distinguishable outcomes. Simply regard the dice as balls and the integers $1, \ldots, 6$ as cells. If the dice are balanced, however, the distinguishable outcomes will not be equally likely.

b If nine indistinguishable dice are tossed, then there are $\binom{8}{5} = 56$ distinguishable outcomes for which each of the integers $1, \ldots, 6$ appears on at least one die. ////

Theorems 1.6.1 and 1.6.2 find application in statistical mechanics.[1] Consider a region of space which contains k particles, such as electrons or photons, and imagine the region subdivided into n subregions (cells). If the particles are regarded as distinguishable, and if every arrangement of particles in cells (with repetition allowed) is equally likely, then the particles are said to obey *Maxwell-Boltzmann statistics*. Although Maxwell-Boltzmann statistics certainly seems to be a reasonable assumption, it applies to no known kind of particle. If the particles are indistinguishable, and if the $\binom{n + k - 1}{n - 1}$ distinguishable arrangements of particles in cells (with repetition allowed) are equally likely, then the particles are said to obey *Bose-Einstein statistics*. Photons obey Bose-Einstein statistics. Finally, if the particles are indistinguishable, if no two may occupy the same cell and if the $\binom{n}{k}$ distinguishable arrangements are equally likely, then the particles are said to obey *Fermi-Dirac statistics*. This model applies to electrons, protons, and neutrons.

EXAMPLE 1.6.2

a If the particles obey Bose-Einstein statistics and $k \geq n$, then the probability that every cell is occupied is $\binom{k - 1}{n - 1} \Big/ \binom{n + k - 1}{n - 1}$.

b Consider a subregion which contains $m < n$ cells. If the particles obey Bose-Einstein statistics, then the probability that the subregion will contain all the particles is $\binom{m + k - 1}{m - 1} \Big/ \binom{n + k - 1}{n - 1}$.

[1] See, for example, Constant (1958), chaps. 5 and 6.

c If the particles obey Fermi-Dirac statistics, then the probability that the subregion contains exactly *r* particles is $\binom{m}{r}\binom{n-m}{k-r}\Big/\binom{n}{k}$, $r = 0, \ldots, k$, by Theorem 1.4.2.   ////

More applications of Theorem 1.6.2 will be found in the problems at the end of this chapter.

1.7 THE GENERALIZED BINOMIAL THEOREM

We shall have occasion to sum certain series and to approximate certain functions. In this section we discuss a tool for performing these operations, namely, Taylor's theorem, which the reader has probably encountered in a calculus course.[1] Taylor's theorem states the following. Let *f* be a function which is defined on an interval (a,b) and has *k* derivatives there; if $x_0 \in (a,b)$, then

$$f(x) = f(x_0) + \sum_{j=1}^{k-1} \frac{1}{j!} f^j(x_0)(x - x_0)^j + \frac{1}{k!} f^k(x_1)(x - x_0)^k \tag{7.1}$$

for $x \in (a,b)$, where x_1 lies between *x* and x_0 and f^j denotes the *j*th derivative of *f*, $j = 1, \ldots, k$. That is, *f* can be approximated by a polynomial in a neighborhood of any given point x_0.

EXAMPLE 1.7.1
a Taking $k = 1$ in (7.1) yields the mean-value theorem, namely,

$$f(x) - f(x_0) = f'(x_1)(x - x_0)$$

where x_1 lies between *x* and x_0.

b Taking $k = 3$ in (7.1) yields the quadratic approximation

$$f(x) - f(x_0) = f'(x_0)(x - x_0) + \tfrac{1}{2}f''(x_0)(x - x_0)^2 + r(x)$$

where the remainder term *r* is defined by $r(x) = \tfrac{1}{6}f'''(x_1)(x - x_0)^3$.   ////

EXAMPLE 1.7.2
Consider the function *f*, defined by $f(x) = \log(1 - x)$ for $-\infty < x < 1$. The first two derivatives of *f* are $f'(x) = -1/(1 - x)$ and $f''(x) = -1/(1 - x)^2$, so that we can expand *f* is a Taylor series about $x_0 = 0$ as

$$\log(1 - x) = -x - r(x)$$

---

[1] See, for example, Thomas (1972), pp. 150–151, for an elementary treatment or Rudin (1964), pp. 95–96, for more detailed treatment.

where $r(x) = \frac{1}{2}(1 - x_1)^{-2}x^2$ with $|x_1| \leq |x|$. Observe also that for $x > 0$, $0 \leq r(x) \leq \frac{1}{2}x^2(1 - x)^{-2}$.

We may apply this observation to estimate the value of the product

$$p_{n,k} = \prod_{i=1}^{k-1}\left(1 - \frac{i}{n}\right)$$

which we encountered in Theorem 1.3.2. Indeed, we have

$$\log p_{n,k} = \sum_{i=1}^{k-1} \log\left(1 - \frac{i}{n}\right) = -\sum_{i=1}^{k-1}\frac{i}{n} - R = -\frac{k(k-1)}{2n} - R$$

where

$$0 \leq R \leq \frac{1}{2}\sum_{i=1}^{k-1}\left(\frac{i}{n}\right)^2\left(1 - \frac{i}{n}\right)^{-2}$$

$$\leq \left(1 - \frac{k}{n}\right)^{-2}\frac{k(k-1)(2k-1)}{12n^2}$$

Here we have used the result of Problem 1.61 to evaluate the summation of i and the summation of i^2.

In the birthday problem of Example 1.3.4b, where $n = 365$ and $k = 25$, we find that $\log p_{n,k} = -0.8219 - R$, where $0 \leq R \leq 0.0212$. That is, $\exp(-0.8431) \leq p_{n,k} \leq \exp(-0.8219)$.

An even better estimate of $p_{n,k}$ can be obtained by taking an additional term in the Taylor series expansion of $\log(1 - x)$.   ////

It is clear from (7.1) that if f has derivatives of all orders, and if

$$\lim \frac{1}{n!} f^n(x_1)(x - x_0)^n = 0$$

as $n \to \infty$ for every $x \in (a,b)$, then we can write f as a power series

$$f(x) = \sum_{k=0}^{\infty} \alpha_k(x - x_0)^k \qquad (7.2)$$

for $x \in (a,b)$, where $\alpha_0 = f(x_0)$ and $\alpha_k = f^k(x_0)/k!$ for $k = 1, 2, \ldots$. We shall call (7.1) and (7.2) *the finite and infinite Taylor series expansions of f about x_0*, respectively. Equation (7.2) is especially useful in the evaluation of infinite series.

EXAMPLE 1.7.3

a Let $f(x) = e^x$ for $-\infty < x < \infty$. Then $f^j(x) = e^x$ for all x and all $j \geq 0$. Let us expand f in an infinite Taylor series about $x_0 = 0$. Observe first that

$f^j(0) = e^0 = 1$ for all $j \geq 0$. Moreover, if $|x_1| \leq |x|$, then $|f^n(x_1)x^n/n!| \leq |x^n|e^{|x|}/n!$, which tends to zero as $n \to \infty$ for any x. Therefore,

$$e^x = \sum_{j=0}^{\infty} \frac{1}{j!} x^j \qquad (7.3)$$

for all x, $-\infty < x < \infty$.

b Similarly, if $f(x) = 1/(1 - x)$ for $-1 < x < 1$, then $f^j(x) = j!/(1 - x)^{j+1}$ for $j = 0, 1, 2, \ldots$. In particular, $f^j(0) = j!$ for $j \geq 0$, and the expansion

$$\frac{1}{1 - x} = \sum_{j=0}^{\infty} x^j \qquad (7.4a)$$

for $-1 < x < 1$ can be deduced from Taylor's theorem.

c A useful extension of part *b* is the following: for $-1 < x < 1$ and $r \geq 0$

$$\sum_{j=r}^{\infty} x^j = x^r \sum_{j=r}^{\infty} x^{j-r} = x^r \sum_{k=0}^{\infty} x^k = \frac{x^r}{1 - x} \qquad (7.4b)$$

////

Equations (7.3) and (7.4a) are known as the *exponential* and *geometric series*, respectively. We shall encounter them again from time to time.

Another useful Taylor series expansion requires the generalization of the binomial coefficients. If α is any real number, let $(\alpha)_0 = 1$ and define

$$(\alpha)_k = \alpha(\alpha - 1) \cdots (\alpha - k + 1) \qquad k \geq 1 \qquad (7.5a)$$

$$\binom{\alpha}{k} = \frac{(\alpha)_k}{k!} \qquad k = 0, 1, 2, \ldots \qquad (7.5b)$$

Then, for any real α, the Taylor series expansion of the function $f(x) = (1 + x)^\alpha$ about the point $x_0 = 0$ is[1]

$$(1 + x)^\alpha = \sum_{k=0}^{\infty} \binom{\alpha}{k} x^k \qquad -1 < x < 1 \qquad (7.6)$$

Equation (7.4a) is a special case. That the right side of (7.6) is the formal Taylor series expansion of $(1 + x)^\alpha$ is easily verified by differentiation.

The numbers $\binom{\alpha}{k}$ defined in (7.5b) are known as *generalized binomial coefficients*, and (7.6) is known as the *generalized binomial theorem*.

---

[1] For a proof that the series converges and is equal to $(1 + x)^\alpha$ for $-1 < x < 1$, see Apostol (1957), pp. 420–421.

1.8   STIRLING'S FORMULA

We have seen that many interesting probabilities can be expressed in terms of the notation $n! = n(n - 1) \cdots 1$. It is clear that for large values of n the exact computation of $n!$ is a formidable task. In this section we shall give an approximation to $n!$ which is valid when n is large. The result is known as *Stirling's formula*.

In the statement of Stirling's formula, we shall use the following notation. If $a_1, a_2, \ldots$ and $b_1, b_2, \ldots$ are two infinite sequences of positive real numbers, then we shall write $a_n \sim b_n$ if and only if $\lim a_n b_n^{-1} = 1$ as $n \to \infty$, and in this case we shall say that a_n is *asymptotic to* b_n. This notation is useful in cases where both a_n and b_n tend to zero or infinity as $n \to \infty$.

Stirling's formula may now be stated as follows.

Theorem 1.8.1 $n! \sim \sqrt{2\pi} \, n^{n+\frac{1}{2}} e^{-n}$ as $n \to \infty$.

In fact, it is possible to give sharp inequalities which relate $n!$ and $\sqrt{2\pi} \, n^{n+\frac{1}{2}} e^{-n}$.

Theorem 1.8.2 $\sqrt{2\pi} \, n^{n+\frac{1}{2}} e^{-n} < n! < \sqrt{2\pi} \, n^{n+\frac{1}{2}} e^{-n}(1 + 1/(12n - 1))$ *for every* $n \geq 1$.

We defer the proof of Theorem 1.8.1 to Section 5.4.1, and we omit the proof of Theorem 1.8.2.[1]

Thus, the relative error incurred by using Stirling's formula,

$$\frac{\sqrt{2\pi} \, n^{n+\frac{1}{2}} e^{-n} - n!}{n!}$$

is positive and at most $1/(12n - 1)$. For $n \geq 9$, this is less than 0.01.

EXAMPLE 1.8.1   If a fair coin is tossed $2n$ times, the probability that exactly n heads will result is $\binom{2n}{n} 4^{-n}$ by Theorem 1.5.1. By Stirling's formula, we have

$$\binom{2n}{n} 2^{-2n} = \frac{(2n)!}{n! \, 2 2^{2n}}$$

$$\sim \frac{\sqrt{2\pi}(2n)^{2n+\frac{1}{2}} e^{-2n}}{(\sqrt{2\pi} \, n^{n+\frac{1}{2}} e^{-n})^2 2^{2n}} = \frac{1}{\sqrt{\pi n}} \qquad (8.1)$$

as $n \to \infty$. For example, the probability that 100 tosses of a fair coin will produce exactly 50 heads is approximately 0.08.

[1] For a proof of Theorem 1.8.1, see Feller (1968), pp. 52–54.

It is interesting to remark that the last line in (8.1) tends to zero as $n \to \infty$. That is, in many tosses of a fair coin, we should *not* expect the coin to turn up heads exactly half of the time. ////

REFERENCES

References are given in full in Appendix D.

The history of the theory of probability is discussed by Todhunter (1865) and David (1962). A series of articles in *Biometrika*, starting in 1955, treat aspects of the more recent history of probability theory.

A more extensive treatment of combinatorial analysis will be found in Riorden (1958). Chapters 2 and 3 of Feller (1968) contain some additional combinatorial analysis and some additional applications of combinatorial analysis to probability theory.

PROBLEMS

1.1 Give a careful definition of an appropriate sample space for each of the following games.

(*a*) A balanced coin is tossed twice; a balanced coin is tossed three times.

(*b*) A balanced die is rolled three times.

(*c*) Two distinct cards are selected sequentially from a standard deck of 52 cards.

(*d*) One card is selected from each of two standard decks.

In each case the sample space should be so selected that the outcomes may be assumed equally likely.

1.2 Give the number of possible outcomes for each of the games described in Problem 1.1.

1.3 If two distinguishable, balanced dice are rolled, what is the probability that the sum of spots on the two dice will be 5? What is the probability that the difference (larger less smaller) will be 2?

1.4 If a balanced coin is tossed three times, what is the probability (*a*) that there will be 2 or more consecutive heads; (*b*) that there will be at least 2 heads?

1.5 If a man has 3 hats, 4 shirts, 4 pairs of trousers, and 2 pairs of shoes, in how many ways can he dress?

1.6 How many 4-letter words can be formed from the English alphabet if we allow any string of 4 letters as a word and regard words as identical if and only if they list the same letters in the same order?

1.7 How many 4-letter words can be formed from the English alphabet if we require:

(*a*) The second letter to be a vowel?

(*b*) Exactly one vowel?

(*c*) At least one vowel?

Here, by definition, a vowel is any of the letters a, e, i, o, or u.

1.8 (*a*) How many 7-digit telephone numbers can be formed? (*b*) Of these, how many contain distinct digits?

1.9   A certain electronic device contains 100 circuits, each of which may be either open or closed. The state of the system is defined to be the vector $(x_1, \ldots, x_{100})$, where $x_i = 1$ or 0 accordingly as the ith circuit is open or closed, $i = 1, \ldots, 100$. How many states are there?

1.10   The ABC Pizza Parlor lists 10 items such as mushrooms or pepperoni,... which may be added to a pizza. If a customer wants 2 additional items, how many choices has he?

1.11   A certain questionnaire lists 10 questions with the possible answers yes or no to each question:

(a) In how many ways can the questionnaire be answered?

(b) In how many ways can the questionnaire be answered with 5 yeses and 5 noes?

1.12   In Problem 1.11, suppose that each questionnaire may be answered by yes, no, or no opinion. In how many ways can the questionnaire be answered with 4 yeses, 4 noes, and 2 no opinion?

1.13   A medical researcher wishes to compare two new drugs and has 20 indistinguishable mice with which to experiment. In how many ways can the 20 mice be divided into two groups of 10?

1.14   In how many ways can a committee of size 4 be chosen from a group of 10:

(a) If all committee members are to have the same status?

(b) There is to be a chairman and 3 others of equal status?

1.15   In how many ways can poker hands be dealt to (a) 2 distinguishable people; (b) 3 distinguishable people?

1.16   Show that $\dbinom{n-1}{k-1} + \dbinom{n-1}{k} = \dbinom{n}{k}$ for $1 \le k \le n$. Interpret your result in terms of combinations.

1.17   Use Problem 1.16 to prove the binomial theorem by mathematical induction.

1.18   Derive the following identities from the binomial theorem:

$$\binom{n}{0} + \binom{n}{1} + \cdots + \binom{n}{n} = 2^n$$

$$\binom{n}{0} - \binom{n}{1} + \cdots \pm \binom{n}{n} = 0$$

$$\binom{n}{1} + 2\binom{n}{2} + \cdots + n\binom{n}{n} = n2^{n-1}$$

1.19   How many subsets are there in a set of n elements? *Hint:* Part (a) of Problem 1.18.

1.20   If two cards are drawn sequentially without replacement from a standard deck, what is the probability that they are (a) both aces; (b) both spades? What is the probability that they are (c) of the same denomination; (d) of the same suit?

1.21   Let an ordered, random sample of size 5 be drawn from a standard deck of 52 cards. What is the probability that the third card drawn will be (a) an ace; (b) a spade?

1.22   (a) What is the probability that all 7 digits of a telephone number will be distinct?

(b) What is the probability that the last 4 digits will be distinct? (Assume all telephone numbers to be equally likely.)

1.23 (*a*) If cards are selected from each of 5 well-shuffled decks, what is the probability that the 5 cards are all different? (*b*) What is the probability that the 5 cards are ·of different denominations?

1.24 Every day the teacher selects one of her 10 pupils to stay after school and clean the blackboard. Johnny, who was selected twice during the first week of school, feels that the teacher is persecuting him. Is it "unusual" that a student should be selected twice during the same 5-day week?

1.25 If a balanced die is rolled 7 times, what is the probability that each face will appear at least once?

1.26 What is the probability that a bridge hand will contain (*a*) exactly 2 aces; (*b*) at least 2 aces?

1.27 What is the probability that a bridge hand will contain (*a*) 8 spades; (*b*) 8 cards of the same suit?

1.28 What is the probability that a bridge hand will contain one each of the 13 denominations?

1.29 (*a*) What is the probability that a bridge hand will contain 4 spades, 3 hearts, 3 diamonds, and 3 clubs? (*b*) What is the probability that a bridge hand will contain 4 cards of one suit and 3 each of the other three suits?

1.30 (*a*) What is the probability that a bridge hand will contain no aces? (*b*) What is the probability that a bridge hand will contain no hearts?

1.31 What is the probability that a poker hand will contain exactly 2 cards of one denomination (a pair) and cards of three different denominations?

1.32 If a committee of size 3 is selected from a group of 6 Democrats and 4 Republicans, what is the probability that the committee will contain (*a*) two Democrats and one Republican; (*b*) more Democrats than Republicans?

1.33 The Senate Committee on Randomness consists of 6 members of party A and 4 members of party B, but the chairman is a member of party B. Recently, the chairman appointed a subcommittee of size 3 which consisted of 2 members of party B and 1 of of party A. The chairman claims to have selected the subcommittee by lot from the 10 committee members. The leader of party A, however, claims that the composition of the subcommittee proves bias beyond a reasonable doubt. Is the leader of party A justified in his claim?

1.34 Sebastian, a magician, claims to have extrasensory perception. In order to test this claim, he is asked to identify the 4 red cards out of 4 red and 4 black cards which are laid face down on the table. Sebastian correctly identifies 3 of the red cards and incorrectly selects 1 of the black cards. Thereafter, he claims to have proved his point. What is the probability that Sebastian would have correctly identified at least 3 of the red cards if he were, in fact, guessing? (Regard the 4 cards selected by Sebastian as an unordered random sample of size 4.)

1.35 A box contains 8 good and 2 defective items. If 5 items are selected at random from the box, what is the probability of finding (*a*) at least 1 of the defective items; (*b*) both defective items?

1.36 In Example 1.4.2, suppose that the lot size is 50 and that it is unprofitable to market lots containing more than 2 defectives. How should k be chosen for the probability of marketing a bad lot to be at most 0.1?

1.37 In Example 1.4.2, show that $q(k,m)$ is a decreasing function of m. *Hint:* Compute $q(k, m + 1) - q(k,m)$.

1.38 Compute and graph the hypergeometric probabilities p_r as a function of r, for
(a) $m = k = 4$ and $n = 8$;
(b) $k = 4$ and $m = n - m = 8$.

1.39 Compute and graph the binomial probabilities $\binom{k}{r} 2^{-k}$ as a function of r for
(a) $k = 4$;
(b) $k = 6$;
(c) $k = 8$.

1.40 If a balanced die is rolled 5 times, what is the probability that exactly 2 of the rolls will produce either 1 or 6 spots?

1.41 Let an ordered random sample be drawn without replacement from a standard deck.
(a) If the sample size is $k = 5$, what is the probability that the sample will contain exactly 2 spades?
(b) What is the probability that the first spade will appear on the fifth draw?

1.42 Repeat Problem 1.41 for sampling with replacement.

1.43 If an ordered random sample of size 5 is drawn without replacement from a standard deck, what is the probability that the second spade will appear on the fifth draw?

1.44 A box contains 6 fuses, 2 of which are defective. If the fuses are inspected in a random order, what is the probability of finding the first defective fuse (a) on the third test; (b) on or before the third test; (c) after the third test?

1.45 Repeat Problem 1.44 with first defective fuse replaced by second defective fuse.

1.46 Which is more probable: obtaining at least 1 six in 6 tosses of a fair die or obtaining at least 2 sixes in 12 tosses of a fair die?

1.47 Let a sample of size $k = 4$ be drawn from an urn which contains 4 red balls and 4 white balls. Is it more probable that all balls drawn will be red if the sampling is with replacement or without replacement?

1.48 Sebastian, a magician, calls heads or tails before each of four tosses of a fair coin. If he is in fact guessing, what is the probability that Sebastian will correctly call (a) all 4; (b) at least 3 of the tosses? Compare your answers with the answer to Problem 1.34.

1.49 If 4 balls are placed in 4 cells according to Bose-Einstein statistics, what is the probability that the first cell will contain (a) exactly 1 ball; (b) exactly 2 balls; (c) at least 1 ball?

1.50 If 6 balls are placed in 4 cells according to Bose-Einstein statistics, what is the probability (a) that every cell is occupied; (b) that at least 3 cells are occupied?

1.51 Repeat Problems 1.49 and 1.50 for Fermi-Dirac statistics.

1.52 If k particles are placed in n cells according to Bose-Einstein statistics, what is the probability that a given subregion, consisting of m cells say, will contain exactly r particles?

1.53   Let k indistinguishable balls be placed in n cells according to Bose-Einstein statistics, and suppose that the cells are labeled by the integers $1, \ldots, n$. What is the probability that the index of the largest occupied cell is m, where $m < n$?

1.54   Write out a proof of Theorem 1.6.1 in the terminology of Section 1.6.

1.55   Derive the following identity for $-1 < x < 1$:

$$\sum_{n=1}^{\infty} \frac{1}{n} x^n = -\log (1 - x)$$

1.56   Find the infinite Taylor series expansions of

$$\cosh x = \frac{e^x + e^{-x}}{2} \quad \text{and} \quad \sinh x = \frac{e^x - e^{-x}}{2}$$

about $x_0 = 0$.

1.57   Show that $e^x \leq 1 + x$ for every x, $-\infty < x < \infty$. *Hint:* Use Example 1.7.1*b*.

1.58   Show that $|\log (1 + x) - x| \leq x^2$ for $-\frac{1}{2} < x < \frac{1}{2}$.

1.59   Show that $\binom{2n}{n} = \binom{-\frac{1}{2}}{n} (-4)^n$ for positive integers $n = 1, 2, \ldots$.

1.60   Evaluate the series $\sum_{n=0}^{\infty} \binom{2n}{n} x^n$ for $-\frac{1}{4} < x < \frac{1}{4}$.

1.61   Show that $\sum_{i=1}^{k} i = \frac{1}{2}k(k + 1)$ and that $\sum_{i=1}^{k} i^2 = \frac{1}{6}k(k + 1)(2k + 1)$ for $k \geq 1$.

1.62   Show that $\exp\left[-k(k - 1)/2(n - k)\right] \leq (n)_k n^{-k} \leq 1$ for $0 \leq k < n$.

1.63   Use Problem 1.62 to derive the following comparison between the hypergeometric and binomial probabilities:

$$\exp\left[-\frac{1}{2} \frac{r(r - 1)}{2(m - r)} - \frac{1}{2} \frac{(k - r)(k - r - 1)}{n - m - k + r} \right] \binom{k}{r} p^r q^{k-r}$$

$$\leq \binom{m}{r} \binom{n - m}{k - r} \binom{n}{k}^{-1}$$

$$\leq \exp\left[\frac{k(k - 1)}{2(n - k)} \right] \binom{k}{r} p^r q^{k-r}$$

1.64   Use Stirling's formula to estimate the number of bridge hands $\binom{52}{13}$.

1.65   Use Stirling's formula to estimate the number of ways that a bridge deck can be partitioned into 4 distinct hands $\binom{52}{13,13,13,13}$.

1.66   An ordered sample of size n is drawn at random and with replacement from an urn containing n distinct balls. Use Stirling's formula to estimate the probability that all n balls are drawn for $n = 10$, 15, and 20.

AXIOMATIC PROBABILITY

2.1 PROBABILITY, FREQUENCY, AND DEGREE OF BELIEF

The classical model presented in Chapter 1 is not flexible enough to encompass many examples which are interesting from both the mathematical and practical points of view. In particular, it cannot be used to describe experiments for which there are infinitely many possible outcomes. In this chapter, we shall develop a more general and flexible model which starts with axioms stating how probabilities should behave and allows various interpretations of the results derived from them. In this section we shall attempt to motivate these axioms and to elucidate various interpretations of the elements of our model.

We begin by examining the meaning of the term "probability" and such related terms as "chance" and "likely." Actually, they may have several meanings, two of which will be of special interest to us. First, they are used by all of us to express our *subjective opinion* or *degree of belief*. For example, statements such as "it will probably rain tomorrow," "he will probably be late," and "the chances that the Mets will win the pennant are about 1 in 3" all express the subjective opinion or degree of belief of the speaker. On the other hand, the term "probability" often denotes *frequency of occurrence*. For example, if a scientist were to report that the probability of curing a particular type of cancer in mice is 0.6, he might well mean that a large number of mice had been treated and of those approximately 60 percent had been cured. The two usages are not mutually exclusive, since one's subjective opinion may be based on past experience with frequencies, but they are distinct and warrant separate consideration.

In considering the two usages, it will be convenient to have some uniform terminology which will apply to both. Thus, consider a variable X whose exact value is unknown to us, and suppose that we can specify a set S in which X must lie. The variable X may represent the outcome of some experiment or game of chance, or it may simply represent some aspect of nature about which we are uncertain. As in the previous chapter, we shall call S the *sample space* and refer to subsets $A, B, \ldots$ of S as *events*. Further, we shall say that the event A *occurs* if and only if $X \in A$.

EXAMPLE 2.1.1

a Games of chance (as in Chapter 1)   Let X denote the number of spots which appear when two fair dice are cast, or let X denote the poker hand dealt a particular player.

b Sampling experiments (as in Chapter 1)   Let X denote the number of defectives found when a lot of manufactured items is examined.

c Scientific experiments   Let X denote the number of particles emitted from a given radioactive substance during a given time interval; or let X denote the number of mice which contract cancer when a group of mice is exposed to cigarette smoke.

d Engineering problems   Let X denote the demand on electricity in New York City on a given day; or let X denote the maximum weight on the George Washington Bridge during a given year.

e Actuarial problems   Let X denote the length of life anticipated for a given man who has just applied for life insurance.

f Uncertainty   Let X denote the exact date of Noah's birth; or let X denote next week's closing Dow-Jones industrial average.   ////

Let us first consider the *frequentistic interpretation* of the term "probability." Here we require X to be the outcome of some game or experiment which may be repeated as often as desired under the same set of relevant experimental conditions. If the game or experiment is so repeated, say n times, and if A is an event, then we can compute the relative frequencies

$$f_n(A) = \frac{1}{n} \times \text{ number of repetitions on which } A \text{ occurs}$$

with which A occurs. That is, $f_n(A)$ is the ratio of the number of times A occurs to the total number of repetitions of the experiment. Now, it is an *empirical fact* that for many types of games and experiments, the relative frequencies $f_n(A)$ tend to stabilize as n increases. That is, they act as if they were approaching limits as $n \to \infty$. The frequentistic interpretation of "probability" defines the probability of A to be

$$P(A) = \lim_{n \to \infty} f_n(A) \qquad (1.1)$$

where the existence of the limit is assumed. (The existence of the limit cannot be proved, for we are not dealing with a purely mathematical subject.)

Thus, according to the frequentistic interpretation of "probability," the probability of an event is determined by the event and the set of experimental conditions. It is independent of the observer and can be determined to an increasing degree of accuracy by simply repeating the experiment to which the event refers often enough and computing the sequence of relative frequencies. For this reason, the frequentistic interpretation of "probability" is sometimes called the *objective interpretation*.

EXAMPLE 2.1.2 A coin is tossed 10,000 times, producing the results shown in Table 2. From the frequentistic point of view, these results are consistent with the hypothesis that the probability of heads on any given toss is $\frac{1}{2}$. ////

Now suppose that we have two events A and B, and suppose that A and B are disjoint; that is, $AB = \emptyset$. Then

$$f_n(A \cup B) = f_n(A) + f_n(B)$$

for every $n = 1, 2, \ldots$. Thus, letting $n \to \infty$, we find that

$$P(A \cup B) = P(A) + P(B) \qquad (1.2)$$

That is, if probabilities are defined by (1.1), they must satisfy condition (1.2) whenever A and B are disjoint events.

In Chapter 1, we used the term "equally likely" without giving a precise definition.

Table 2

No. of tosses	No. of heads	Relative frequency
100	46	0.460
500	239	0.478
1000	495	0.495
5000	2529	0.506
10000	5049	0.505

We can now give such a definition from the point of view of the frequentistic interpretation. If S is a finite set, then the outcomes $s \in S$ are equally likely if the events
$\{s\}$ will occur with approximately the same relative frequency after many repetitions
of the game or experiment under consideration. That is, the outcomes are equally
likely if $f_n(\{s\})$ all converge to the same limit $P(\{s\}) = c$ for all $s \in S$. Equation (1.2)
then requires[1] that $P(A) = |A|/|S|$ for $A \subset S$. Thus, the model of Chapter 1 is
applicable, and the results of Chapter 1 now admit the following frequentistic interpretation. If the outcomes $s \in S$ are equally likely, and if A is any event whose
probability was computed to be $P(A) = p$ in Chapter 1, then the relative frequency
$f_n(A)$ with which A will occur will be approximately p after many repetitions of the
game or experiment under consideration.

Let us now consider the *subjective interpretation* of the term "probability."
Here a problem presents itself immediately, for most subjective probability statements
are qualitative (for example, "it will probably rain tomorrow"), not quantitative.
If we wish to fit a subjective interpretation into a mathematical theory of probability,
we shall need a method for quantifying subjective probability statements. One way
to do so is to relate them to betting odds, and this is the approach we shall follow.
Let A be an event, and let G denote the following gamble:

1   One pays p units to play.
2   One receives 1 unit if A occurs and nothing if A does not occur.

Equivalently, the gamble can be described by saying that one wins $1 - p$ units if A
occurs and one loses p units if A does not occur. We shall say that G offers *odds* of
$1 - p$ to p on the occurrence of A.[2]

Let us agree to say that a person regards the gamble G as *fair* if he is indifferent
to the two sides to G. That is, the person regards G as fair if and only if he would as
soon win $1 - p$ units if A occurs and lose p units if A does not occur as win p units
if A does not occur and lose $1 - p$ units if A does occur.

We now adopt the following definition of subjective probability. If there is a
unique value of p, $0 \leq p \leq 1$, for which a person regards the gamble G as fair, then
we shall say that that person's subjective probability for A is $P(A) = p$. Observe that
subjective probabilities are determined by the observer and are influenced by the
event itself only insofar as the observer is well informed about it. *Two different
people may assign different subjective probabilities to the same event, even if they have
access to the same information.*

---

[1] Here we anticipate the result of Theorem 2.3.3; see Example 2.3.5.
[2] The units here should be taken to be amounts of money small compared to one's
total resources. We wish to avoid, for example, the possibility that the loss of a
unit would result in bankruptcy.

Now suppose that a person has two events A and B to which he has assigned subjective probabilities $P(A) = p$ and $P(B) = q$, and suppose also that A and B are disjoint. Then by hypothesis he regards both the following bets as fair.

1   One pays p units to play and receives 1 unit if and only if A occurs.
2   One pays q units to play and receives 1 unit if and only if B occurs.

If he were to place both bets, he would pay $p + q$ units to play and since A and B are disjoint, he would receive 1 unit if either A or B occurred (and nothing otherwise). Since the new bet is formed by placing two fair bets, it seems reasonable that he should consider it to be fair. That is, it seems reasonable that he should assign subjective probability

$$P(A \cup B) = p + q = P(A) + P(B) \qquad (1.3)$$

to the event $A \cup B$. We say that a person's subjective probabilities are *consistent* if and only if they satisfy (1.3) whenever A and B are disjoint. It can be shown (see Problems 2.4 to 2.6) that a person with inconsistent subjective probabilities can be led to accept bets on the conjunction of which he must certainly lose money. Thus, we restrict our attention to consistent subjective probabilities.

The subjective meaning of the term "equally likely" should now be clear. If S is a finite set, one regards the outcomes $s \in S$ as equally likely if and only if one assigns the same subjective probability $P(\{s\}) = c$ to each event $s \in S$. As above, Equation (1.3) then requires that one assign subjective probability $P(A) = |A|/|S|$ to each event $A \subset S$, so that the model of Chapter 1 is applicable. Moreover, the results of Chapter 1 now admit the following subjective interpretation. If A is an event whose probability was computed to be $P(A) = p$ in Chapter 1, and if one regards the outcomes of the game to which A refers as equally likely, then, in order to be consistent in one's beliefs, one must assign subjective probability $P(A) = p$ to A.

2.2   A MATHEMATICAL MODEL

In this section we present a mathematical model which is general enough to encompass the two interpretations of probability presented in Section 2.1 and sufficiently flexible to allow the derivation of a useful mathematical theory. Our model will consist of the following basic elements:

1   A nonempty set S called the *sample space*.
2   A class $\mathscr{S}$ of subsets of S, the elements of which will be called *events*.
3   For every event $A \in \mathscr{S}$ a real number $P(A)$ which we call the *probability* of A.

That is, we require a real-valued function P which is defined on the class $\mathscr{S}$ of events.

The sample space S, events $A \subset S$, and probability P can all be interpreted as in the previous section. That is, S can be regarded as the set of possible outcomes of some game or experiment; an event A is said to occur if and only if the outcome of the game or experiment is an element of A; and $P(A)$ can be regarded as either the frequentistic or subjective probability of the event A.

In many examples the class $\mathscr{S}$ will consist of all subsets of S, but in others $\mathscr{S}$ will be a proper subclass of the class of all subsets of S. Henceforth, we shall say that a subset $A \subset S$ is an *event* if and only if $A \in \mathscr{S}$. We shall have to perform certain set-theoretic operations with events, such as the formation of complements, unions, and intersections, and we shall require the class $\mathscr{S}$ to be closed with respect to these operations.

We impose three requirements on $\mathscr{S}$:

1 The sample space S and the empty set $\varnothing$ must be events. We shall call S the *sure event* and $\varnothing$ the *impossible event*.

2 If A is an event, then the complement $A' = S - A$ is also an event. We shall call A' *the event that A does not occur*.

3 If $A_1, A_2, \ldots$ is a finite or infinite sequence of events, then the union $\cup A_i$ and intersection $\cap A_i$ are also events. We shall call the union (intersection) *the event that A_i occurs for some i (for all i)*.

A class $\mathscr{S}$ of subsets of S will be called a σ *algebra* of subsets of S if and only if it satisfies conditions 1, 2, and 3.

EXAMPLE 2.2.1

a The class of all subsets of a nonempty set S is a σ algebra since conditions 1 to 3 are trivially satisfied in this case.

b If S is an interval of real numbers, then there is a smallest σ algebra of subsets of S which contains all subintervals of S (see Problems 2.21 and 2.22). This σ algebra is known as *the class of Borel sets*, and its elements are known as *Borel sets*.

The relevant properties of the class of Borel sets are the following:

1 Every subinterval of S is a Borel set.

2 The class of Borel sets is closed with respect to the formation of complements and the formation of unions and intersections of finite or infinite sequences of its members.　　　　　　　　　////

Probability theory has developed its own name for several set-theoretic relations between events. We shall say that events $A_1, A_2, \ldots$ are *mutually exclusive* if they are disjoint, that is, $A_i A_j = \varnothing$ for $i \neq j$. We shall say that events $A_1, A_2, \ldots$ are *exhaustive* if their union is the entire sample space S, that is, if $\cup A_i = S$. Finally, we shall say that the event A *implies* the event B if A is a subset of B, $A \subset B$.

Let us also record the *De Morgan laws*: if $A_1, A_2, \ldots$ are events, then

$$(\cup A_i)' = \cap A_i' \quad \text{and} \quad (\cap A_i)' = \cup A_i'$$

See Appendix A for a derivation.

Let us now consider the function P. What properties might we reasonably demand of P? First, we wish probabilities to be numbers between 0 and 1, and we wish certainly to imply a probability of 1. Thus, we shall require

$$0 \leq P(A) \leq 1 \quad \text{and} \quad P(S) = 1 \qquad (2.1)$$

for $A \in \mathscr{S}$. Moreover, we saw in Section 2.1 that within either the subjective or frequentistic interpretation of probability we should have

$$P(A \cup B) = P(A) + P(B) \qquad (2.2)$$

whenever A and B are mutually exclusive events. Thus, we shall require conditions (2.1) and (2.2).

Conditions (2.1) and (2.2) work splendidly if S is a finite set, as in Chapter 1, but they do not lead to a sufficiently rich mathematical theory if S is infinite. We are therefore led to introduce the following strengthened version of (2.2): if $A_1, A_2, \ldots$ is an infinite sequence of mutually exclusive events, then

$$P\left(\bigcup_{i=1}^{\infty} A_i\right) = \sum_{i=1}^{\infty} P(A_i) \qquad (2.3)$$

Condition (2.3) does imply (2.2) in general and is equivalent to (2.2) if S is a finite set (see Problems 2.17 and 2.18). In any case, we shall adopt (2.3) as an axiom. Accordingly, we define a *probability measure* to be a function P which is defined on a σ algebra $\mathscr{S}$ and satisfies conditions (2.1), (2.2), and (2.3).

We can now define our mathematical model for probability. We define a *probability space* to be ordered triple $(S, \mathscr{S}, P)$, where S is a nonempty set, $\mathscr{S}$ is a σ algebra of subsets of S, and P is a probability measure defined on $\mathscr{S}$. A probability space may be regarded as a model for an experiment or game of chance with the convention that S represents the set of possible outcomes of the experiment or game, $\mathscr{S}$ represents the class of observable events, and, for each $A \in \mathscr{S}$, $P(A)$ is the probability that the event A will occur. Probability spaces form the basis for the theory of probability to be presented in this book.

EXAMPLE 2.2.2   Discrete probability spaces   Let $S = \{s_1, s_2, \ldots\}$ be a finite or countably infinite[1] set, and let f be a real-valued function which is defined on S and satisfies[2]

$$f(s) \geq 0 \qquad \text{for all } s \in S \qquad \text{and} \qquad \sum_S f(s) = 1 \qquad (2.4)$$

Then we can define a function P on the class $\mathscr{S}$ of all subsets of S by letting

$$P(A) = \sum_A f(s) \qquad (2.5)$$

for every $A \subset S$. We have $P(A) \geq 0$ and $P(A) \leq P(S)$ for every A since $f(s) \geq 0$ for every s, and we have $P(S) = 1$ by (2.4). Thus, condition (2.1) is satisfied. Moreover, if A and B are disjoint, then

$$\begin{aligned} P(A \cup B) &= \sum_{A \cup B} f(s) \\ &= \sum_A f(s) + \sum_B f(s) = P(A) + P(B) \end{aligned}$$

Thus, condition (2.2) is satisfied, and similarly, condition (2.3) is also satisfied. Thus, P is a probability measure, and $(S, \mathscr{S}, P)$ is a probability space.

Taking $A = \{s\}$ in (2.5) yields $P(\{s\}) = f(s)$ for $s \in S$. Thus, $f(s)$ gives the probability that the outcome of the game or experiment under consideration will be s.

////

EXAMPLE 2.2.3

a   If S is a finite set, and if $f(s) = 1/|S|$ for all $s \in S$, then (2.5) yields $P(A) = |A|/|S|$ for $A \subset S$. Thus, the classical model of Chapter 1 is a special case of Example 2.2.2.

b   Consider an experiment in which a coin is tossed until a head appears and the total number of tosses is recorded. We can describe the outcome of the experiment by a positive integer (the number of tosses required), and we therefore take S to be the set of all positive integers $S = \{1, 2, \ldots\}$. Moreover, in Example 5.1.1*a* we showed that the probability that the first head appears on the sth toss is simply 2^{-s}. Let $f(s) = 2^{-s}$. Then

$$\sum_S f(s) = \sum_{s=1}^{\infty} 2^{-s} = 1$$

---

[1] A set is called *countably infinite* if there is a one-to-one correspondence between S and the set of positive integers $Z = \{1, 2, \ldots\}$.

[2] The notation $\Sigma_S f(s)$ means that the numbers $f(s)$, $s \in S$, are added. This may be a finite sum if S is finite or an infinite series if S is countably infinite.

by Equation (7.4) of Chapter 1, so that condition (2.4) is satisfied. We now define a probability measure by (2.5) to obtain a probability space to represent the experiment. If, for example, we wish to compute the probability that an even number of tosses will be required, we find the probability of the event $A = \{2,4,\ldots\}$. By (2.5) and (7.4) of Chapter 1, this is

$$P(A) = \sum_A 2^{-s} = \sum_{k=1}^{\infty} 2^{-2k} = \tfrac{1}{4}(1 - \tfrac{1}{4})^{-1} = \tfrac{1}{3} \qquad ////$$

More examples of discrete probability spaces will be found in the problems at the end of this chapter. Let us now consider an example of a different nature.

EXAMPLE 2.2.4   Absolutely continuous probability spaces   Let S be a finite or infinite interval of real numbers, and let f be a real-valued function defined on S for which

$$f(s) \geq 0 \qquad \text{for all } s \in S \qquad \text{and} \qquad \int_S f(s) \, ds = 1 \qquad (2.6)$$

By analogy with (2.5), it seems natural to define a probability measure P by

$$P(A) = \int_A f(s) \, ds \qquad (2.7)$$

The problem here is the class $\mathscr{S}$ of events. It is simply not true that the integral on the right side of (2.7) will exist as a proper or improper integral for every $A \subset S$. It is possible, however, to define a probability measure P on the class $\mathscr{S}$ of Borel sets of S (Example 2.2.1b) in such a manner that (2.7) holds whenever A is a subinterval of S. Moreover, the probability measure P is uniquely determined by (2.7) and conditions (2.1), (2.2), and (2.3).

Probability spaces $(S,\mathscr{S},P)$ for which S is an interval, $\mathscr{S}$ is the class of Borel sets of S, and P is of the form (2.7) are called *absolutely continuous*. For such spaces, the probability of a subinterval A of S is given by (2.7), and the probability of more complicated events must be deduced from (2.7) and the axioms of probability (2.1), (2.2), and (2.3). $\qquad ////$

EXAMPLE 2.2.5   Consider an experiment in which a number is drawn from the unit interval $S = [0,1]$ in such a manner that the probability that the number lies in a subinterval of S is equal to the length of the subinterval. Taking $f(s) = 1$, $0 \leq s \leq 1$, in (2.7) yields $P(A) = $ length of A, so that the above discussion guarantees the existence of a probability space to represent our experiment. Let us compute, for example, the probability that the number drawn will be a rational number. That is, let us compute $P(R^{\#})$, where $R^{\#}$ denotes the set of rational numbers in S. We write[1]

$$R^{\#} = \{r_1, r_2, \ldots\} = \bigcup_{n=1}^{\infty} A_n$$

where $A_n = \{r_n\}$ is the set whose only element is r_n, $n = 1, 2, \ldots$. Now each A_n is an interval of length 0, so that

$$P(A_n) = \int_{A_n} ds = r_n - r_n = 0$$

for $n = 1, 2, \ldots$. It now follows from (2.3) that

$$P(R^{\#}) = \sum_{n=1}^{\infty} P(A_n) = \sum_{n=1}^{\infty} 0 = 0$$

That is, the probability that the number drawn will be a rational number is zero.   ////

2.3   SOME ELEMENTARY CONSEQUENCES OF THE FIRST TWO AXIOMS

In this section we shall develop some elementary consequences of the axioms presented in Section 2.2. We shall assume throughout that S is a nonempty set, that $\mathscr{S}$ is a σ algebra of subsets of S, and that P is a function defined on $\mathscr{S}$ which satisfies conditions (2.1) and (2.2). That is, we suppose

$$0 \leq P(A) \leq 1 \quad \text{and} \quad P(S) = 1 \quad (2.1)$$

for all $A \in \mathscr{S}$, and

$$P(A \cup B) = P(A) + P(B) \quad (2.2)$$

whenever A and B are disjoint. Although most later applications of our results will be to the case that $(S, \mathscr{S}, P)$ is a probability space, we make no use of condition (2.3) here, and we shall not assume it. We shall continue to refer to elements of $\mathscr{S}$ as events.

Theorem 2.3.1   *Let A and B be events. If $A \subset B$, then*

$$P(B - A) = P(B) - P(A) \quad (3.1)$$

In particular, $P(A) \leq P(B)$.

PROOF   If $A \subset B$, then we may write $B = A \cup (B - A)$. Since $A(B - A) = \varnothing$, we obtain $P(B) = P(A) + P(B - A)$ by condition (2.2). Equation (3.1) follows immediately. The final assertion of the theorem follows from the fact that $P(B - A) \geq 0$ by (2.1).   ////

[1] The set of rational numbers is countably infinite; see, for example, Rudin (1964), p. 26.

Corollary 2.3.1 *If A is any event, then*

$$P(A') = 1 - P(A) \qquad (3.2)$$

In particular, $P(\varnothing) = 0$.

PROOF   Taking $B = S$ in Equation (3.1), we obtain $P(A') = P(S - A) = P(S) - P(A) = 1 - P(A)$ by (2.1). This establishes (3.2). The final assertion of the corollary follows, since $P(\varnothing) = P(S') = 1 - P(S) = 0$.   ////

The final assertion of Theorem 2.3.1 can be paraphrased as follows: if the event A implies the event B, then the probability of A is less than or equal to the probability of B. We shall see later that this simple remark can be extremely useful. Equations (3.1) and (3.2) are also very useful. We illustrate with some examples.

EXAMPLE 2.3.1

a   If a poker hand is selected at random from a standard deck of 52 cards, what is the probability that the hand will contain at least 1 ace? Let A be the event that the hand contains at least 1 ace. Then A' is the event that the hand contains no aces, so that $P(A') = \binom{4}{0}\binom{48}{5} / \binom{52}{5} = \binom{48}{5} / \binom{52}{5}$ by Theorem 1.4.1. Therefore $P(A) = 1 - P(A') = 1 - \binom{48}{5} / \binom{52}{5}$ by (3.2).

b   What is the probability that the highest denomination in a randomly selected poker hand is a queen (aces high)? Let B be the event that the highest denomination is at most a queen, and let A be the event that the highest denomination is at most a jack. Then the event that the highest denomination is a queen is $C = B - A$. Since A implies B, we have $P(C) = P(B) - P(A)$ by (3.1), and so it will suffice to compute $P(A)$ and $P(B)$. To compute $P(B)$, regard aces and kings as red balls and 2s, 3s, . . . , queens as white balls. Then B is the event that no red balls are drawn in a sample of size 5, so that $P(B) = \binom{44}{5} / \binom{52}{5}$. Similarly, $P(A) = \binom{40}{5} / \binom{52}{5}$. Thus,

$$P(C) = \left[\binom{44}{5} - \binom{40}{5} \right] / \binom{52}{5} \qquad ////$$

Given any two events A and B, we define their *symmetric difference* to be the event that either A occurs or B occurs but not both. That is, we define their symmetric difference to be $A \vartriangle B = (A \cup B) - AB$.

Theorem 2.3.2   *If A and B are events, then*

$$P(A \cup B) = P(A) + P(B) - P(AB) \qquad (3.3)$$

$$P(A \triangle B) = P(A) + P(B) - 2P(AB) \qquad (3.4)$$

PROOF   We may write $A \cup B = A \cup (B - A) = A \cup (B - AB)$. That is, either A or B occurs if and only if either A occurs or B occurs but AB does not. Therefore, since A and $B - AB$ are mutually exclusive, and since AB implies B, we have

$$P(A \cup B) = P(A) + P(B - AB)$$
$$= P(A) + P(B) - P(AB)$$

by Equations (2.2) and (3.1). This establishes (3.3), from which (3.4) follows since $P(A \triangle B) = P(A \cup B) - P(AB)$ by (3.1). $\qquad ////$

EXAMPLE 2.3.2   Of the entering freshman class at a given university, 22 percent take a mathematics course, 29 percent take a science course, and 15 percent take both. If a student is selected at random from the freshman class, what is the probability that he takes either a mathematics course or a science course? What is the probability that he takes either a mathematics course or a science course but not both? Let A be the event that the randomly selected freshman takes a mathematics course, and let B be the event that he takes a science course. Then, we are given that $P(A) = 0.22$, $P(B) = 0.29$, and $P(AB) = 0.15$, and we require the probability of $A \cup B$ and $A \triangle B$. These are $P(A \cup B) = 0.22 + 0.29 - 0.15 = 0.36$ and $P(A \triangle B) = 0.21$ by Equations (3.3) and (3.4) respectively. $\qquad ////$

We present an extension of Equations (3.3) and (3.4) in the next section and conclude this section with an extension of Equation (2.2).

Theorem 2.3.3   *Let $A_1, A_2, \ldots, A_n$ be any events. If $A_1, \ldots, A_n$ are mutually exclusive, then*

$$P\left(\bigcup_{i=1}^{n} A_i\right) = \sum_{i=1}^{n} P(A_i) \qquad (3.5)$$

In any case (even if $A_1, \ldots, A_n$ are not mutually exclusive), we have

$$P\left(\bigcup_{i=1}^{n} A_i\right) \le \sum_{i=1}^{n} P(A_i) \qquad (3.6)$$

PROOF   We shall prove (3.5) by induction on n. If $n = 1$, then (3.5) is trivially true. Now, suppose that (3.5) is true when $n = m \geq 1$, and consider the case that $n = m + 1$. In this case, we find that the events

$$A = \bigcup_{i=1}^{m} A_i \quad \text{and} \quad B = A_{m+1}$$

are mutually exclusive. Therefore,

$$P\left(\bigcup_{i=1}^{m+1} A_i\right) = P(A \cup B) = P(A) + P(B)$$

$$= \sum_{i=1}^{m} P(A_i) + P(A_{m+1}) = \sum_{i=1}^{m+1} P(A_i)$$

by Equation (2.2) and the induction hypothesis. Expression (3.6) can be established by a similar argument which uses Equation (3.3) in place of Equation (2.2).   ////

EXAMPLE 2.3.3   Let an unordered sample of size k be drawn from an urn which contains m red balls and $n - m$ white balls. What is the probability that at least r of the balls drawn will be red? For $j = 0, \ldots, k$, let E_j be the event that exactly j of the balls drawn will be red. Then $E_0, \ldots, E_k$ are mutually exclusive, and

$$P(E_j) = \frac{\binom{m}{j}\binom{n-m}{k-j}}{\binom{n}{k}}$$

for $j = 0, \ldots, k$ by Theorem 1.4.1. Moreover, the event that at least r of the balls drawn are red is simply

$$L_r = \bigcup_{j=r}^{k} E_j$$

Therefore,

$$P(L_r) = \sum_{j=r}^{k} \frac{\binom{m}{j}\binom{n-m}{k-j}}{\binom{n}{k}}$$

by Equation (3.5). The probability that at most r red balls will be drawn is

$$P(M_r) = \sum_{j=0}^{r} \frac{\binom{m}{j}\binom{n-m}{k-j}}{\binom{n}{k}}$$ ////

EXAMPLE 2.3.4 Consider a lottery in which 100,000 tickets are sold, of which 5 win prizes. If a man buys 10 tickets, what is the probability that he will win at least 1 prize? Regard the 10 tickets as an ordered sample without replacement from the 100,000 tickets, and let A_k be the event that the kth ticket wins a prize. Then, $P(A_k) = 0.00005$, $k = 1, \ldots, 10$, and the event that the man wins at least 1 prize is $A = \bigcup_{k=1}^{10} A_k$. Since the events $A_1, \ldots, A_{10}$ are not mutually exclusive, Equation (3.5) is not applicable. However, (3.6) is applicable and yields $P(A) \leq \sum_{k=1}^{10} P(A_k) = 0.0005$.

The exact probability can also be computed. Indeed, by (3.2) and Theorem 1.5.1, we have $P(A) = 1 - P(A') = 1 - (99,995)_{10}/(100,000)_{10}$. Thus, we have both a simple upper bound for the probability in question and a rather complicated expression for its exact value. ////

EXAMPLE 2.3.5 If S is a finite set, if $\mathscr{S}$ is the class of all subsets of S, and if $P(\{s\}) = c$ is the same for all $s \in S$, then $P(A) = |A|/|S|$ for all $A \subset S$. Indeed, if $A \subset S$, then $A = \bigcup_A \{s\}$, where the union extends over all distinct $s \in A$, so that $P(A) = \sum_A P(\{s\}) = c|A|$. Taking $A = S$, we now find that $1 = P(S) = c|S|$, or $c = 1/|S|$. The assertion follows. ////

There are a number of interesting combinatorial identities which follow from Theorem 2.3.3; we list two of them in the following example.

EXAMPLE 2.3.6

a Let $E_0, E_1, \ldots, E_k$ be as described in Example 2.3.3. Then $E_0, E_1, \ldots, E_k$ are both mutually exclusive and <u>exhaustive</u>. Therefore,

$$1 = P(S) = P\left(\bigcup_{j=0}^{k} E_j\right) = \sum_{j=0}^{k} \frac{\binom{m}{j}\binom{n-m}{k-j}}{\binom{n}{k}}$$

That is,

$$\sum_{j=0}^{k} \binom{m}{j}\binom{n-m}{k-j} = \binom{n}{k}$$

b Similarly, if an ordered sample of size n is drawn without replacement from an urn containing $m \geq 1$ red balls and $n - m$ white balls, then at least 1 red ball must be drawn. Let F_k be the event that the first red ball is drawn on the kth draw, $k = 1, \ldots, n$. Then $F_1, \ldots, F_n$ are mutually exclusive and exhaustive, and $P(F_k) = m(n - m)_{k-1}/(n)_k = m(n - m)_{k-1}(n - k)!/n!$, $k = 1, \ldots, n$ by Example 1.5.1. Therefore, we have

$$\sum_{k=1}^{n} m(n - m)_{k-1}(n - k)! = n! \qquad ////$$

2.4 COMBINATIONS OF EVENTS[1]

In Theorem 2.3.3 we showed that the probability of the union of n events is always less than or equal to the sum of their probabilities, with equality if the events are mutually exclusive. In this section we shall develop an exact expression for the probability of the union of n arbitrary events. We shall use the following notation. Let $A_1, A_2, \ldots, A_n$ be any events, and for every subset J of $\{1, \ldots, n\}$ let

$$B_J = \bigcap_{i \in J} A_i$$

Thus, B_J is the event that A_i occurs for every $i \in J$, with no restrictions placed on the occurrence of A_i for $i \notin J$. Further, let

$$S_k = \sum_{|J|=k} P(B_J) \qquad (4.1)$$

where the summation extends over all subsets J of size k, $k = 1, \ldots, n$. Thus,

$$S_1 = \sum_{i=1}^{n} P(A_i)$$

$$S_2 = \sum_{i=2}^{n} \sum_{j=1}^{i-1} P(A_i A_j)$$

and so forth. The formula we shall develop is given in the following theorem.

Theorem 2.4.1 *Let $A_1, \ldots, A_n$ be any n events, and let $A = A_1 \cup \cdots \cup A_n$ be the event that at least one of $A_1, \ldots, A_n$ occurs. Then*

$$P(A) = \sum_{k=1}^{n} (-1)^{k-1} S_k \qquad (4.2)$$

Theorem 2.4.1 can be proved by straightforward mathematical induction on n using Equation (3.3), which Theorem 2.4.1 generalizes. We shall give the details below, but first we shall consider some examples.

[1] This section treats a special topic and may be omitted without loss of continuity.

Most applications of Equation (4.2) will be to cases in which the events $A_1, \ldots,$ A_n are symmetric in the sense that

$$P(B_J) = P(A_1 A_2 \cdots A_k) \qquad (4.3)$$

for all subsets J of size k, $k = 1, \ldots, n$. In this case S_k simplifies to

$$S_k = \binom{n}{k} P(A_1 \cdots A_k) \qquad (4.1a)$$

since there are $\binom{n}{k}$ summands in Equation (4.1).

EXAMPLE 2.4.1 Matching A computer prepares monthly bills for its n customers and addresses an envelope for each. A programming error then causes it to put the bills into the envelopes at random. What is the probability that it places at least one bill in the correct envelope? Let us number the bills and envelopes in such a manner that each bill receives the same number as the envelope addressed for it. We may then describe the outcome of the experiment by a permutation $x = (x_1, \ldots, x_n)$ of the integers $1, \ldots, n$, where x_k denotes the number of the envelope into which the kth bill is placed, $k = 1, \ldots, n$. We may therefore take our sample space S to be the set of all such permutations, and we interpret the phrase "at random" to mean that all $n!$ outcomes $x = (x_1, \ldots, x_n) \in S$ are equally likely. The event that the kth letter is correctly placed is then $A_k = \{x \in S: x_k = k\}$ for $k = 1, \ldots, n$, and we require the probability of the union $B = \bigcup_{k=1}^{n} A_k$. It is easy to see that the symmetry condition (4.3) is satisfied. Moreover, $P(A_1 \cdots A_k) = (n - k)!/n!$, because $A_1 \cdots A_k$ specifies that $x_i = i$, $i = 1, \ldots, k$, and allows $x_{k+1}, \ldots, x_n$ to be permuted in any order. It now follows that $S_k = \binom{n}{k} (n - k)!/n! = 1/k!$. Therefore, by Equation (4.2)

$$\binom{n}{k} = \frac{n!}{k! \, (n-k)!}$$

$$P\left(\bigcup_{i=1}^{n} A_i\right) = \sum_{k=1}^{n} (-1)^{k-1} \frac{1}{k!} = 1 - \sum_{k=0}^{n} (-1)^k \frac{1}{k!}$$

The last sum, however, is simply the first n terms in the infinite Taylor series expansion of e^x for $x = -1$, so that

$$P\left(\bigcup_{i=1}^{n} A_i\right) \approx 1 - e^{-1} = 0.632 \qquad (4.4)$$

for large n. In fact, the approximation (4.4) is valid to two decimal places provided only that $n \geq 5$. ////

EXAMPLE 2.4.2 The coupon collector's problem A manufacturer gives away coupons of t different types with his product and gives a prize to anyone who collects at least one of all t types. If a man collects n coupons, what is the probability that he will collect at least one of all t types? If the t types of coupons are distributed in equal numbers, and if there are a large number of coupons, we can rephrase the question as follows.

If a t-sided, balanced die is tossed n times, what is the probability that each of the t faces appears at least once? Let A_i be the event that the ith face does not appear at least once. Then we require the probability that the event $A = \bigcup_{i=1}^{t} A_i$ does not occur, that is, $1 - P(A)$. We can compute $P(A)$ from Equation (4.2). Indeed, it is again easily verified that the symmetry condition (4.3) is satisfied, and

$$P(A_1 \cdots A_k) = \left(1 - \frac{k}{t}\right)^n$$

for $k = 1, \ldots, t$, since $A_1 \cdots A_k$ requires each of the n tosses to result in one of $t - k$ specified faces (see Lemma 1.5.1). Therefore,

$$P(A) = \sum_{k=1}^{t} (-1)^{k-1} \binom{t}{k} \left(1 - \frac{k}{t}\right)^n \qquad (4.5)$$

by Equation (4.2). Equation (4.5) does not simplify but is amenable to computation. We list some typical values in Table 3 for $t = 6$.

PROOF of Theorem 2.4.1   We shall prove Theorem 2.4.1 by induction on n. If $n = 1$, then (4.2) is trivial. Thus, suppose that (4.2) is true for $n \le m$, and consider the case that $n = m + 1$. If $A_1, \ldots, A_{m+1}$ are any $m + 1$ events, then

$$P\left(\bigcup_{i=1}^{m+1} A_i\right) = P\left(\bigcup_{i=1}^{m} A_i\right) + P(A_{m+1}) - P\left(\bigcup_{i=1}^{m} A_i A_{m+1}\right) \qquad (4.6)$$

by (3.3). Moreover, the first and last terms on the right side of (4.6) are probabilities of the union of m events, so that

$$P\left(\bigcup_{i=1}^{m} A_i\right) = \sum_{k=1}^{m} (-1)^{k-1} \left[\sum_{|J|=k} P\left(\bigcap_{i \in J} A_i\right)\right] \qquad (4.7a)$$

$$P\left(\bigcup_{i=1}^{m} A_i A_{m+1}\right) = \sum_{k=1}^{m} (-1)^{k-1} \left[\sum_{|J|=k} P\left(A_{m+1} \bigcap_{i \in J} A_i\right)\right] \qquad (4.7b)$$

Table 3

n	8	12	16	20	24
$1 - P(A)$	0.114	0.438	0.698	0.848	0.925

////

by the induction hypothesis. Finally, if (4.7a) and (4.7b) are substituted into (4.6), and if the $(k + 1)$st term in (4.7a) is grouped with the kth term in (4.7b), Equation (4.2) is obtained. Theorem 2.4.1 now follows by mathematical induction.   ////

In closing, we mention the following extension of Theorem 2.4.1. Let $A_1, \ldots, A_n$ be any n events; for $r = 1, \ldots, n$ let L_r be the event that at least r of the events $A_1, \ldots, A_n$ occur; and let E be the event that exactly r of the events $A_1, \ldots, A_n$ occur. Thus,

$$L_r = \bigcup_{|J| \geq r} B_J \quad \text{and} \quad E_r = L_r - L_{r+1}$$

where B_J is as in Equation (4.1).

Theorem 2.4.2   *For $r = 1, \ldots, n$,*

$$P(L_r) = \sum_{k=r}^{n} (-1)^{k-r} \binom{k-1}{r-1} S_k \qquad (4.8)$$

$$P(E_r) = \sum_{k=r}^{n} (-1)^{k-r} \binom{k}{r} S_k \qquad (4.9)$$

where S_k is defined by (4.1).

Equation (4.8) can be established by an inductive argument which is similar to that given in the proof of Theorem 2.4.1. Equation (4.9) can then be obtained from the identity $P(E_r) = P(L_r) - P(L_{r+1})$. We omit the details. Another, simpler proof of Theorem 2.4.1 is sketched in Problem 8.66.

2.5   EQUIVALENTS OF THE THIRD AXIOM[1]

We shall now turn our attention to the third axiom (2.3) and develop several useful equivalents to it. We say that an infinite sequence of events $A_1, A_2, \ldots$ is *increasing* if and only if $A_1 \subset A_2 \subset \cdots \subset A_n \subset A_{n+1} \subset \cdots$ for every $n = 1, 2, \ldots$. That is, $A_1, A_2, \ldots$ is increasing if and only if the occurrence of A_n implies the occurrence of A_{n+1} for every $n = 1, 2, \ldots$. We define the *limit of an increasing sequence $A_1, A_2, \ldots$* to be the union

$$A = \bigcup_{n=1}^{\infty} A_n$$

and write $A = \lim A_n$. Thus, $A = \lim A_n$ occurs if and only if A_n occurs for some $n = n_0$, in which case A_k occurs for all $k \geq n_0$. Similarly, we say that an infinite sequence of events $A_1, A_2, \ldots$ is *decreasing* if and only if $A_1 \supset A_2 \supset \cdots \supset A_n \supset$

---

[1] This section treats a special topic and may be omitted without loss of continuity.

$A_{n+1} \supset \cdots$ for every $n = 1, 2, \ldots$, and we define the *limit of decreasing sequence* of events to be the intersection

$$A = \bigcap_{n=1}^{\infty} A_n$$

In this case $A = \lim A_n$ occurs if and only if A_n occurs for every $n = 1, 2, \ldots$. We observe that a sequence $A_1, A_2, \ldots$ is increasing (decreasing) if and only if $A'_1, A'_2, \ldots$ is decreasing (increasing) and that in either case

$$(\lim A_n)' = \lim A'_n \qquad (5.1)$$

EXAMPLE 2.5.1 Let $S = (0,1)$ be the open unit interval, and for $n = 1, 2, \ldots$ let $A_n = (1/n, 1)$ be the open interval from $1/n$ to 1. Then, since $1/(n + 1) < 1/n$, $n \geq 1$, $A_1, A_2, \ldots$ is an increasing sequence of events, and since $1/n \to 0$ as $n \to \infty$,

$$\lim A_n = \bigcup_{n=1}^{\infty} A_n = (0,1)$$

Similarly, letting $B_n = (0, 1/n)$, $n \geq 1$, we find that $B_1, B_2, \ldots$ is a decreasing sequence of events with limit $\bigcap_{n=1}^{\infty} B_n = \varnothing$, since there are no real numbers x with $0 < x < 1/n$ for $n = 1, 2, \ldots$. ////

Now suppose that S is a nonempty set, that $\mathscr{S}$ is a σ algebra of subsets of S, and that P is a function defined on $\mathscr{S}$ which satisfies (2.1) and (2.2), and consequently (3.1) to (3.8). Then we say that P is *continuous from below* (*from above*) if and only if

$$P(\lim A_n) = \lim P(A_n) \qquad (5.2)$$

as $n \to \infty$, whenever $A_1, A_2, \ldots$ is an increasing (decreasing) sequence of events. The main result of this section is that the third axiom is equivalent to continuity as defined in (5.2).

Theorem 2.5.1 *Let S be a nonempty set, let $\mathscr{S}$ be a σ algebra of subsets of S, and let P be a function on $\mathscr{S}$ which satisfies (2.1) and (2.2). Then, the following are equivalent:*

(*i*) *Equation (2.3) holds.*
(*ii*) *P is continuous from below.*
(*iii*) *P is continuous from above.*

In particular, (i), (ii), and (iii) all hold if P is a probability measure.

PROOF We shall show that $(i) \Rightarrow (ii) \Rightarrow (iii) \Rightarrow (i)$. We begin with the proof that $(i) \Rightarrow (ii)$. Suppose that P is a probability measure, and let $A_1, A_2, \ldots$ be an increasing sequence of events. Then we can define a new sequence B_1, $B_2, \ldots$ by

$$B_1 = A_1 \quad \text{and} \quad B_n = A_n - A_{n-1}$$

for $n = 2, 3, \ldots$. Then $B_1, B_2, \ldots$ are mutually exclusive,

$$A_n = \bigcup_{k=1}^{n} B_k \qquad (5.3)$$

for $n = 1, 2, \ldots$, and

$$\lim A_n = \bigcup_{n=1}^{\infty} A_n = \bigcup_{n=1}^{\infty} \bigcup_{k=1}^{n} B_k = \bigcup_{n=1}^{\infty} B_n$$

Therefore, by (2.3),

$$P(\lim A_n) = \sum_{k=1}^{\infty} P(B_k) = \lim_{n \to \infty} \sum_{k=1}^{n} P(B_k) = \lim_{n \to \infty} P(A_n)$$

Here, the first equality follows from (2.3), the second from the definition of an infinite sum, and the third from (5.3).

The proof that $(ii) \Rightarrow (iii)$ is now trivial. Indeed, if $A_1, A_2, \ldots$ is a decreasing sequence of events, then $A_1', A_2', \ldots$ is an increasing sequence of events and (5.1) holds. Thus, if P is continuous from below, then

$$P(\lim A_n) = 1 - P(\lim A_n') = 1 - \lim P(A_n') = \lim P(A_n)$$

as $n \to \infty$, so that P is also continuous from above.

Finally, we must show that $(iii) \Rightarrow (i)$. Let P be continuous from below, and let $A_1, A_2, \ldots$ be a sequence of mutually exclusive events with union

$$A = \bigcup_{n=1}^{\infty} A_n$$

For $n = 1, 2, \ldots$ define

$$B_n = \bigcup_{k=1}^{n} A_k$$

Then, $B_1, B_2, \ldots$ is an increasing sequence of events with limit $\lim B_n = A$, so that $C_n = A - B_n$ forms a decreasing sequence of events with limit $\lim C_n = \emptyset$. Now, for every $n = 1, 2, \ldots$, we have

$$P(A) = P(B_n) + P(C_n) = \sum_{k=1}^{n} P(A_k) + P(C_n) \qquad (5.4)$$

by (2.2) and (3.5). Finally, since P is assumed to be continuous from above, we must have $\lim P(C_n) = P(\lim C_n) = P(\emptyset) = 0$ as $n \to \infty$. Thus, letting $n \to \infty$ in (5.4), we find

$$P(A) = \sum_{k=1}^{\infty} P(A_k)$$

as required by (2.3). This completes the proof of Theorem 2.5.1. ////

Interest in Theorem 2.5.1 derives from two facts. First, it shows that the third axiom (2.3), which was not as well motivated as (2.1) and (2.2), is equivalent to requiring probabilities to be continuous in the sense of Equation (5.2). The reader may find the continuity assumption more palatable than (2.3) as originally stated. Also, it shows that probability measures are continuous in the sense of (5.2), and this fact will be useful to us later on.

REFERENCES

A concise discussion of various interpretations of probability is given by De Finetti (1968). A more extensive discussion of these interpretations will be found in Smokler and Kyburg (1964). A different approach to the quantification of subjective probabilities is given by DeGroot (1970), who also gives further references.

Readers familiar with measure theory may wish to consult a more advanced text for further information on the mathematical foundations of probability. Neveu (1965) and Tucker (1967) are recommended.

PROBLEMS

2.1 Define an appropriate sample space for each of the following experiments. It is no longer necessary that the outcomes be assumed equally likely:

(a) A loaded die is rolled twice.

(b) A die is rolled until an ace appears. $S = \{1, 2, 3, \cdots - \}$

(c) A fair coin is tossed until two heads have appeared. $S = \{2, 3, 4, 5, \cdots \}$

(d) You wish to guess the year of Noah's birth.

2.2 Define appropriate sample spaces for the following experiments:

(a) The length of time required for a radioactive substance to register 25 emissions is observed.

(b) The weight of a randomly selected man is recorded.

(c) The annual precipitation in Seattle is recorded.

(d) The value of IBM stock is recorded each day for a week.

(e) The number of traffic accidents in a particular city on a particular day is recorded.

2.3 Try to assess your subjective probability that it will rain tomorrow.

NOTE: Problems 2.4 to 2.6 show that a person with inconsistent subjective probabilities will be willing to place bets on which he will certainly lose money.

2.4 Let A be an event, and let p and q denote your subjective probabilities for A and A', respectively. If $p + q \neq 1$, then would you regard as fair two bets the combination of which would force you to lose? *Hint:* If $p + q > 1$, what happens if you bet in favor of both A and A'?

2.5 Extend the result of Problem 2.4 to the case of the disjoint events whose union is S.

2.6 If A and B are disjoint events and your subject probability for $A \cup B$ is not $P(A) + P(B)$, then would you be willing to place bets the combination of which would force you to lose? *Hint:* Let $C = A'B'$, and show that either $P(C) + P(C') \neq 1$ or $P(A) + P(B) + P(C) \neq 1$; then apply Problems 2.5 and 2.6.

2.7 A person is selected at random from the population of a given city. Let A be the event that the person is male; let B be the event that the person is under 30 years of age; and let C be the event that the person speaks a foreign language. Describe in symbols:

(a) A male who is under 30 and does not speak a foreign language.

(b) A female who either is under 30 or speaks a foreign language.

(c) A person who is either under 30 or female but not both.

(d) A male who either is under 30 or speaks a foreign language but not both.

2.8 Let A, B, and C be as Problem 2.7. Describe in words the following events:

(a) $A(B \cup C)$ (b) $A \cup BC$ (c) $A - BC$

(d) $A - (B \cup C)$ (e) $(A \cup B) - AB$ (f) $(A \cup B \cup C) - ABC$

(g) $(A \cup B \cup C) - (AB \cup BC \cup AC)$

(h) $AB - C$ (i) $(A \cup B) - C$ (j) $(A - B) - C$

2.9 Consider a die which is loaded in such a manner that the probability that k spots will appear when the die is rolled is proportional to k. If the die is rolled once, what is the probability that an even number of spots will appear?

2.10 If the probability that a telephone exchange will make exactly k wrong connections during a 24-hour day is proportional to $1/k!$ for $k = 0, 1, 2, \ldots$, what is the probability (a) that there will be no wrong connections; (b) that there will be at most 2 wrong connections?

2.11 Let $S = \{1,2,\ldots\}$ be the set of positive integers, and let $f(s) = 1/s(s + 1)$ for $s \in S$. Show that $f(1) + f(2) + \cdots = 1$.

2.12 Let S and f be as Problem 2.11 and define P as in Example 2.2.2. Find the probability of the events $A = \{1,2,3,4\}$ and $B = \{10,11,\ldots\}$.

2.13 In Problem 2.12 find the probability of the event $A = \{2,4,6,\ldots\}$ that an even integer is selected. *Hint:* Integrate the Taylor series expansion of $-\frac{1}{2} \log (1 - x^2)$.

2.14 Let $S = [0,1]$ denote the unit interval, and let a point s be selected at random from S as in Example 2.2.5. What is the probability (a) that the first decimal in the decimal expansion of s is 1; (b) that it is at most 5?

2.15 In Problem 2.14 replace first decimal by second decimal.

2.16 Let $S = (0,\infty)$, and let probabilities be assigned as in Example 2.2.4 with $f(s) = e^{-s}$ for $s \in S$. Let A be the set of $s \in S$ which differ from a positive integer by at most $\frac{1}{4}$. Find $P(A)$. *Hint:* Let A_n be the set of s for which $|s - n| \leq \frac{1}{4}$; compute $P(A_n)$; and sum.

2.17 Show that condition (2.3) implies condition (2.2).

2.18 Show that condition (2.2) and (2.3) are equivalent if S is a finite set.

2.19 Let a point be chosen at random from the unit interval $S = [0,1]$, as in Example 2.2.5, and let the event A be defined as follows. First, we define A_1 to be the interval $(\frac{1}{3},\frac{2}{3})$. Next, we define A_2 to be the union of the intervals $(\frac{1}{9},\frac{2}{9})$ and $(\frac{7}{9},\frac{8}{9})$. In general we define A_n, $n \geq 3$, to be the union of the middle thirds of the 2^{n-1} intervals which are comprised by $(A_1 \cup \cdots \cup A_{n-1})'$. Finally, we let $A = \bigcup_{k=1}^{\infty} A_n$ be the union of the A_n. The complement of A, $C = S - A$, is known as the *Cantor set* and has many interesting properties. Show that $P(A) = 1$ and consequently that $P(C) = 0$. *Hint:* $P(A_n)$ is easily computed, and the A_n are disjoint.

2.20 In Problem 2.19, replace middle third by middle fourth throughout (also in the definition of A_1 and A_2). Compute $P(A)$ in this case.

2.21 Show that if A is an index set and for each $\alpha \in A$ $\mathscr{S}_\alpha$ is a σ algebra of subsets of a nonempty set S, then $\bigcap_A \mathscr{S}_\alpha$ is again a σ algebra of subsets of S.

2.22 Let S be an interval of real numbers, and let $\mathscr{S}$ be the intersection of all those σ algebras of subsets of S which contain all subintervals of S. Show that $\mathscr{S}$ is a σ algebra and that if $\mathscr{T}$ is any other σ algebra which contains all subintervals of S, then $\mathscr{S} \subset \mathscr{T}$.

2.23 Some of the requirements in the definition of a σ algebra are redundant. Show that $\mathscr{S}$ is a σ algebra if $S \in \mathscr{S}$; if $A \in \mathscr{S}$ implies $A' \in \mathscr{S}$; and if $A_k \in \mathscr{S}$ for $k = 1, 2, \ldots$ implies $\bigcup_{k=1}^{\infty} A_k \in \mathscr{S}$.

2.24 If an unordered random sample of size 10 is drawn from a lot of manufactured items, of which 10 are defective and 90 are nondefective, what is the probability that the sample will contain (*a*) at least 1 defective; (*b*) at least 2 defectives?

2.25 If a fair coin is tossed until a head appears, what is the probability that between 3 and 8 tosses (inclusive) will be required?

2.26 If cards are drawn sequentially from a standard deck until a spade appears, what is the probability that between 3 and 8 draws will be required?

2.27 What is the probability that the smallest denomination in a poker hand will be a 4?

2.28 If a balanced die is rolled twice, what is the probability that the largest number of spots to appear will be j, $j = 1, \ldots, 6$?

2.29 If a balanced die is rolled n times, what is the probability that the largest number of spots to appear will be j for $j = 1, 2, \ldots, 6$?

2.30 What is the probability that a randomly selected poker hand will contain at least 2 cards of at least one denomination?

2.31 What is the probability that a poker hand will contain:
(*a*) Exactly 2 aces or exactly 2 kings or both?
(*b*) Exactly 2 aces or exactly 2 kings but not both?

2.32 An official in the Internal Revenue Service believes:
(*a*) That 40 percent of all taxpayers fail to list all their taxable income.
(*b*) That 36 percent list more deductions than they actually have.
(*c*) That 22 percent do both.
If he is consistent in his beliefs, what percentage of taxpayers does he believe cheat by either method (*a*) or method (*b*)?

2.33 If A, B, and C are events, derive a formula for $P(A \cup B \cup C)$ in terms of the probabilities of intersections of A, B, and C.

2.34 What is the probability that a randomly selected bridge hand will contain at least seven cards of the same suit?

2.35 What is the probability that a randomly selected poker hand will contain at least three cards of the same denomination?

2.36 If an unordered random sample of size 10 is selected from a group of 55 Democrats and 45 Republicans, what is the probability that the sample will contain more Democrats than Republicans?

2.37 In Problem 2.36 what is the probability that the sample will contain between 4 and 8 Democrats (inclusive)?

2.38 Let A_1, A_2, ... be any infinite sequence of events, and let $B_1 = A_1$ and $B_k = A_k - (A_1 \cup \cdots \cup A_{k-1})$ for $k \geq 2$. Show that B_1, B_2, ... are mutually exclusive and $\bigcup_{k=1}^{\infty} A_k = \bigcup_{k=1}^{\infty} B_k$.

2.39 Let A_1, A_2, ... be any infinite sequence of events with union $A = \bigcup_{k=1}^{\infty} A_k$. Show that $P(A) \leq \sum_{k=1}^{\infty} P(A_k)$.

2.40 Obtain an upper bound on the probability that a randomly selected bridge hand will contain a void (no cards of at least one suit).

2.41 Compute the probability that a randomly selected bridge hand will contain a void.

2.42 Compute the probability that a randomly selected bridge hand will contain exactly 6 cards of at least one suit.

2.43 If a man randomly selects 4 socks from a drawer which contains 4 distinguishable pairs of socks, what is the probability that he will select at least 1 pair?

2.44 Cards labeled $1, 2, \ldots, n$ are turned over in a random order. Let A_k be the event that the card labeled k is turned over on the kth turn. What is the probability that at least one of $A_1, \ldots, A_n$ occur?

2.45 In Problem 2.44 let $p_n(j)$ be the probability that exactly j of $A_1, \ldots, A_n$ will occur. Show that $p_n(j) \to 1/ej!$ for $j = 0, 1, 2, \ldots$ as $n \to \infty$.

2.46 Show that if A_1, A_2, ... is any infinite sequence of events, then $P(\bigcup_{k=1}^{\infty} A_k) = \lim P(\bigcup_{k=1}^{n} A_k)$ as $n \to \infty$ and $P(\bigcap_{k=1}^{\infty} A_k) = \lim P(\bigcap_{k=1}^{n} A_k)$ as $n \to \infty$.

3

CONDITIONAL PROBABILITY AND INDEPENDENCE

3.1 CONDITIONAL PROBABILITY

Let $(S,\mathscr{S},P)$ be a probability space, and let B be an event with positive probability. Thus, $(S,\mathscr{S},P)$ may be thought of as a model for some experiment or game of chance and B as an event with a positive chance of occurring. Now suppose that we learn that B has in fact occurred. Then our original assignment of probabilities, represented in the model by P, may no longer be appropriate. Indeed, since we now know that B has occurred, we know that it is impossible for B' to occur, although we may have originally assigned positive probability to B'. The question which we propose to answer in this section is therefore: How should our probabilities change in the light of new information?

From the frequentistic point of view, the answer is quite simple. *Our new probabilities should represent limiting relative frequencies of events among exactly those trials on which B occurs.* That is, if the game or experiment under consideration is repeated n times, as in Section 2.1, and if n_A denotes the number of times that A

occurs during the n repetitions of the game or experiment, then the relative frequency of A among those trials on which B occurs is

$$\frac{n_{AB}}{n_B} = \frac{n_{AB}/n}{n_B/n}$$

In the frequentistic interpretation of probability, the latter quantity is (for large n) approximately $P(AB)/P(B)$, which therefore seems to be a reasonable candidate for our new probability.

We are also led to the ratio $P(AB)/P(B)$ from the subjective point of view. Thus, consider the following game: if B occurs, then

1 One pays q units to play.
2 One receives 1 unit if A occurs and nothing if A does not occur.

If B does not occur, no bets are placed. How can q be chosen in such a manner that the above bet is fair? If one has already assigned subjective probabilities to the events A, B, AB, and $A'B$ in a consistent manner, this question has an easy answer. Since one wins $1 - q$ units with probability $P(AB)$ and loses q units with probability $P(A'B)$, the intuitive notion of fairness requires that $(1 - q)P(AB) = qP(A'B)$. This may also be written

$$P(AB) = q[P(AB) + P(A'B)] = qP(B)$$

where we have used the consistency in the final step. Solving for q now yields $q = P(AB)/P(B)$, which therefore seems to be a reasonable candidate for our new probability for A from the subjective point of view too.

We have motivated the following definition: if A and B are events for which $P(B) > 0$, then we define the *conditional probability of A given B* to be

$$P(A \mid B) = \frac{P(AB)}{P(B)} \qquad (1.1)$$

Before we proceed to examples, let us remark that our original probabilities $P(A)$ may also be regarded as conditional probabilities given the sample space S. Indeed, taking $B = S$ in Equation (1.1) yields $AS = A$ and $P(S) = 1$, so that $P(A \mid S) = P(A)$. This remark admits the following interpretation: our original probabilities are conditional probabilities given our initial store of information about the problem at hand; our new probabilities $P(A \mid B)$, where $B \neq S$, are conditional given some additional information.

EXAMPLE 3.1.1

a If $A \subset B$, then $AB = A$, so that $P(A \mid B) = P(A)/P(B)$. In this case the new probability for A is larger than the original probability for A except in the trivial case that $P(B) = 1$. In particular, $P(B \mid B) = 1$.

b If $A \subset B'$, then $AB = \emptyset$, so that $P(A \mid B) = 0$.   ////

EXAMPLE 3.1.2

a If an ordered random sample of size $k = 2$ is drawn from an urn which contains m red balls and $n - m$ white balls, what is the conditional probability that the second ball drawn will be red (event A) given that the first ball drawn is red (event B)? We have $P(B) = m/n$ and $P(AB) = m(m - 1)/n(n - 1)$. Consequently, $P(A \mid B) = P(AB)/P(B) = (m - 1)/(n - 1)$. A similar calculation will show that $P(A \mid B') = m/(n - 1)$. In either case, the conditional probability of drawing a red ball on the second draw, given the outcome of the first draw, is proportional to the number of red balls in the urn at the time of the second draw.

b If the sampling had been with replacement in part a, we would have found that $P(B) = m/n$, $P(AB) = m^2/n^2$ and consequently that $P(A \mid B) = m/n$. Similarly, $P(A \mid B') = m/n$. Again, the conditional probability of drawing a red ball on the second draw is proportional to the number of red balls in the urn at the time of the second draw.   ////

EXAMPLE 3.1.3

a What is the probability that a randomly selected poker hand contains exactly 3 aces (event A), given that it contains at least 2 aces (event B)? Since A implies B, we have $AB = A$ is the event that the hand contains exactly 3 aces. Thus $P(AB) = \binom{4}{3}\binom{48}{2} / \binom{52}{5}$ by Theorem 1.4.1. The event B occurs if the hand contains 2, 3, or 4 aces, so that

$$P(B) = \frac{\binom{4}{2}\binom{48}{3} + \binom{4}{3}\binom{48}{2} + \binom{4}{4}\binom{48}{1}}{\binom{52}{5}}$$

by Theorems 1.4.1 and 2.3.3. Therefore,

$P(A \mid B) = P(AB)/P(B)$

$$= \binom{4}{3}\binom{48}{2} \Big/ \left[\binom{4}{2}\binom{48}{3} + \binom{4}{3}\binom{48}{2} + \binom{4}{4}\binom{48}{1}\right]$$

$$= 0.0416$$

b What is the probability that a randomly selected poker hand will contain exactly 2 kings (event A), given that it contains exactly 2 aces (event B)? We have

$$P(B) = \binom{4}{2}\binom{48}{3}\bigg/\binom{52}{5} \quad \text{and} \quad P(AB) = \binom{4}{2}\binom{4}{2}\binom{44}{1}\bigg/\binom{52}{5}$$

by Theorems 1.4.1 and 1.4.2. Thus, $P(A \mid B) = \binom{4}{2}\binom{44}{1}\bigg/\binom{48}{3}$. Observe that this is also the probability that a sample of size 3 from a deck with no aces will contain exactly 2 kings. ////

In the classical model, where $P(A) = |A|/|S|$ for $A \subset S$, conditional probabilities assume a particularly simple form. Indeed, if B is a nonempty subset of S, then

$$P(A \mid B) = \frac{|AB|/|S|}{|B|/|S|} = \frac{|AB|}{|B|} \tag{1.2}$$

for every $A \subset S$. *Thus, if we originally regard the outcomes in S as equally likely and we learn that B has occurred, then we regard the outcomes in B as equally likely.* In effect, we reduce our sample space from S to B. That is, we compute probabilities as if B were the sample space.

This remark can be extremely useful in the computation of conditional probabilities in sampling experiments. Indeed, it says that given that the sample has some particular property (event B), all remaining possible samples are equally likely. Examples 3.1.2 and 3.1.3b are special cases.

EXAMPLE 3.1.4

a If an unordered random sample of size k is drawn from an urn containing r red balls, b black balls, and w white balls, what is the conditional probability that the sample will contain exactly i white balls (event A) given that it contains exactly j red balls (event B)? Given that the sample contains exactly j red balls, we may regard the remaining $k - j$ balls in the sample as a new sample from an urn containing b black balls, w white balls, and no red balls. That is, we form a new sample space consisting of all outcomes in B. The conditional probability of A is then just the probability that a sample of size $k - j$ from an urn containing b black balls and w white balls will contain exactly i white balls; that is,

$$P(A \mid B) = \frac{\binom{b}{i}\binom{w}{k-j-i}}{\binom{b+w}{k-j}}$$

by Theorem 1.4.1. Example 3.1.3b is a special case.

Similar remarks apply to ordered samples.

b Let an ordered random sample of size $k = k_1 + k_2$ be drawn from an urn, where k_1 and k_2 are positive integers. Then, given the outcome of the first k_1 draws, all possible k_2-tuples of balls are equally likely to be drawn from the urn on the remaining k_2 draws. Parts *a* and *b* of Example 3.1.2 are a special case.

c A committee of size $k = 5$ is to be selected sequentially from a group of 6 Democrats and 4 Republicans. Given that the first 2 committee members to be selected were Democrats, what is the conditional probability that the committee will consist of 3 Democrats and 2 Republicans? We require the probability that a sample of size 3 from a group of 4 Democrats and 4 Republicans consists of 1 Democrat and 2 Republicans. The answer is therefore $\binom{3}{1} 4(4)_2/(8)_3 = \frac{3}{7}$.

$////$

The simple idea embodied in (1.2) generalizes. Thus, consider a probability space $(S, \mathscr{S}, P)$ and an event B with positive probability $P(B)$. Define a new class of events $\mathscr{S}_B$ to be all $A \in \mathscr{S}$ for which $A \subset B$, and define a function P_B on $\mathscr{S}_B$ by

$$P_B(A) = P(A \mid B) \qquad A \in \mathscr{S}_B$$

That $\mathscr{S}_B$ is, in fact, a σ algebra is left as an exercise.

Theorem 3.1.1 $(B, \mathscr{S}_B, P_B)$ *is a probability space.*

PROOF We have to show that P_B satisfies (2.1), (2.2), and (2.3) of Chapter 2. Now $A \in \mathscr{S}_B$ implies $A \subset B$, in which case $P_B(A) = P(A)/P(B)$. Thus, $P_B(B) = 1$. To establish (2.2), let A_1 and A_2 be disjoint elements of $\mathscr{S}_B$. Then

$$P_B(A_1 \cup A_2) = \frac{P(A_1 \cup A_2)}{P(B)} = \frac{P(A_1) + P(A_2)}{P(B)}$$

$$= P_B(A_1) + P_B(A_2)$$

where we have used the fact that P is a probability measure in the second step; (2.3) can be established in a similar manner.

$////$

Theorem 3.1.1 admits an interpretation which is similar to that of Equation (1.2). That is, *given that B has occurred, we may regard B as the new sample space for our experiment provided that we change our assignment of probabilities from P to P_B.* However, Theorem 3.1.1 does provide additional information. *Theorem 3.1.1 says that we are allowed to use all the results of Chapter 2 in computing conditional probabilities because those results are valid for any probability space.*

EXAMPLE 3.1.5   Bridge

a Given that a bridge player has 7 spades, what is the conditional probability that his partner has at least 1 spade? Given that one player has 7 spades and 6 nonspades, we may regard his partner's hand as a sample of size 13 from a deck containing 6 spades and 33 nonspades. Therefore, the conditional probability that his partner has no spades is $\binom{33}{13}\Big/\binom{39}{13} = 0.0706$, and the probability that his partner has at least 1 spade is $1 - 0.0706 = 0.9294$.

b If North and South have exactly 8 trumps in their combined hands, what is the conditional probability that the remaining 5 trumps are split 3 and 2 between East and West? We require that conditional probability that the number of trumps in East's hand be either 2 or 3. Given that North and South together have exactly 8 trumps, we may regard East's hand as a sample of size 13 from a deck containing 5 trumps and 21 nontrumps. The conditional probability that East has either 2 or 3 trumps is therefore

$$\frac{\binom{5}{2}\binom{21}{11} + \binom{5}{3}\binom{21}{10}}{\binom{26}{13}} = 0.678 \qquad ////$$

Theorem 3.1.1 opens an interesting possibility, namely, iterating the operation of conditioning. That nothing really new is thereby obtained is the content of our next theorem.

Theorem 3.1.2 *Let A, B, and C be events for which A and C are subsets of B and $P(C) > 0$. Then*

$$P_B(A \mid C) = P(A \mid C)$$

PROOF   Observe first that $P(B) \geq P(C) > 0$, so that P_B is well defined. Now, by definition, $P_B(A \mid C) = P_B(AC)/P_B(C)$, which may be written

$$\frac{P(ABC)/P(B)}{P(BC)/P(B)} = \frac{P(AC)}{P(C)} = P(A \mid C) \qquad ////$$

3.2   BAYES' THEOREM

Conditional probabilities are not only interesting as new probabilities given some additional information; they may also be used as tools in the computation of unconditional probabilities. For example, the formula

$$P(AB) = P(A \mid B)P(B) \qquad (2.1)$$

follows immediately from the definition of conditional probability and allows one to compute $P(AB)$ from knowledge of $P(A \mid B)$ and $P(B)$. It is useful since we can sometimes compute $P(A \mid B)$ by regarding B as the sample space for a new experiment. Moreover, since any event A can be written $A = AB \cup AB'$ with $AB \cap AB' \subset BB' = \varnothing$, we have $P(A) = P(AB) + P(AB')$, which, by Equation (2.1), can be written

$$P(A) = P(A \mid B)P(B) + P(A \mid B')P(B') \qquad (2.2)$$

provided that $0 < P(B) < 1$. Finally, if $P(A) > 0$, we may use Equations (2.1) and (2.2) to compute $P(B \mid A)$. Thus

$$P(B \mid A) = \frac{P(AB)}{P(A)} = \frac{P(A \mid B)P(B)}{P(A \mid B)P(B) + P(A \mid B')P(B')} \qquad (2.3)$$

Equation (2.3) is a special case of *Bayes' theorem*, discussed below. Let us first consider some examples.

EXAMPLE 3.2.1 Let urn I contain 4 red balls and 2 white balls, and let urn II contain 3 balls of each color. If a ball is drawn at random from urn I and transferred to urn II and then a ball is drawn at random from urn II, what is the probability that the second ball drawn will be red? Let A be the event that the ball drawn from urn II is red, and let B be the event that the ball transferred is red. Then, $P(A \mid B) = \frac{4}{7}$ since there will be 4 red balls and 3 white balls in urn II at the time of the second drawing if B occurs. Similarly $P(A \mid B') = \frac{3}{7}$. Since $P(B) = \frac{2}{3}$, we have

$$P(A) = (\tfrac{4}{7})(\tfrac{2}{3}) + (\tfrac{3}{7})(\tfrac{1}{3}) = \tfrac{11}{21}$$

by Equation (2.2). Now suppose that we observe the color of the second ball to be red but do not observe the color of the transferred ball. Then we can compute the conditional probability that the ball transferred was red from Equation (2.3). Indeed, we have $P(B \mid A) = (\tfrac{4}{7})(\tfrac{2}{3})/(\tfrac{11}{21}) = \tfrac{8}{11}$.　　　　////

EXAMPLE 3.2.2 In a certain community, it is found that 60 percent of all property owners oppose an increase in the property tax while 80 percent of non-property owners favor it. If 65 percent of all registered voters are property owners, what proportion of registered voters favor the tax increase? Let A be the event that a randomly selected voter favors the tax increase. Then we require $P(A)$. Let B be the event that a randomly selected voter is a property owner. Then, we are given that $P(A \mid B) = 0.40$, $P(A \mid B') = 0.80$, and $P(B) = 0.65$. By Equation (2.2) we find $P(A) = (0.40) \times (0.65) + (0.80)(0.35) = 0.54$. That is, 54 percent of the registered voters favor the tax increase.

What percentage of those registered voters who favor the tax increase are property owners? We require $P(B \mid A)$. By Equation (2.3), this is simply $P(B \mid A) = (0.40)(0.65)/0.54 = 0.4815$. ////

Equations (2.1), (2.2), and (2.3) can all be generalized. We begin with the generalization of (2.1).

Theorem 3.2.1 *Let* $A_1, \ldots, A_n$ *be events, and let* $B_k = A_1 \cap \cdots \cap A_k$ *for* $k = 1, \ldots, n$. *If* $P(B_{n-1}) > 0$, *then*

$$P(B_n) = P(B_1) \prod_{k=2}^{n} P(B_k \mid B_{k-1})$$

PROOF   We observe that B_k implies $B_{k-1}(B_k \subset B_{k-1})$, so that

$$P(B_k \mid B_{k-1}) = P(B_k)/P(B_{k-1})$$

for $k = 2, \ldots, n$. Therefore,

$$\cancel{P(B_1)} \frac{\cancel{P(B_2)}}{\cancel{P(B_1)}} \cdots \frac{P(B_n)}{\cancel{P(B_{n-1})}} = P(B_n)$$

as asserted. ////

EXAMPLE 3.2.3   A certain communication system, or *channel*, is designed to transmit either the symbol 0 or the symbol 1. There are 4 relays, each of which may malfunction. In fact, each relay changes a received 1 to a transmitted 0 with probability 0.1 and changes a received 0 to a transmitted 1 with probability 0.2.

$$\text{Source} \rightarrow \boxed{1} \rightarrow \boxed{2} \rightarrow \boxed{3} \rightarrow \boxed{4} \rightarrow \text{receiver}$$

If a 1 is sent, what is the probability that a 1 is transmitted by every relay? Let A_i be the event that the ith relay transmits a 1, and let $B_k = A_1 \cap \cdots \cap A_k$ for $k = 1, \ldots, 4$. Then we are given that $P(B_1) = P(B_k \mid B_{k-1}) = 0.9$ for $k = 2, 3, 4$. Thus, $P(A_1 A_2 A_3 A_4) = P(B_4) = (0.9)^4 = 0.6561$.

If a 1 is sent, what is the probability that a 1 will be received? A 1 will be received if and only if zero, two, or four of the relays malfunction. The probability that none of the relays malfunction has just been computed. The probability that relays 1 and 2 malfunction while relays 3 and 4 operate correctly is $(0.1)(0.2)(0.9)^2$ by a similar argument. This is also the probability that any two specified relays malfunction while the other two operate correctly, so the probability that exactly two of the relays malfunction is $\binom{4}{2}(0.1)(0.2)(0.9)^2 = 0.0972$. Finally, the probability that all four relays malfunction is $(0.1)^2(0.2)^2 = 0.0004$. Thus, the probability that a 1 is received is $0.6561 + 0.0972 + 0.0004 = 0.7537$. ////

Let us now consider the generalization of Equations (2.2) and (2.3).

Theorem 3.2.2 *Let $B_1, B_2, \ldots$ be a finite or infinite sequence of mutually exclusive, exhaustive events and let $P(B_i) > 0$ for all i. If A is any event, then*

$$P(A) = \sum_i P(A \mid B_i)P(B_i) \qquad (2.4)$$

where the summation extends over all i. If $P(A) > 0$, then

$$P(B_j \mid A) = \frac{P(A \mid B_j)P(B_j)}{\sum_i P(A \mid B_i)P(B_i)} \qquad (2.5)$$

for every j.

PROOF Equations (2.2) and (2.3) are, of course, special cases of (2.4) and (2.5) with $B_1 = B$ and $B_2 = B'$. To prove (2.4) observe that since $\bigcup_i B_i = S$, by assumption, we can write $A = \bigcup_i AB_i$. Moreover, since B_i are mutually exclusive, we must have

$$P(A) = \sum_i P(AB_i) = \sum_i P(A \mid B_i)P(B_i)$$

by Equation (2.1). This establishes (2.4). Equation (2.5) then follows from the definition $P(B_j \mid A) = P(AB_j)/P(A)$ on writing $P(AB_j) = P(A \mid B_j)P(B_j)$ and substituting (2.4) for $P(A)$. ////

Equation (2.5) is known as *Bayes' theorem* after the seventeenth-century clergyman, Thomas Bayes.

Both Equations (2.4) and (2.5) are useful in describing experiments which proceed in two stages and have the property that the chance mechanism of the second stage is determined by the outcome of the first stage of experimentation. For example, Example 3.2.1 is of this nature. There the composition of the urn from which the ball was drawn on the second stage was determined by the outcome of the first stage. We shall call such experiments *compound experiments*. In applications of Theorem 3.2.2 to compound experiments, one usually lets the B_i represent the possible outcomes of the first stage of experimentation and $P(A \mid B_i)$ describe the chance mechanism of the second stage under the hypothesis that B_i occurred on the first stage, as in Example 3.2.1.

The B_i may also be thought of as possible states of nature, in which case $P(A \mid B_i)$ is to be interpreted as the probability of A under the hypothesis that nature is in state B_i. With this interpretation the $P(B_i)$ are typically subjective probabilities which represent our opinion about nature prior to any experimentation and are known as *prior probabilities*. The conditional probabilities $P(B_j \mid A)$ may then be thought of as describing our new opinion about nature after some experiment has been performed and the event A observed to occur; for this reason they are known as *posterior probabilities*. Thus, Bayes' theorem may be thought of as an algorithm for changing one's mind in the light of experimental evidence, and it is this interpretation from which Bayes' theorem derives its fame. Of course, one must be able and willing to express one's opinion in terms of subjective probabilities in order to use this algorithm. The reluctance of the scientific community to accept the subjective interpretation of probability has slowed the acceptance of the latter interpretation of Bayes' theorem. However, it has gained considerable ground during the last decade and now forms the basis for an analytical theory of decision making which will be discussed in Section 10.5.

EXAMPLE 3.2.4 If the probability that a family will have exactly n children is 2^{-n} for $n = 1, 2, \ldots$, and if all 2^n permutations of the sexes of the n children are equally likely, what is the probability that a family will have no boys? Let A be the event that a family has no boys, and let B_n be the event that it has exactly n children. Then, we are given that $P(B_n) = 2^{-n}$, and clearly, $P(A \mid B_n) = 2^{-n}$ for every $n = 1, 2, \ldots$. (Given B_n, A requires n girls.) By Theorem 3.2.1, we therefore have

$$P(A) = \sum_{n=1}^{\infty} P(A \mid B_n)P(B_n)$$

$$= \sum_{n=1}^{\infty} 2^{-n} \times 2^{-n} = \frac{1}{4} \sum_{n=0}^{\infty} \left(\frac{1}{4}\right)^n = \left(\frac{1}{4}\right)\left(\frac{4}{3}\right) = \frac{1}{3}$$

The conditional probability that a family will have n children, given that it has no boys, is $P(B_n \mid A) = P(A \mid B_n)P(B_n)/P(A) = 3 \times 4^{-n}$ for $n = 1, 2, \ldots$.   ////

EXAMPLE 3.2.5 Traffic accidents A certain state groups its licensed drivers according to age into the following categories: (1) 16 to 25; (2) 26 to 45; (3) 46 to 65; and (4) over 65. Table 4 lists, for each group, the proportion of licensed drivers who belong to the group and the proportion of drivers in the group who had accidents.

Table 4

Group	Size	Accident rate
1	0.151	0.098
2	0.356	0.044
3	0.338	0.056
4	0.155	0.086

What proportion of licensed drivers had accidents? What proportion of those licensed drivers who had accidents were over 65? Let A be the event that a randomly selected licensed driver has an accident, and let B_k be the event that a randomly selected licensed driver falls into group k, $k = 1, 2, 3, 4$. We require $P(A)$ and $P(B_4 \mid A)$, respectively. Now, $P(B_k)$ and $P(A \mid B_k)$ are given by the columns entitled size and accident rate, respectively. Thus, $P(A) = P(A \mid B_1)P(B_1) + \cdots + P(A \mid B_4)P(B_4) = (0.098)(0.151) + \cdots + (0.086)(0.155) = 0.06272$, and $P(B_4 \mid A) = P(A \mid B_4)P(B_4)/P(A) = (0.086)(0.155)/0.06272 = 0.2125$.   ////

3.3 INDEPENDENCE

It seems natural to ask the following question: For what events A and B is it true that $P(A \mid B) = P(A)$? That is, for what events A and B is it true that the occurrence of B provides no information about the chance that A will occur? The answer is easily derived. We shall have $P(A \mid B) = P(A)$ if and only if $P(AB)/P(B) = P(A)$. That is, $P(A \mid B) = P(A)$ if and only if

$$P(AB) = P(A)P(B) \qquad (3.1)$$

We therefore define two events A and B to be *independent* if and only if (3.1) holds, and we expressly allow the possibility that $P(B) = 0$ in (3.1). The definition of independence is then symmetric in A and B.

The intuitive meaning of independence should be clear. A and B are independent if and only if the occurrence of B does not affect the chance that A will occur, and conversely. The importance of the notion of independence derives from the fact that many naturally occurring phenomena operate independently, that is, in such a manner that the outcome of one does not affect that of the other(s).

EXAMPLE 3.3.1
  a If A and B are disjoint, then $P(AB) = P(\emptyset) = 0$, so that A and B cannot be independent unless either $P(A) = 0$ or $P(B) = 0$.
  b If $A \subset B$, then $AB = A$, so that $P(AB) = P(A)$. In this case A and B cannot be independent unless $P(B) = 1$.
  c The empty set $\emptyset$ and the sample space S are independent of every other event. Indeed, $P(AS) = P(A) = P(A)P(S)$, and $P(A\emptyset) = P(\emptyset) = 0 = P(\emptyset)P(A)$ for every $A \subset S$.   ////

EXAMPLE 3.3.2 Let an ordered random sample of size $k = 2$ be drawn from an urn which contains $m \geq 1$ red balls and $n - m$ white balls. Further, let B be the event that a red ball is drawn on the first draw, and let A be the event that a red ball is drawn on the second draw. If the sampling is with replacement, then $P(A \mid B) = m/n = P(A)$ by Theorem 1.3.1 and Example 3.1.2, so that A and B are independent. If the sampling is without replacement, then $P(A \mid B) = (m - 1)/(n - 1) \neq m/n = P(A)$. Thus, A and B are not independent if the sampling is without replacement.

This result is clearly in accord with our intuitive notion of independence. Indeed, when sampling with replacement, the color of the ball drawn on the first draw does not affect the supply of red balls in the urn at the time of the second draw, while it does when the sampling is without replacement.   ////

EXAMPLE 3.3.3

a Let two distinguishable, balanced dice be rolled in such a manner that all possible outcomes are equally likely. Then the event A that only one spot appears on the first die and the event B that only one spot appears on the second die are independent. In fact, $P(A) = P(B) = \frac{1}{6}$, while $P(AB) = \frac{1}{36}$. This is a special case of Example 3.3.2.

b Let a card be selected at random from a deck of 52 cards. Let A be the event that a heart is drawn, and let B be the event that a face card (jack, queen, or king) is drawn. Then $P(A) = 13/52 = \frac{1}{4}$, $P(B) = 12/52 = \frac{3}{13}$, since there are 13 hearts and 12 face cards. Moreover, AB is the event that the jack of hearts, queen of hearts, or king of hearts is drawn, so that $P(AB) = 3/52 = P(A)P(B)$. Therefore, A and B are independent.

c Let a point be selected from the unit square $S = \{(x,y): 0 \le x \le 1$ and $0 \le y \le 1\}$ in such a manner that the probability that the point falls into a subregion $C \subset S$ is equal to the area of C. Let $A = \{(x,y): 0.25 \le x \le 0.75\}$, and let $B = \{(x,y): 0.25 \le y \le 0.75\}$ (see Figure 1*a*). Then A is a rectangle of height 0.5 and length 1, so that $P(A) = 0.5$, and similarly, $P(B) = 0.5$. Moreover, $AB = \{(x,y): 0.25 \le x \le 0.75, 0.25 \le y \le 0.75\}$ is a square whose sides are of length 0.5, so that $P(AB) = 0.25$. Therefore, A and B are independent.

d Let a point s be chosen from the unit interval $S = [0,1]$ in such a manner that the probability that the point belongs to a subinterval $I \subset S$ is the length of I, as in Example 2.2.5. Let us write s in its decimal expansion as $s = .s_1 s_2, \ldots,$ where s_k are integers between 0 and 9 inclusive. For example, if $s = \frac{1}{8}$, then $s_1 = 1, s_2 = 2, s_3 = 5$, and $s_k = 0$ for $k \ge 4$. Let A be the event that $s_1 = 0$, and let B be the event that $s_2 = 0$. Then A is the interval $[0,0.1)$, so that $P(A) = 0.1$; and B is the union $[0,0.01) \cup [0.10,0.11) \cup \cdots \cup [0.80,0.81) \cup [0.90,0.91)$, so that $P(B) = 0.1$ also (see Figure 1*b*). Finally, AB is the interval $[0,0.01)$, so that $P(AB) = 0.01 = P(A)P(B)$. Therefore, A and B are independent.   ////

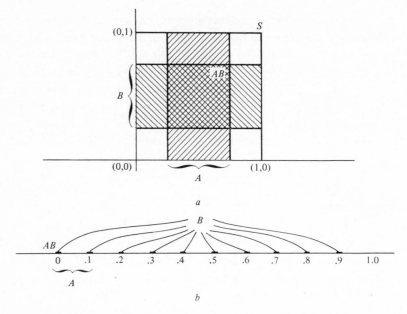

FIGURE 1.
(*a*) The unit square; (*b*) the unit interval.

There are two seemingly natural ways to extend the definition of independence from two events to several, say $A_1, \ldots, A_n$, where $n \geq 2$ is a positive integer. One is to require that

$$P(A_i A_j) = P(A_i)P(A_j) \qquad (3.2)$$

whenever $i \neq j$. The other is to require that

$$P\left(\bigcap_{i \in J} A_i\right) = \prod_{i \in J} P(A_i) \qquad (3.3)$$

for every nonempty subset $J \subset \{1, \ldots, n\}$. The two notions are not equivalent (see Example 3.3.5), and it is the second (3.3) which is the more useful. We therefore define events $A_1, \ldots, A_n$ to be *pairwise independent* if and only if (3.2) holds and to be *mutually independent* if and only if (3.3) holds. Thus, independence of either type is symmetric in the events $A_1, \ldots, A_n$ and has the property that subcollections of independent events are again independent. Most of the examples of independence which we shall encounter will be examples of mutual independence. Accordingly, we shall use the unqualified term "independence" to mean mutual independence.

EXAMPLE 3.3.4 Let an ordered random sample of size n be drawn with replacement from an urn containing r red balls and b black balls, and let A_i be the event that a red ball is drawn on the ith draw, $i = 1, \ldots, n$. Then, $A_1, \ldots, A_n$ are independent. Indeed, both sides of (3.3) are equal to $r^k/(r + b)^k$, where $k = |J|$.   ////

EXAMPLE 3.3.5 We shall give an example of events which are pairwise independent but not mutually independent. Let an integer be chosen from the set $S = \{1,2,3,4\}$ in such a manner that each integer is equally likely to be chosen. Further, let $A_i = \{i,4\}$ be the event that either i or 4 is chosen, $i = 1, 2, 3$. Then, $P(A_i) = 2/4 = \frac{1}{2}$, $i = 1, 2, 3$, and $P(A_iA_j) = P(\{4\}) = \frac{1}{4} = P(A_i)P(A_j)$ whenever $i \neq j$. Therefore, A_1, A_2, A_3 are pairwise independent. However, $P(A_1A_2A_3) = P(\{4\}) = \frac{1}{4} \neq \frac{1}{8} = P(A_1)P(A_2)P(A_3)$, so that A_1, A_2, and A_3 are not mutually independent.

////

EXAMPLE 3.3.6 Stratified sampling Consider a population of n individuals of which an unknown number m favor a particular political candidate or issue. We suppose that the population is divided into *strata*, by which we mean disjoint subclasses. For example, the population might be divided into urban dwellers, suburban dwellers, and country dwellers; or it might be divided according to the age of its members; or it might be divided according to the income of its members, etc. We suppose that there are a total of t different strata and that there are a total of n_i members in the ith strata, of which m_i favor the political candidate or issue in question. Thus, $n = n_1 + \cdots + n_t$ and $m = m_1 + \cdots + m_t$.

Suppose next that we are allowed to sample k of the members of the population in order to learn about m. Then two possible sampling schemes present themselves. First, we might take a random sample (without replacement) of size k from the entire population. In this case the probability that the sample will contain exactly r people who favor the candidate or issue is $\binom{m}{r}\binom{n-m}{k-r}\Big/\binom{n}{k}$ by Theorem 1.4.1.

We might also divide the sample size k into groups of sizes $k_1, \ldots, k_t$, where $k_1 + \cdots + k_t = k$, and take a random sample of size k_i from the ith stratum for $i = 1, \ldots, t$. We suppose that the t different samples are so taken that the sample selected from the ith stratum does not affect that selected from the other strata. In this case we may assume that the outcomes of the t different sampling experiments are independent, and we compute the probability that the sample from the ith stratum contains exactly r_t who favor the candidate or issue for all $i = 1, \ldots, t$ to be

$$\frac{\binom{m_1}{r_1}\binom{n_1 - m_1}{k_1 - r_1}}{\binom{n_1}{k_1}} \cdots \frac{\binom{m_t}{r_t}\binom{n_t - m_t}{k_t - r_t}}{\binom{n_t}{k_t}}$$

The second of the two sampling schemes described above is known as *stratified sampling*. We continue our study of stratified sampling in Example 8.6.1, where we show that if $k_1, \ldots, k_t$ are chosen proportional to $n_1, \ldots, n_t$, then stratified sampling is more informative than simple random sampling.   ////

3.4 SOME PROPERTIES OF INDEPENDENCE

Certain set-theoretic operations preserve independence. We give some of them in the following theorems.

Theorem 3.4.1 *If $A_1, \ldots, A_n$ are independent events, then so are the following collections of events:*

(i) *$B_1, \ldots, B_n$, where B_i is either A_i or A'_i, $i = 1, \ldots, n$.*
(ii) *$C_1, \ldots, C_k$, where $k \leq n$ and $C_1, \ldots, C_k$ are formed by taking unions or intersections of disjoint subcollections of $A_1, \ldots, A_n$.*

PROOF   In order to keep the notation in hand, we shall prove only (i) for $n = 2$ and (ii) for $k = 2$. The proofs for general n and k are conceptually no more difficult but notationally fearsome.

It is enough to prove (i) in the special case that $B_1 = A_1$ and $B_2 = A'_2$, for we can exchange A_1 and A_2 or A_1 and A'_2. If $B_1 = A_1$ and $B_2 = A'_2$, then

$$P(B_1 B_2) = P(A_1 - A_1 A_2) = P(A_1) - P(A_1 A_2)$$
$$= P(A_1) - P(A_1)P(A_2) = P(A_1)[1 - P(A_2)]$$
$$= P(B_1)P(B_2)$$

as asserted.

Let us first prove (ii) in the special case that both C_1 and C_2 are formed by taking intersections of disjoint subcollections of $A_1, \ldots, A_n$. In this case we may assume (by relabeling $A_1, \ldots, A_n$ if necessary) that

$$C_1 = \bigcap_{i=1}^{r} A_i \quad \text{and} \quad C_2 = \bigcap_{i=t}^{n} A_i$$

where $1 \leq r < t \leq n$. In this case the result is obvious, for

$$P(C_1 C_2) = P(A_1 \cap \cdots \cap A_r \cap A_t \cap \cdots \cap A_n)$$
$$= \left[\prod_{i=1}^{r} P(A_i) \right] \left[\prod_{i=t}^{n} P(A_i) \right] = P(C_1)P(C_2)$$

Now suppose that C_1 is as above and that $C_2 = \bigcup_{i=t}^{n} A_i$. Let $B_i = A_i$, $i = 1, \ldots, r$, and let $B_i = A_i'$, $i = r + 1, \ldots, n$. Then, $B_1, \ldots, B_n$ are independent by (i), so that

$$D_1 = \bigcap_{i=1}^{r} B_i \quad \text{and} \quad D_2 = \bigcap_{i=t}^{n} B_i$$

are independent by what has just been shown. Finally, if D_1 and D_2 are independent, then so are $C_1 = D_1$ and $C_2 = D_2'$, again by (i). The remaining cases under (ii) may be handled similarly to complete the proof.   ////

Theorem 3.4.2 *Let $A_1, \ldots, A_m$ be mutually exclusive events, and let $B_1, \ldots, B_n$ be mutually exclusive events. If A_i and B_j are independent for every choice of i and j, then*

$$A = \bigcup_{i=1}^{m} A_i \quad \text{and} \quad B = \bigcup_{j=1}^{n} B_j$$

are independent.

PROOF

$$AB = \bigcup_{i=1}^{m} \bigcup_{j=1}^{n} A_i B_j$$

and the events $C_{ij} = A_i B_j$ are mutually exclusive. Therefore,

$$P(AB) = \sum_{i=1}^{m} \sum_{j=1}^{n} P(A_i B_j) = \sum_{i=1}^{m} \sum_{j=1}^{n} P(A_i)P(B_j)$$

$$= \left[\sum_{i=1}^{m} P(A_i) \right]\left[\sum_{j=1}^{n} P(B_j) \right] = P(A)P(B)$$

as asserted.[1]   ////

In the presence of independence, many calculations simplify. We illustrate with some examples.

EXAMPLE 3.4.1

a Three missiles are fired at a target. If each missile has probability 0.6 of hitting the target, what is the probability that at least one of the missiles hits

---

[1] If $a_1, \ldots, a_m$ and $b_1, \ldots, b_n$ are real numbers, then

$$\left(\sum_{i=1}^{m} a_i\right)\left(\sum_{j=1}^{n} b_j\right) = \sum_{i=1}^{m} \sum_{j=1}^{n} a_i b_j$$

This is easily established by mathematical induction.

the target? Let A_i be the event that the ith missile hits the target, $i = 1, 2, 3$. Then, the event that at least one of the missiles hits the target is $B = A_1 \cup A_2 \cup A_3 = (A_1' \cap A_2' \cap A_3')'$. Therefore, assuming that A_1, A_2, and A_3 are independent, we have $P(B) = 1 - P(A_1' \cap A_2' \cap A_3') = 1 - P(A_1')P(A_2') \times P(A_3') = 1 - 0.4^3 = 0.936$.

b Suppose that n individuals work independently on a problem. If each has probability p of solving the problem, what is the probability that they all solve the problem? What is the probability that at least one of them solves the problem? Let A_i be the event that the ith individual solves the problem, so that $P(A_i) = p$ for $i = 1, \ldots, n$. The event that all n individuals solve the problem is $A = \bigcap_{i=1}^{n} A_i$, so that $P(A) = \prod_{i=1}^{n} P(A_i) = p^n$. The event that at least one of the individuals solves the problem is $L = \bigcup_{i=1}^{n} A_i = (\bigcap_{i=1}^{n} A_i')'$. Thus, $P(L) = 1 - P(\bigcap_{i=1}^{n} A_i') = 1 - (1 - p)^n$.

In the special case that $p = 0.5$ and $n = 4$, we have $P(A) = 0.0625$ and $P(L) = 0.9375$. ////

3.5 REPEATED TRIALS: PRODUCT SPACES[1]

Suppose that we have n experiments or games $E_1, \ldots, E_n$, where $n \geq 2$ is an integer. Suppose also that each experiment E_i may be described by a probability space $(S_i, \mathscr{S}_i, P_i)$, $i = 1, \ldots, n$. Finally, suppose that we perform all n experiments, either one at a time or simultaneously, in such a manner that the outcome of one experiment does not affect the outcomes of the others. Is it possible to describe the new experiment, formed by performing all $E_1, \ldots, E_n$? That is, is it possible to define a probability space to represent the new experiment in such a manner that events depending on different experiments are independent? The answer is yes, as we shall show in this section.

Before we give the details of the construction, let us remark that the problem posed above contains the following problem as a special case. Suppose that we have one experiment E_0, which is described by a probability space $(S_0, \mathscr{S}_0, P_0)$, and suppose that experiment E_0 is repeated n times. Is it possible to define a probability space which describes the new experiment and has the property that events depending on different trials (repetitions) are independent? The answer is again yes, since the second question is a special case of the first with $E_i = E_0$, $i = 1, \ldots, n$.

For simplicity, we shall give the construction only in the special case that the probability spaces $(S_1, \mathscr{S}_1, P_1), \ldots, (S_n, \mathscr{S}_n, P_n)$ are all discrete. That is, we consider

---

[1] This section may be omitted without loss of continuity.

only the case that each S_i is either a finite or countably infinite set and each $\mathscr{S}_i$ consists of all subsets of S_i, $i = 1, \ldots, n$. Let S denote the cartesian product

$$S = S_1 \times S_2 \times \cdots \times S_n$$

Thus, S consists of all ordered n-tuples $(s_1, \ldots, s_n)$ with $s_i \in S_i$, $i = 1, \ldots, n$. We shall use S as the sample space for the new experiment with the convention that s_i denotes the outcome of experiment E_i, $i = 1, \ldots, n$. Further, we shall let $\mathscr{S}$ be the class of all subsets of S, and we define a function P on $\mathscr{S}$ by

$$f(\{s\}) = \prod_{i=1}^{n} P_i(\{s_i\})$$

for $s = (s_1, \ldots, s_n) \in S$ and

$$P(A) = \sum_{s \in A} f(\{s\})$$

for $A \subset S$. Since

$$\sum_{s \in S} f(\{s\}) = \prod_{i=1}^{n} \left[\sum_{s_i \in S_i} P_i(\{s_i\}) \right] 1^n = 1$$

it follows from Example 2.2.2 that $(S, \mathscr{S}, P)$ is a probability space.

We shall say that an event $B \subset S$ *depends only on the ith trial* if and only if there is a subset $A \subset S_i$ for which

$$B = \{(s_1, \ldots, s_n) \in S : s_i \in A\} \qquad (5.1)$$

Moreover, if $A \subset S_i$ and $B \subset S$ are related by (5.1), we shall refer to B as *the event that A occurs on the ith trial*.

Lemma 3.5.1 *For $i = 1, \ldots, n$, let $A_i \subset S_i$, and let B_i be the event that A_i occurs on the ith trial. Then,*

$$\bigcap_{i=1}^{n} B_i = A_1 \times A_2 \times \cdots \times A_n$$

That is, the intersection of $B_1, \ldots, B_n$ is the cartesian product of $A_1, \ldots, A_n$.

PROOF B_i is the set of $(s_1, \ldots, s_n) \in S$ for which $s_i \in A_i$. Thus, both $B_1 \cap B_2 \cap \cdots \cap B_n$ and $A_1 \times A_2 \times \cdots \times A_n$ may be described as the set of $s = (s_1, \ldots, s_n)$ for which $s_i \in A_i$ for all $i = 1, \ldots, n$.   ////

Lemma 3.5.2 *Let $A_i \subset S_i$, $i = 1, \ldots, n$, and let $A = A_1 \times A_2 \times \cdots \times A_n$. Then*

$$P(A) = \prod_{i=1}^{n} P_i(A_i)$$

PROOF   By definition of P, we have

$$P(A) = \sum_{s \in A} \left[\prod_{i=1}^{n} P_i(\{s_i\}) \right]$$

and the latter sum is easily seen to be

$$\prod_{i=1}^{n} \left[\sum_{s_i \in A_i} P_i(\{s_i\}) \right] = \prod_{i=1}^{n} P_i(A_i) \qquad \text{////}$$

Theorem 3.5.1   *For $i = 1, \ldots, n$, let $A_i \subset S_i$, and let B_i be the event that A_i occurs on the ith trial. Then*

$$P(B_i) = P_i(A_i) \qquad i = 1, \ldots, n \tag{5.2}$$

and $B_1, \ldots, B_n$ are mutually independent.

PROOF   Let us first prove (5.2). For each i, $B_i = S_1 \cdots S_{i-1} \times A_i \times S_{i+1} \cdots S_n$. Therefore,

$$P(B_i) = P_i(A_i) \prod_{j \neq 1} P_j(S_j) = P_i(A_i)$$

as asserted. To establish the mutual independence of $B_1, \ldots, B_n$, let J be a subset of $\{1, \ldots, n\}$. Then we may write

$$\bigcap_{i \in J} B_i = C_1 \times C_2 \times \cdots \times C_n$$

where $C_i = A_i$ if $i \in J$ and $C_i = S_i$ otherwise. Therefore,

$$P\left(\bigcap_{i \in J} B_i \right) = \prod_{i=1}^{n} P_i(C_i) = \prod_{i \in J} P_i(A_i) = \prod_{i \in J} P(B_i)$$

by Lemma 3.5.2 and Equation (5.2). The mutual independence follows.   ////

PROBLEMS

3.1   From an urn containing 5 red and 5 white balls, an unordered random sample of size 5 is drawn. Given that there are at least 2 red balls in the sample, find the conditional probability that there are exactly 3 red balls in the sample.

3.2   If in Problem 3.1 an ordered random sample had been drawn without replacement, what is the conditional probability that the sample contains exactly 3 red balls:
   (*a*) Given that the first 2 balls drawn are red?
   (*b*) Given that the first and last balls drawn are red?

3.3   If a balanced coin is tossed 5 times, what is the conditional probability of getting exactly 3 heads:
   (*a*) Given that there are at least 2 heads?
   (*b*) Given that the first and last tosses resulted in heads?

3.4 If a balanced die is rolled 6 times, what is the conditional probability of getting exactly 2 sixes:

(a) Given exactly 2 aces?

(b) Given at least 2 aces?

3.5 If two balanced dice are rolled, find the conditional probability that the sum of spots will be 7, given that it is odd.

3.6 In poker a flush consists of 5 cards of the same suit. Given that all the cards in a randomly selected poker hand are red (either hearts or diamonds), what is the conditional probability that the hand is a flush?

3.7 If George Gambler is dealt 4 spades and 1 nonspade and then discards the nonspade in order to draw another card, what is the conditional probability that he will successfully complete a flush?

3.8 In Problem 3.7, suppose that George had been dealt 3 spades and 2 nonspades. If he discards the 2 nonspades and draws 2 new cards, what is the conditional probability that he will complete his flush?

3.9 In bridge, suppose that North and South have 9 trumps in their combined hands but do not have the king of trumps. What is the conditional probability that the king is unguarded, that is, appears with no other trump in either East's hand or West's?

3.10 In Problem 3.9, suppose also that South has the ace of trumps. What is the conditional probability that the king is either unguarded or in West's hand (so that it can be finessed)?

3.11 A box contains three drawers. In one drawer there are 2 gold coins; in another there are 1 gold coin and 1 silver coin; and in the third drawer there are 2 silver coins. A drawer is selected at random, and then 1 coin is selected at random from that drawer. Given that the coin selected is gold, what is the conditional probability that the remaining coin in the opened drawer is also gold?

3.12 Voter registration in a certain city revealed the tabulated statistics. If a person is selected at random from the registered voters of this city, what is the conditional probability that the person will be male given that the person is a Democrat?

	Male, %	Female, %
Democrat	20	25
Independent	10	15
Republican	15	15

3.13 In Problem 3.12 what is the conditional probability that the person will be a Democrat given that the person is male?

3.14 Let $(S, \mathscr{S}, P)$ be a probability space, and let B be an event with positive probability. Define Q on $\mathscr{S}$ by $Q(A) = P(A \mid B)$. Show that Q is a probability measure on $\mathscr{S}$.

3.15 Show that in Problem 3.14 if C is an event with $Q(C) > 0$, then $Q(A \mid C) = P(A \mid BC)$ for all $A \in \mathscr{S}$.

3.16 A fair coin is tossed 5 times. Given that the first toss resulted in heads and that the 5 tosses produced at least 2 heads, what is the conditional probability that the 5 tosses resulted in exactly 2 heads?

3.17 A fair coin is tossed 10 times. Given that the 10 tosses produced exactly 5 heads, what is the conditional probability that:

(*a*) The first toss resulted in heads?

(*b*) Exactly 3 of the first 5 tosses resulted in heads?

3.18 A university finds that 75 percent of its graduating seniors scored above 80 on the entrance examination, while only 25 percent of those who fail to graduate score above 80. They also find that half of entering freshmen graduate. What is the conditional probability that a freshman will graduate given:

(*a*) That he scored above 80 on the entrance examination?

(*b*) Given that he scored 80 or below?

3.19 Suppose that 10 percent of the licensed drivers in a given state are incompetent. Suppose also that a diagnostic test is available, which is 90 percent effective in the following sense. If a driver is incompetent, the probability that the test will so indicate is 0.9; and if a driver is not incompetent, the probability that the test will so indicate is also 0.9. Given that the test indicates that a particular driver is incompetent, what is the conditional probability that the driver is in fact incompetent?

3.20 George Gambler always plays by the following strategy. If he is dealt a flush (five cards of the same suit), he plays it. If he is dealt 4 cards of one suit and 1 of another, he discards the nonmatching card and draws another. Otherwise, he does not attempt a flush. What is the probability that he will either be dealt or successfully attempt a flush?

3.21 Percy Paranoid is virtually certain that the coin he has is unfair. In fact, he attributes subjective probability 0.9 to the event that the coin has probability 0.75 of turning up heads and only probability 0.1 to the event that the coin is balanced (has probability 0.5 of turning up heads). If 4 independent tosses of the coin produce 2 heads and 2 tails, how should Percy modify his subjective probabilities?

3.22 Consider two urns. Urn I contains 4 red balls and 2 white balls, and urn II contains 3 balls of each color. If 2 balls are drawn from urn I without replacement and transferred to urn II and then a ball is drawn from urn II, what is the probability that the ball drawn from urn II will be red? Given that the ball drawn from urn II was red, what is the conditional probability that (*a*) 0, (*b*) 1, (*c*) 2 red balls were transferred?

3.23 In Problem 3.22 suppose that 2 balls are drawn without replacement from urn II. Given that both are red, what is the conditional probability that (*a*) 0, (*b*) 1, (*c*) 2 red balls were transferred?

3.24 In Example 3.2.4, (*a*) find the probability that a family has exactly k boys. (*b*) Find the conditional probability that a family has n children given that it has exactly k boys.

3.25 In Example 3.2.5, what is the conditional probability:

(*a*) That a driver will have an accident given that he is under 46 years of age?

(*b*) That a driver is under 45 years of age given that he has an accident?

NOTE: Problems 3.26 to 3.29 sketch an application of conditional probability to mathematical learning theory; see Estes (1959). Each day an experimental animal is exposed to a certain set of stimuli designed to elicit a particular response. Let A_k be the event that the animal makes the desired response on the kth day, and suppose that $P(A_{k+1} \mid A_k) = \beta$ and $P(A_{k+1} \mid A_k') = \alpha$, where $0 < \alpha < \beta \le 1$.

− 3.26 Let $p_k = P(A_k)$. Show that $p_{k+1} = \alpha + (\beta - \alpha)p_k$.

−3.27 If $\beta = 1$ and $p_1 = 0$, show that $p_k = 1 - (1 - \alpha)^{k-1}$.

− 3.28 Show that $\lim p_k = \alpha/(1 + \alpha - \beta)$ as $k \to \infty$.

3.29 If $\alpha = 0.05$, $\beta = 0.9$, and $p_1 = 0$, find the probability that the animal will make the desired response on days 11 and 12.

NOTE: Problems 3.30 to 3.34 develop properties of *Polya's urn scheme*, which may be described as follows. Balls are drawn sequentially from an urn which initially contains $r \ge 1$ red balls and $w \ge 1$ white balls. After each drawing, the ball drawn is put back into the urn along with $t \ge 1$ balls of the same color.

3.30 Show that the probability of drawing red balls on the first k draws is

$$\frac{(r/t + k - 1)_k}{(r/t + w/t + k - 1)_k}$$

3.31 Show that the probability of drawing red balls on the first k draws and white balls on the next j draws is

$$p = \frac{(r/t + k - 1)_k(w/t + j - 1)_j}{(r/t + w/t + n - 1)_n}$$

where $n = k + j$.

3.32 Show that the probability of drawing exactly k red balls on the first $n = k + j$ draws is $\binom{n}{k} p$, where p is as in Problem 3.31.

3.33 Show that the unconditional probability of drawing a red ball on the second draw is $r/(r + w)$.

3.34 Show that the probability of drawing a red ball on the nth draw is $r/(r + w)$ for every $n = 1, 2, \ldots$.

− 3.35 Three missiles are fired at a target. If their probabilities of hitting the target are 0.4, 0.5, and 0.6, respectively, and if the missiles are fired independently, what is the probability:

(a) That all three missiles hit the target?

(b) That at least one of the three hits the target?

3.36 In Problem 3.35 find the probability that (a) exactly 1; (b) exactly 2 of the missiles hit the target.

3.37   A die is loaded in such a manner that the probability that exactly k spots will appear when it is rolled is proportional to k. If two independent tosses of the die are made, what is the probability that the sum of spots will be 7?

3.38   In Problem 3.37 what is the probability that the same number of spots will appear on both dice?

3.39   Peter and Paul each toss a fair coin until a head has appeared:
(a) What is the probability that they will require the same number of tosses?
(b) What is the probability that Peter will require more tosses than Paul?
Assume the outcomes of all tosses to be independent.

3.40   In Problem 3.39 what is the probability that Peter will require at least twice as many tosses as Paul?

3.41   Two opinion pollers take independent random samples of size $k = 5$ without replacement from a population of 5 Democrats and 5 Republicans. What is the probability that the two samples will contain *exactly* the same number of Democrats?

3.42   If A is independent of A, what can be said about $P(A)$?

3.43   Show, directly from the definition, that if A, B, and C are independent, then so are A, B', and C'.

3.44   Let two balanced dice be rolled. Which of the following pairs of events are independent?
(a) A is the event that at most 2 spots appear on the first die, and B is the event that at least 2 appear on the second die.
(b) A is the event that the total number of spots on the two dice is odd, and B is the event that the total number of spots exceeds 7.

3.45   Let four cards be drawn without replacement from a standard bridge deck. Which of the following pairs of events are independent?
(a) A is the event that there are exactly 2 hearts, and B is the event that there is at least 1 spade.
(b) A is the event that there are at least 2 spades, and B is the event that there is at least 1 ace.

3.46   The independence or dependence of events depends not only on the events themselves but also on the probability function P. For example, consider one roll of a die, and let A be the event that either 1 or 6 spots appear, and let B be the event that an odd number of spots appear. Then, A and B are independent if the die is balanced, but they are not independent if the die is loaded in such a manner that the probability of obtaining k spots is proportional to k.

3.47   A string of Christmas tree lights is connected in series, so that if any one of the lights malfunctions, none of the lights glow. If there are 20 lights and each malfunctions with probability $p = 0.1$, what is the probability that all 20 lights will glow? Assume independence.

3.48   A fair die is rolled repeatedly. If 1 or 6 spots appear on the first roll, you win. If k spots appear on the first roll, where $2 \leq k \leq 5$, the die is rolled until 1, k, or 6 spots appear. If k spots appear before 1 or 6, then you win. Otherwise, you lose. Calculate the probability that you win. *Hint:* Let A_n be the event that you win after exactly n rolls and find the probability of $A_1 \cup A_2 \cup \cdots$.

3.49 Let S and T be finite or countably infinite sets, let P_0 be a probability measure on $\mathscr{S}$, the class of all subsets of S, and for each $s \in S$ let Q_s be a probability measure on $\mathscr{T}$, the class of all subsets of T.
(a) Define P on the class of subset of $S \times T$ by $P(B) = \sum_B Q_s(\{t\})P_0(\{s\})$, where the summation extends over all $(s,t) \in B$.
(b) Show that P is a probability measure.

3.50 As a continuation of Problem 3.49, for $A \subset S$, show that $P(A \times T) = P_0(A)$. Show also, that if $P_0(\{s\}) > 0$, then $P(S \times B \mid \{s\} \times T) = Q_s(B)$ for $B \subset T$.

4

THE BINOMIAL AND RELATED PROBABILITIES

4.1 THE BINOMIAL PROBABILITIES

In this section we shall consider independent events $A_1, \ldots, A_n$ with the same probability $P(A_i) = p$, $i = 1, \ldots, n$. One context in which such events arise is that of independent trials of the same experiment. Thus, let $(S_0, \mathscr{S}_0, P_0)$ be a probability space, and imagine the experiment to which $(S_0, \mathscr{S}_0, P_0)$ refers to be repeated n times, where n is positive integer. Further, let $A \in \mathscr{S}_0$ be an event which refers to the basic experiment, and let A_i be the event that A occurs on the ith trial (repetition). Then, as explained in Section 3.5, $A_1, \ldots, A_n$ are mutually independent events with the same probability $P(A_i) = P_0(A)$, $i = 1, \ldots, n$.

We shall now give a formula for the probability that *exactly k of the events $A_1, \ldots, A_n$ will occur.*

Theorem 4.1.1 *Let $A_1, \ldots, A_n$ be independent events with common probability $P(A_i) = p$, $i = 1, \ldots, n$. Then the probability that exactly k of $A_1, \ldots, A_n$ will occur is*

$$b(k;n,p) = \binom{n}{k} p^k q^{n-k} \qquad (1.1)$$

for $k = 0, \ldots, n$, where $q = 1 - p$.

PROOF   For any fixed subset $J \subset \{1, \ldots, n\}$, let

$$B_J = \left(\bigcap_{i \in J} A_i \right) \cap \left(\bigcap_{i \notin J} A_i' \right)$$

be the event that A_i occurs for $i \in J$ and A_i does not occur for $i \notin J$. If there are k elements in J, then

$$P(B_J) = \prod_{i \in J} P(A_i) \prod_{i \notin J} P(A_i') = p^k q^{n-k}$$

by the independence of $A_1, \ldots, A_n$, since $P(A_i) = p$ and $P(A_i') = 1 - p = q$, $i = 1, \ldots, n$. Now the event that exactly k of $A_1, \ldots, A_n$ occur is simply

$$E_k = \bigcup_{|J|=k} B_J$$

where the union extends over all subsets J of size k. Since the events B_J are mutually exclusive, and since there are $\binom{n}{k}$ subsets of size k, it now follows that

$$P(E_k) = \sum_{|J|=k} P(B_J) = \binom{n}{k} p^k q^{n-k}$$

as asserted.   ////

Equation (1.1) is one of the most important formulas in all of probability theory. Its right side defines the _binomial probabilities_, which are tabulated in Appendix Table C.1 for selected values of n and p.[1] As explained above, it applies to independent repetitions of any fixed experiment.

EXAMPLE 4.1.1   Theorem 4.1.1 contains Equation (5.4) of Chapter 1 as a special case. Indeed, if an ordered random sample of size n is drawn with replacement from an urn containing r red balls and w white balls, and if we let A_i be the event that a red ball is drawn on the ith draw, $i = 1, \ldots, n$, then $A_1, \ldots, A_n$ are independent with common probability $P(A_i) = p = r/(r + w)$, the proportion of red balls in the urn (Example 3.3.4). Note that $q = w/(r + w)$. Hence, the probability that exactly k red balls will be drawn is

$$\binom{n}{k} \left(\frac{r}{r + w} \right)^k \left(\frac{w}{r + w} \right)^{n-k}$$

which is (5.4) of Chapter 1 in a different notation.   ////

[1] More extensive tables will be found in Beyer (1966) or Selby (1965).

EXAMPLE 4.1.2

a   If a fair coin is tossed n times, the probability that exactly k heads will result is $b(k;n,\tfrac{1}{2}) = \binom{n}{k} 2^{-n}$. For the special case that $n = 8$, these probabilities are given in Table 5. The remaining values can be obtained by the symmetry $b(k;n,\tfrac{1}{2}) = b(n - k;n,\tfrac{1}{2})$, and a graph will be found in Figure 4.

b   If a pair of balanced dice are rolled n times, then the probability that exactly k of the rolls will produce a total of exactly 7 spots is $b(k;n,\tfrac{1}{6})$ because the probability that a total of 7 spots will result from one roll is $\tfrac{1}{6}$.

c   If a bridge player plays 8 hands during an evening, what is the probability that he will get no aces on exactly 4 of the hands? The probability that he will get no aces on a single hand is $p = \binom{48}{13} \Big/ \binom{52}{13} = 0.3038$. Thus, if the hands are dealt independently of one another, the probability of getting no aces on exactly 4 hands is $b(4;8,p)$. By linear interpolation in Table 1 we find $b(4;8,p) \approx 0.14$.

d   Suppose that the probability of curing a given disease in experimental animals with a given treatment is $p = 0.7$. If the treatment is administered independently to $n = 10$ such animals, then the probability that exactly 7 will be cured is $b(7;10,0.7) = 0.267$.   ////

For later reference, we observe the symmetry

$$b(k;n,p) = b(n - k, n; q) \qquad (1.2)$$

which was used above in a special case.

Some additional properties of $b(k;n,p)$ can be deduced from the identity

$$b(k;n, p) = \frac{(n - k + 1)p}{kq} b(k - 1; n, p) \qquad (1.3)$$

which holds for $0 < p < 1$ and $k = 1, \ldots, n$. To establish (1.3) observe that

$$b(k;n, p) = \binom{n}{k} p^k q^{n-k}$$

$$= \frac{n - k + 1}{k} \binom{n}{k - 1} p^k q^{n-k}$$

$$= \frac{n - k + 1}{k} \frac{p}{q} b(k - 1; n, p)$$

for $k = 1, \ldots, n$ and $0 < p < 1$.

Table 5

k	0	1	2	3	4
$b(k;8,\tfrac{1}{2}) =$	0.0039	0.0313	0.1094	0.2188	0.2734

Since $(n - k + 1)p > kq$ if and only if $k < (n + 1)p$, it follows from (1.3) that $b(k - 1; n, p) < b(k;n,p)$ for $k < (n + 1)p$. That is, $b(k;n,p)$ is an increasing function of k on the interval $0 \leq k < (n + 1)p$. Similarly, $b(k;n,p)$ is a decreasing function of k on the interval $(n + 1)p < k \leq n$. In particular, $b(k;n,p)$ is maximized by taking $k = [(n + 1)p]$, the greatest integer which is less than or equal to $(n + 1)p$. A more complete picture of the behavior of the binomial probabilities will be given in Section 4.5.

In the context of Theorem 4.1.1, it is of interest to ask for the probability that at least k or at most k of the events $A_1, \ldots, A_n$ occur. Letting E_k denote the event that exactly k of $A_1, \ldots, A_n$ occur, the latter events are

$$L_k = \bigcup_{j=k}^{n} E_j \quad \text{and} \quad M_k = \bigcup_{j=0}^{k} E_j$$

respectively. Since the events $E_0, \ldots, E_n$ are mutually exclusive, we now have the following corollary.

Corollary 4.1.1 *Let $A_1, \ldots, A_n$ be independent with common probability $P(A_i) = p$, $i = 1, \ldots, n$. Then*

$$P(L_k) = \sum_{j=k}^{n} b(j;n, p) \qquad (1.4a)$$

$$P(M_k) = \sum_{j=0}^{k} b(j;n, p) \qquad (1.4b)$$

for $k = 0, \ldots, n$.

EXAMPLE 4.1.3

a If a fair coin is tossed 20 times, the probability of obtaining exactly 10 heads is $b(10;20,0.5) = 0.1762$. The probability of obtaining at least 10 heads is $b(10;20,0.5) + \cdots + b(20;20,0.5) = 0.5881$.

b If the probability of curing a certain type of disease in experimental animals with a particular treatment is $p = 0.7$, and if the treatment is administered independently to 10 such animals, then the probability that at least 7 will be cured is $b(7;10,0.7) + \cdots + b(10;10,0.7) = 0.6496$.   ////

Theorem 4.1.1 can be generalized. Thus, consider a probability space $(S_0, \mathscr{S}_0, P_0)$, and let $A_1, \ldots, A_k$, $k \geq 2$, be mutually exclusive and exhaustive events. Further, denote the probability of A_i by p_i, so that $p_i \geq 0$, $i = 1, \ldots, k$ and $p_1 + \cdots + p_k = 1$. Now imagine the experiment to which $(S_0, \mathscr{S}_0, P_0)$ refers repeated independently n times, where n is a positive integer, and let $n_1, \ldots, n_k$ be integers for which

$$n_i \geq 0 \qquad i = 1, \ldots, k$$

$$n_1 + \cdots + n_k = n \qquad (1.5)$$

Then we can compute the probability that A_i occurs exactly n_i times during the n trials, $i = 1, \ldots, k$.

Theorem 4.1.2   *The probability that A_i occurs exactly n_i times, $i = 1, \ldots, k$, is*

$$m(n_1, \ldots, n_k; p) = \binom{n}{n_1, \ldots, n_k} p_1^{n_1} \cdots p_k^{n_k} \qquad (1.6)$$

for all $n_1, \ldots, n_k$ which satisfy (1.5). Here p denotes the vector $p = (p_1, \ldots, p_k)$, and

$$\binom{n}{n_1, \ldots, n_k} = \frac{n!}{n_1! \cdots n_k!}$$

denotes the multinomial coefficient.

PROOF   The proof of Theorem 4.1.2 is similar to that of Theorem 4.1.1, which it generalizes. Let A_{ij} be the event that A_i occurs on the jth trial, $i = 1, \ldots, k, j = 1, \ldots, n$, and for each partition $\alpha = (\alpha_1, \ldots, \alpha_k)$ of the integers $\{1, \ldots, n\}$, let

$$B_\alpha = \bigcap_{i=1}^{k} \bigcap_{j \in \alpha_i} A_{ij}$$

be the event that A_i occurs on trials $j \in \alpha_i$, $i = 1, \ldots, k$. Then, by independence,

$$P(B_\alpha) = \prod_{i=1}^{k} \prod_{j \in \alpha_i} P(A_{ij}) = \prod_{i=1}^{k} p_i^{r_i}$$

where $r_i = |\alpha_i|$ denotes the number of elements in α_i, $i = 1, \ldots, k$. Now, the event that A_i occurs exactly n_i times, $i = 1, \ldots, k$, is simply $C = \bigcup_\alpha B_\alpha$, where the union extends over all α for which $|\alpha_i| = n_i$, $i = 1, \ldots, k$. Therefore, since the events B_α are mutually exclusive, and since there are $\binom{n}{n_1, \ldots, n_k}$ such α, by Theorem 1.2.3, it follows that

$$P(C) = \binom{n}{n_1, \ldots, n_k} \prod_{i=1}^{k} p_i^{n_i}$$

as asserted.   ////

The probabilities (1.6) are known as the *multinomial probabilities*.

EXAMPLE 4.1.4

a   If a balanced die is tossed 12 times, the probability that each face appears exactly twice is

$$\binom{12}{2, \ldots, 2} 6^{-12} = \left(\frac{12!}{2^6}\right) 6^{-12} = 0.0034$$

(Let A_i be the event that exactly i spots appear on a single toss, $i = 1, \ldots, 6$, and observe that $p_1 = p_2 = \cdots = p_6 = \frac{1}{6}$.)

b In an evening of bridge, South plays 6 hands. What is the probability that South will have exactly 2 aces in exactly 2 hands, exactly 1 ace in exactly 2 hands, and no aces on exactly 2 hands? On a single hand, the probability that South receives exactly i aces is

$$p_i = \frac{\binom{4}{i}\binom{48}{13 - i}}{\binom{52}{13}}$$

for $i = 0, \ldots, 4$ by Theorem 1.4.1. The desired probability is therefore

$$\binom{6}{2,2,2,0,0} p_0{}^2 p_1{}^2 p_2{}^2 p_3{}^0 p_4{}^0 = \binom{6}{2,2,2} p_0{}^2 p_1{}^2 p_2{}^2 \qquad ////$$

4.2 THE NEGATIVE BINOMIAL PROBABILITIES

In this section we continue our study of independent trials of an experiment. Thus, let $(S_0, \mathscr{S}_0, P_0)$ be a probability space, and imagine the experiment to which $(S_0, \mathscr{S}_0, P_0)$ refers to be repeated n times, where n is a positive integer. As in the previous section, let $A \in \mathscr{S}_0$, and let A_i be the event that A occurs on the ith trial, $i = 1, \ldots, n$, so that $A_1, \ldots, A_n$ are mutually independent with common probability $p = P_0(A)$.

We shall compute the probability that A occurs for the rth time on the kth trial for any integers r and k with $1 \leq r \leq k \leq n$. For $r = 1$, this is easy. Indeed, the event that A occurs for the first time on the kth trial is simply $B_k = A'_1 \cap \cdots \cap A'_{k-1} \cap A_k$, so that $P(B_k) = P(A'_1) \cdots P(A'_{k-1})P(A_k) = pq^{k-1}$, where $q = 1 - p$. For $r > 1$, we have.

Theorem 4.2.1 *The probability that A occurs for the rth time on the kth trial is*

$$a(k;r,p) = \binom{k - 1}{r - 1} p^r q^{k-r} \qquad (2.1)$$

for $1 \leq r \leq k \leq n$. In particular, the probability that A occurs for the first time on the kth trial is

$$a(k;p) = a(k;1,p) = pq^{k-1} \qquad (2.2)$$

for $k = 1, \ldots, n$.

PROOF A will occur for the rth time on the kth trial if and only if A_k occurs and exactly $r - 1$ of $A_1, \ldots, A_{k-1}$ occur. Let B be the event that exactly $r - 1$ of $A_1, \ldots, A_{k-1}$ occur. Then

$$P(B) = b(r - 1; k - 1, p) = \binom{k - 1}{r - 1} p^{r-1} q^{k-r}$$

by Theorem 4.1.1. Moreover, A_k and B are independent, since B depends only on $A_1, \ldots, A_{k-1}$ by Theorems 3.4.1 and 3.4.2. Therefore,

$$P(BA_k) = P(B)P(A_k) = pP(B) = \binom{k - 1}{r - 1} p^r q^{k-r}$$

as asserted.   ////

EXAMPLE 4.2.1   In repeated tosses of a fair coin, the probability that the first head appears on the kth toss is 2^{-k}. The probability that the rth head appears on the kth toss is $\binom{k - 1}{r - 1} 2^{-k}$.   ////

EXAMPLE 4.2.2   The world series

a   Suppose that two teams I and II play a series of at most 7 games with the convention that the first team to win 4 games wins the series. Suppose also that the outcomes of the games are independent of each other and that team I has a constant probability p of winning on each game. Let B_k be the event that team I wins the series in exactly k games. Then B_k is the event that team I wins for the fourth time on the kth game, so that $P(B_k) = \binom{k - 1}{3} p^4 q^{k-4}$ for $k = 4, 5, 6, 7$. The event that team I wins the series is then $B = B_4 \cup B_5 \cup B_6 \cup B_7$, so that

$$P(B) = \sum_{k=4}^{7} \binom{k - 1}{3} p^4 q^{k-3}$$

See Table 6.

b   If the teams are evenly matched, $p = 0.5$, what is the probability that the series will require all 7 games? We require the probability that either team I

Table 6

p	0.55	0.60	0.65	0.70	0.75
$P(B)$	0.6083	0.7102	0.8002	0.8740	0.9294

wins for the fourth time on the seventh game or team II wins for the fourth time on the seventh game. The two events are mutually exclusive, and they have the same probability by symmetry. Thus, the answer is $2\binom{6}{3}2^{-7} = 0.3125$. ////

The right sides of (2.1) and (2.2) are independent of n, and hence $a(k;r,p) = \binom{k-1}{r-1}p^r q^{k-r}$ are defined for all $k = r, r+1, \ldots$. These numbers are known as the *negative binomial probabilities* and in the special case that $r = 1$, $a(k;p) = pq^{k-1}$, $k = 1, 2, \ldots$, are known as the *geometric probabilities*. We shall now show that they are, in fact, probabilities.

Lemma 4.2.1 *If $p > 0$, then for any $r = 1, 2, \ldots$, we have*

$$\sum_{k=r}^{\infty}\binom{k-1}{r-1}p^r q^{k-r} = 1 \qquad (2.3)$$

PROOF For fixed $n \geq 1$, consider n trials of an experiment, as in the opening paragraph of this section. Let C_n be the event that A occurs $r - 1$ or fewer times, and for $k = r, \ldots, n$ let B_k be the event that A occurs for the rth time after exactly k trials. Then $C_n, B_r, \ldots, B_n$ are mutually exclusive, exhaustive events, so that $P(C_n) + P(B_r) + \cdots + P(B_n) = 1$. Moreover, $P(B_k) = \binom{k-1}{r-1}p^r q^{k-r}$ for $k = r, \ldots, n$, so that

$$P(C_n) + \sum_{k=r}^{n}\binom{k-1}{r-1}p^r q^{k-r} = 1$$

Therefore, it will suffice to show that $\lim P(C_n) = 0$ as $n \to \infty$. To see this observe that

$$P(C_n) = \sum_{j=0}^{r-1}\binom{n}{j}p^j q^{n-j}$$

by Corollary 4.1.1 and that $\binom{n}{j}p^j q^{n-j} \sim p^j q^{-j}n^j q^n/j!$, which tends to zero as $n \to \infty$ for each fixed j, since $q < 1$. An alternate proof can be based on the generalized binomial theorem of Section 1.7. ////

The geometric probabilities have an interesting property which may be described as *lack of memory*. As in the introduction to this section, let A be an event, and let A_i be the event that A occurs on the ith of n independent trials of the experiment to which

A refers. Further, let us refer to the occurrence of A_i as "success" on the ith trial and to the nonoccurrence of A_i as "failure." Let C_k be the event that there are no successes during the first k trials. Equivalently, C_k may be described as the event that the first success takes place after the kth trial, if at all. We now claim that

$$P(C_{k+j} \mid C_k) = P(C_j) \qquad (2.4)$$

for all positive integers k and j for which $k + j \le n$. To see this, simply observe that $C_k = A_1' \cap \cdots \cap A_k'$, so that $P(C_k) = q^k$ by independence. Since C_{k+j} implies C_k, we now have $P(C_{k+j} \mid C_k) = P(C_{k+j})/P(C_k) = q^{k+j}/q_k = q^j = P(C_j)$, as asserted.

Equation (2.4) can be paraphrased as follows. Given that one has waited for at least k trials without a success, the conditional probability that one has to wait an additional j trials for a success is the same as the probability that one has to wait for j trials before a success at the beginning. That is, the process "forgets" the initial string of k failures. This property is, in fact, characteristic of the geometric probabilities (see Problem 5.12).

4.3 POISSON'S THEOREM: THE LAW OF RARE EVENTS

In this section we shall develop an approximation to the binomial probabilities

$$b(k;n,p) = \binom{n}{k} p^k q^{n-k}$$

which is valid when n is large, p is small, and the product $\beta = np$ is moderate. More precisely, we shall prove the following theorem.

Theorem 4.3.1 *Let* $p_1, p_2, \ldots$ *be a sequence of real numbers for which* $0 < p_n < 1$, $n \ge 1$, $\lim p_n = 0$, *and* $\lim np_n = \beta$, *as* $n \to \infty$, *where* $0 < \beta < \infty$. *Then*

$$\lim b(k;n,p_n) = \frac{1}{k!} \beta^k e^{-\beta}$$

as $n \to \infty$ *for every* $k = 0, 1, 2, \ldots$.

In the proof of Theorem 4.3.1, we shall need the following lemma from analysis.

Lemma 4.3.1 *Let* $x, x_1, x_2, \ldots$ *be a sequence of real numbers. If* $\lim x_n = x$ *as* $n \to \infty$, *then*

$$\lim \left(1 + \frac{x_n}{n} \right)^n = e^x$$

as $n \to \infty$.

PROOF   Since $x_n \to x$, it follows that $x_n/n \to 0$ as $n \to \infty$. Therefore, there is an integer n_0 for which $|x_n/n| \le \frac{1}{2}$ for $n \ge n_0$. For such n, we may use Taylor's theorem (Section 1.7) to write

$$\log\left(1 + \frac{x_n}{n}\right) = \frac{x_n}{n} + \frac{1}{2}(1 + y)^{-2}\left(\frac{x_n}{n}\right)^2$$

where $y = y(x_n, n)$ is an intermediate value and $|y| \le |x_n/n| \le \frac{1}{2}$. Therefore,

$$n\log\left(1 + \frac{x_n}{n}\right) = x_n + \frac{1}{2}(1 + y)^{-2}\left(\frac{x_n^2}{n}\right) \to x$$

as $n \to \infty$. Therefore,

$$\left(1 + \frac{x_n}{n}\right)^n = \exp\left[n\log\left(1 + \frac{x_n}{n}\right)\right] \to e^x$$

as $n \to \infty$.   ////

PROOF of Theorem 4.2.1   Let $\beta_n = np_n$. Then $\beta_n \to \beta$, by assumption, and

$$b(k; n, p_n) = \binom{n}{k} p_n^k (1 - p_n)^{n-k}$$

$$= \frac{1}{k!} \frac{(n)_k}{n^k} \beta_n^k \left(1 - \frac{\beta_n}{n}\right)^n \left(1 - \frac{\beta_n}{n}\right)^{-k} \qquad (3.1)$$

for $k = 0, 1, 2, \ldots$. As $n \to \infty$, the factors on the right side of (3.1) converge to $1/k!$, 1, β^k, $e^{-\beta}$, and 1, respectively. The theorem follows.   ////

The numbers

$$p(k; \beta) = \frac{1}{k!} \beta^k e^{-\beta} \qquad k = 0, 1, 2, \ldots \qquad (3.2)$$

are known as the *Poisson probabilities* and are tabulated in Appendix Table C.2 for various values of k and β. We remark that they are probabilities (sum to 1), because

$$\sum_{k=0}^{\infty} \frac{1}{k!} \beta^k = e^{\beta}$$

by Taylor's theorem (Section 1.7).

The content of Theorem 4.3.1 can now be stated somewhat more informally as follows. If n is large, p is small, and the product $\beta = np$ is moderate, then the binomial probabilities $b(k; n, p)$ can be approximated by the Poisson probabilities $p(k; \beta)$. The requirement that n be large and p be small leads to still another description of the Poisson probabilities. The Poisson probabilities $p(k; \beta)$ give the probability of the occurrence of exactly k of a large number n of improbable events (p small). For this reason, the Poisson probabilities are known as the *law of rare events*.

EXAMPLE 4.3.1 Suppose that a machine on an assembly line has probability $p = 0.01$ of producing a defective item each time it operates. If the machine produces 300 items during a given day, then the probability that exactly 4 of the 300 will be defective is approximately $p(4;3) = 0.168$. The probability that at most 4 of the 300 items will be defective is approximately $p(0;3) + p(1;3) + p(2;3) + p(3;3) + p(4;3) = 0.815$. 　　　　　　　　　　　　　　　　　　　　　　　　　　　////

EXAMPLE 4.3.2 Connections to a wrong number During a 24-hour period a telephone exchange handles a large number of calls, say n calls. There is also a small probability p that each call will be connected to a wrong number. We may therefore expect the probability of exactly k connections to a wrong number to be approximately $p(k;\beta)$, where $\beta = np$. 　　　　　　　　　　　　　////

EXAMPLE 4.3.3 Radioactive decay Consider a radioactive substance which emits radioactive particles at a rate of β per second. That is, suppose that during a long time interval, the average rate of emission is β per second (the number emitted during any particular second will, of course, be random). If there are a total of n particles in the substance, it seems reasonable to assume that each will be emitted with probability approximately $p = (1/n)\beta t$ during a time interval of length t. It also seems reasonable to suppose that the particles are emitted independently of each other. With these assumptions, it follows from Theorem 4.3.1 that the probability of exactly k emissions during a time interval of length t is approximately

$$p(k;\beta t) = \frac{1}{k!}(\beta t)^k e^{-\beta t}$$

for $k = 0, 1, 2, \dots$. The above derivation is incomplete, but the result is true, provided only that t is small when compared to the half-life of the substance. We shall return to this question in Section 7.6. 　　　　　　　　　　　　　　////

4.4 THE NORMAL CURVE

In the next section, we shall develop another approximation to the binomial probabilities $b(k;n,p)$. The new approximation is valid when npq is large and is therefore complementary to the Poisson approximation of Section 4.3.

The approximation involves the function

$$\phi(x) = \frac{1}{\sqrt{2\pi}} e^{-\frac{1}{2}x^2} \qquad -\infty < x < \infty \qquad (4.1)$$

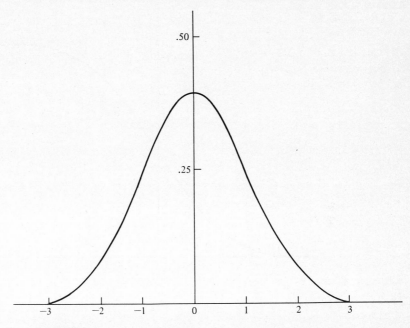

FIGURE 2
The standard normal density function.

which we shall refer to as the *standard normal density function* (Figure 2). Clearly, ϕ is symmetric about zero [$\phi(x) = \phi(-x)$], and ϕ attains its maximum value of $1/\sqrt{2\pi}$ at $x = 0$. Moreover, differentiation shows that $\phi''(x) = (1 - x^2)\phi(x)$, so that ϕ has inflection points at ± 1. Thus, the graph of ϕ is bell-shaped.

We shall need the following lemma.

Lemma 4.4.1 $\int_{-\infty}^{\infty} \phi(x)\,dx = 1$.

PROOF Let

$$I = \int_{-\infty}^{\infty} e^{-\frac{1}{2}x^2}\,dx$$

Then, we must show that $I = \sqrt{2\pi}$ or, equivalently, that $I^2 = 2\pi$. Now

$$I^2 = \iint_{-\infty}^{\infty} \exp\left[-\tfrac{1}{2}(x^2 + y^2)\right]\,dx\,dy$$

Make the change of variables[1] $x = r\cos\theta$, $y = r\sin\theta$. Then, $x^2 + y^2 = r^2$,

---

[1] Multiple integrals are discussed in Section 6.4, and the change-of-variable formula for multiple integrals is discussed in Section 7.4. For an elementary discussion of these concepts see, for example, Thomas (1972), Chap. 15.

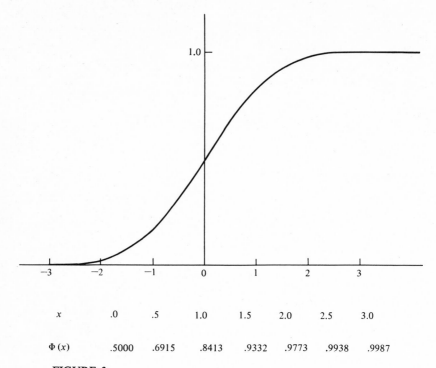

x	.0	.5	1.0	1.5	2.0	2.5	3.0
$\Phi(x)$	.5000	.6915	.8413	.9332	.9773	.9938	.9987

FIGURE 3

The standard normal distribution function.

and $dx\, dy = r\, dr\, d\theta$. Therefore,

$$I^2 = \int_0^\infty \int_0^{2\pi} re^{-\frac{1}{2}r^2}\, d\theta\, dr$$

$$= 2\pi \int_0^\infty re^{-\frac{1}{2}r^2}\, dr = 2\pi e^{-\frac{1}{2}r^2}\big|_0^\infty = 2\pi \qquad ////$$

We shall need a notation for the indefinite integral of ϕ. Let

$$\Phi(x) = \int_{-\infty}^x \phi(y)\, dy \qquad -\infty < x < \infty \qquad (4.2)$$

Φ is known as the *standard normal distribution function*. The expression for Φ cannot be simplified, but Φ has been computed numerically and is tabulated in Appendix Table C.3. We give an abbreviated form of this table in Figure 3.

The value of $\Phi(x)$ for negative values of x can be obtained from the identity

$$\Phi(-x) = 1 - \Phi(x) \qquad -\infty < x < \infty \qquad (4.3)$$

which follows easily from the symmetry of ϕ. In fact, the change of variables $u = -y$ shows

$$\Phi(-x) = \int_{-\infty}^{-x} \phi(y) \, dy = \int_{x}^{\infty} \phi(u) \, du = 1 - \Phi(x)$$

by symmetry of ϕ and Lemma 4.4.1.

Finally, we remark that $\Phi(x)$ approaches 1 quite rapidly as $x \to \infty$. In fact, we have the following inequality.

Lemma 4.4.2 *For $x > 0$, $1 - \Phi(x) < (1/x)\phi(x)$, and $1 - \Phi(x) \sim (1/x)\phi(x)$ as $x \to \infty$.*

PROOF The derivative of $\phi(x)$ is $-x\phi(x)$, and the derivative of $1 - \Phi(x)$ is $-\phi(x)$, so that

$$\phi(x) = \int_{x}^{\infty} y\phi(y) \, dy = x[1 - \Phi(x)] + \int_{x}^{\infty} [1 - \Phi(y)] \, dy \qquad (4.4)$$

for $x > 0$. The second equality follows from integration by parts. Now the second term in the last line of (4.4) is positive, so that $x[1 - \Phi(x)] \leq \phi(x)$ for $x > 0$, as asserted in the lemma. Let us now replace $1 - \Phi(y)$ by its upper bound $y^{-1}\phi(y)$ to obtain

$$\phi(x) \leq x[1 - \Phi(x)] + \int_{x}^{\infty} \frac{1}{y} \phi(y) \, dy$$

$$\leq x[1 - \Phi(x)] + \frac{1}{x} \int_{x}^{\infty} \phi(y) \, dy$$

$$= \left[x + \frac{1}{x}\right][1 - \Phi(x)]$$

for $x > 0$. Thus, $1 - \Phi(x) \sim (1/x)\phi(x)$ as $x \to \infty$. ////

4.5 NORMAL APPROXIMATION

The normal density ϕ can be used to approximate the binomial probabilities $b(k;n,p)$ when n is large. In fact, the following result is true and will be demonstrated in the next section. For fixed p, $0 < p < 1$, let

$$x_{nk} = \frac{k - np}{\sqrt{npq}} \qquad (5.1)$$

and define r_{nk} by

$$\sqrt{npq} \, b(k;n,p) = \phi(x_{nk}) + r_{nk}$$

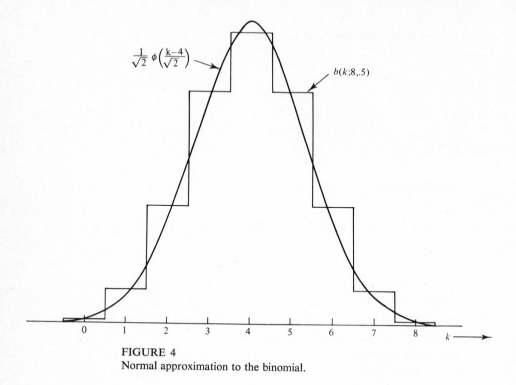

$\frac{1}{\sqrt{2}} \phi\left(\frac{k-4}{\sqrt{2}}\right)$

$b(k;8,.5)$

$k \longrightarrow$

FIGURE 4
Normal approximation to the binomial.

Then the remainder term r_{nk} is negligible when n is large in the sense that

$$\lim\left(\max_k |r_{nk}|\right) = 0 \qquad \text{as } n \to \infty$$

That is, we can approximate the binomial probabilities $b(k;n,p)$ by the simpler expression $\phi(x_{nk})/\sqrt{npq}$ when n is large, and we shall denote the relationship by writing

$$\sqrt{npq}\, b(k;n,p) \approx \phi(x_{nk}) \qquad (5.2)$$

As a corollary, we see that a bar graph of the binomial probabilities $b(k;n,p)$ has the approximate shape of the standard normal density centered at $k = np$ with units of width $1/\sqrt{npq}$ on both the k and $b(k;n,p)$ axes (Figure 4). When $p = 0.5$, the approximation is excellent for values of n as small as $n = 8$.

To state our next result, we shall use the following notation. We shall write $\Pr(\cdot)$ to denote the probability of the event described inside the parentheses. For example, if $A_1, \ldots, A_n$ are independent events with the same probability $P(A_i) = p$, and if X denotes the number of $A_1, \ldots, A_n$ which actually occur, then $\Pr(j \le X \le k)$ will denote the probability of the event that at least j and at most k of the events $A_1, \ldots, A_n$ occur for $0 \le j \le k \le n$. By Theorems 2.3.3 and 4.1.1, this is

$$\Pr(j \le X \le k) = \sum_{i=j}^{k} b(i;n,p)$$

We may now state the following result, which is related to (5.2). Let $A_1, \ldots, A_n$ and X be as above, and let $0 \leq j \leq k \leq n$. Define α and β by

$$\alpha = \frac{j - np - \frac{1}{2}}{\sqrt{npq}} \quad \text{and} \quad \beta = \frac{k - np + \frac{1}{2}}{\sqrt{npq}} \quad (5.3)$$

and define $r'_n = r'_n(j,k)$ by

$$\Pr\,(j \leq X \leq k) = \Phi(\beta) - \Phi(\alpha) + r'_n \quad (5.4)$$

Then the remainder term r'_n is negligible for large values of n. That is, we can approximate $\Pr\,(j \leq X \leq k)$ by the simpler expression $\Phi(\beta) - \Phi(\alpha)$, and we shall denote the relationship by

$$\Pr\,(j \leq X \leq k) \approx \Phi(\beta) - \Phi(\alpha) \quad (5.5)$$

It is hard to overemphasize the power of (5.5), for it gives a simple, effective approximation to complicated sums of binomial probabilities.

Relations (5.2) and (5.5) are known as the *DeMoivre-Laplace local and integral limit theorems*, respectively. We shall prove them in the next section. Relation (5.5) is a special case of the *central-limit theorem*, which we discuss in Section 9.4.

Let us now consider some examples.

EXAMPLE 4.5.1 The approximation (5.5) is generally quite good when p is close to $\frac{1}{2}$. Let X denote the number of heads in n tosses of a fair coin. We give in Table 7 the exact and approximate values of

$$\Pr\,(0 \leq X \leq k) = \sum_{i=0}^{k} b(i\,;n,\tfrac{1}{2})$$

for selected values of n and k.

Table 7   **EXACT AND APPROXIMATE VALUES OF Pr** $(X \leq k)$

$n = 8$ and $p = 0.5$				$n = 16$ and $p = 0.5$			
k	Exact	Approx.	Error	k	Exact	Approx.	Error
0	0.004	0.006	−0.002	0	0.0000	0.0001	−0.0001
1	0.035	0.038	−0.003	1	0.0003	0.0006	−0.0003
2	0.145	0.144	0.001	2	0.0021	0.0030	−0.0009
3	0.363	0.361	0.002	3	0.0106	0.0122	−0.0016
4	0.637	0.638	−0.001	4	0.0384	0.0401	−0.0017
				5	0.1051	0.1056	−0.0005
				6	0.2272	0.2266	0.0006
				7	0.4018	0.4013	0.0005
				8	0.5982	0.5987	−0.0005

By symmetry, the approximation must be just as good in the range $k > 0.5n$ as in the range $k < 0.5n$. Thus, the error (exact − approximate) is uniformly small for all k for n as small as 8. The relative error, (exact − approximate)/exact, will be large for small values of k, however. ·

Conversely, the approximation is generally poor if p is close to 0 or to 1. This is clear from the results of Section 4.3.   ////

EXAMPLE 4.5.2

a   In 400 tosses of a fair coin, what is the probability that the number of heads, X say, will differ from 200 by at most 10? We require

$$\Pr(190 \leq X \leq 210) = \sum_{i=190}^{210} b(i;400,\tfrac{1}{2})$$

Relation (5.5) applies with $n = 400$, $p = \frac{1}{2}$, $j = 190$, and $k = 210$. We find easily that $np = 200$, $\sqrt{npq} = 10$, $\alpha = -1.05$, and $\beta = 1.05$. From Appendix Table C.3, we then find that the desired probability is approximately

$$\Phi(\beta) - \Phi(\alpha) = 2\Phi(1.05) - 1 = 0.706$$

b   In 10,000 births, what is the probability that the proportion of males lies between 0.49 and 0.51? Let A_i be the event that the ith child is male, $i = 1,\ldots,$ 10^4, and suppose that the A_i are independent with probability 0.5. Then we require the probability that $4900 \leq X \leq 5100$, where X is the number of A_i that occur. We have $np = 5000$ and $\sqrt{npq} = 50$, so that

$$\alpha = \frac{4900 - 5000 - 0.5}{50} = -2.01$$

and similarly, $\beta = 2.01$. Thus, the required probability is approximately $\Phi(2.01) - \Phi(-2.01) = 2\Phi(2.01) - 1 = 0.956$.

c   A medical researcher believes that the probability of curing a particular type of disease in laboratory mice with a particular treatment is $p = 0.7$. If he is correct, and if he administers the treatment independently to 100 such mice, what is the probability that at least 65 of them will be cured? Here we have $n = 100$, $p = 0.7$, $j = 65$, and $k = 100$. After some calculation, we find $\Phi(\beta) - \Phi(\alpha) = 0.885$.   ////

We shall now discuss a practical application of (5.5). Consider a coin with unknown probability p of turning up heads. Equivalently, consider a drug which has unknown probability p of curing a disease, or consider a large electorate an unknown proportion p of which favors a particular candidate or issue. We might estimate p

as follows. We toss the coin n times; we count the number of heads which result, X_n say; and we compute the relative frequency of heads $F_n = (1/n)X_n$. According to the frequentistic interpretation of probability, F_n converges to p as $n \to \infty$, so that it seems reasonable to estimate p by F_n. We might therefore ask for preassigned constants $\varepsilon > 0$ and γ, $0 < \gamma < 1$, how large n must be in order that

$$\Pr\left(|F_n - p| \le \varepsilon\right) \ge \gamma \qquad (5.6)$$

The γ then serves as a natural measure of our confidence that, in fact, $|F_n - p| \le \varepsilon$, and ε serves to measure the accuracy of our estimate. For example, if we knew that (5.6) held with $\varepsilon = 0.01$ and $\gamma = 0.99$, then we could be virtually certain that our estimate F_n would be within 0.01 of the unknown p.[1]

Using (5.5), we can find an n such that (5.6) is approximately satisfied.

EXAMPLE 4.5.3 Given $\varepsilon, \gamma, 0 < \varepsilon, \gamma < 1$, how large must n be for (5.6) to be "approximately" satisfied? Let j be the smallest integer which is greater than or equal to $n(p - \varepsilon)$, and let k be the greatest integer which is less than or equal to $n(p + \varepsilon)$. Then

$$\Pr\left(|F_n - p| \le \varepsilon\right) = \sum_{i=j}^{k} b(i;n,p)$$

which is approximately $\Phi(\beta) - \Phi(\alpha)$ with $\alpha = (j - np - \frac{1}{2})/\sqrt{npq}$ and $\beta = (k - np + \frac{1}{2})/\sqrt{npq}$. Let $\sigma^2 = pq$. Then, both β and $-\alpha$ differ from $\sqrt{n}\, \varepsilon\sigma^{-1}$ by at most $1/\sigma\sqrt{n}$. Since Φ is continuous, it follows that $\Phi(\beta) - \Phi(\alpha)$ is approximately

$$\Phi(\sqrt{n}\, \varepsilon\sigma^{-1}) - \Phi(-\sqrt{n}\, \varepsilon\sigma^{-1}) = 2\Phi(\sqrt{n}\, \varepsilon\sigma^{-1}) - 1$$

Thus, if n is so large that $2\Phi(\sqrt{n}\, \varepsilon\sigma^{-1}) - 1 \ge \gamma$, then (5.6) should be approximately satisfied. That is, we must have

$$n \ge \frac{\sigma^2 \Phi^{-1}[(1 + \gamma)/2]^2}{\varepsilon^2}$$

where Φ^{-1} denotes the inverse function to Φ. Finally, since $\sigma^2 = pq = p(1 - p) \le \frac{1}{4}$ for $0 < p < 1$, as is easily verified by differentiation, we see that the latter condition will be satisfied if $n \ge n_0$, where

$$n_0 = \frac{\Phi^{-1}[(1 + \gamma)/2]^2}{4\varepsilon^2} \qquad (5.7)$$

Thus n_0 appears to be the appropriate choice of n. ////

In applications, Table 8 will be useful.

---

[1] Statisticians refer to the interval $[F_n - \varepsilon, F_n + \varepsilon]$ as a *confidence interval* and to γ as the *confidence coefficient*.

EXAMPLE 4.5.4   Suppose that two candidates, A and B, are running for an office. Let p be the proportion of the electorate which favors candidate A. In order to estimate p, an opinion poll is taken. That is, a random sample of size n is selected from the electorate and asked their preference. Let F_n denote the proportion of the sample which favor A. How large should n be chosen in order that $\Pr\,(|F_n - p| \le 0.05) \ge 0.95$, approximately?

If the electorate is large, we may ignore the difference between sampling without replacement and sampling with replacement (see Section 1.5). For sampling with replacement, (5.7) applies with $\varepsilon = 0.05$ and $\gamma = 0.95$ to yield $n_0 = 384$ to the nearest integer. /////

Table 8

γ	0.900	0.950	0.975	0.990	0.995
$\Phi^{-1}(\gamma)$	1.282	1.645	1.960	2.326	2.576

4.6   THE DEMOIVRE-LAPLACE THEOREMS[1]

In this section we shall discuss the proofs of (5.2) and (5.5). Recall that $a_n \sim b_n$ means $a_n b_n^{-1} \to 1$ as $n \to \infty$.

Theorem 4.6.1   *Let $0 < p < 1$, and let k_n be any sequence of integers for which $0 \le k_n \le n$ for $n \ge 1$, and let*

$$x_{nk_n} = \frac{k_n - np}{\sqrt{npq}} \qquad (6.1)$$

If k_n depends on n in such a manner that $\lim n^{-1/6} x_{nk_n} = 0$ as $n \to \infty$, then $\sqrt{npq}\, b(k_n;n,p) \sim \phi(x_{nk_n})$ as $n \to \infty$.

PROOF   To simplify the notation let us write k for k_n, x for x_{nk_n}, and j for $n - k$. Then

$$k = np + x\sqrt{npq} \qquad (6.2a)$$

$$j = nq - x\sqrt{npq} \qquad (6.2b)$$

by definition of x. Moreover, since $n^{-1/6} x \to 0$ as $n \to \infty$, we must also have $n^{-1/2} x \to 0$ as $n \to \infty$, so that $k/n \to p$ and $j/n \to q$ as $n \to \infty$. In particular,

---

[1] This section may be omitted without loss of continuity.

both k and j tend to infinity as $n \to \infty$, so we may apply Stirling's formula (Section 1.8) to deduce that

$$k! \sim \sqrt{2\pi k}\, k^k e^{-k}$$

$$j! \sim \sqrt{2\pi j}\, j^j e^{-j}$$

$$n! \sim \sqrt{2\pi n}\, n^n e^{-n}$$

as $n \to \infty$. If we substitute these relations into the definition of $b(k;n,p)$ and use the fact that $k + j = n$, we find that

$$b(k;n,p) = \frac{n!}{k!\,j!}\, p^k q^j$$

$$\sim \sqrt{\frac{n}{2\pi kj}} \left(\frac{k}{np}\right)^{-k} \left(\frac{j}{nq}\right)^{-j}$$

as $n \to \infty$. Thus,

$$\frac{\sqrt{npq}\; b(k;n,p)}{\phi(x)} = A_n B_n \qquad (6.3)$$

where

$$A_n = \sqrt{\frac{n^2 pq}{kj}} \quad \text{and} \quad B_n = \left(\frac{k}{np}\right)^{-k} \left(\frac{j}{nq}\right)^{-j} e^{\frac{1}{2}x^2}$$

so that it will suffice to show that $\lim A_n = 1$ and $\lim B_n = 1$ as $n \to \infty$. That $A_n \to 1$ as $n \to \infty$ is clear since $k/n \to p$ and $j/n \to q$, as we observed above.

To show that $B_n \to 1$ as $n \to \infty$, write k and j in the form of (6.2) to obtain

$$\log B_n = -k \log \frac{k}{np} - j \log \frac{j}{nq} + \tfrac{1}{2}x^2$$

$$= -(np + x\sqrt{npq}) \log \left(1 + x\sqrt{\frac{q}{np}}\right)$$

$$- (nq - x\sqrt{npq}) \log \left(1 - x\sqrt{\frac{p}{nq}}\right) + \tfrac{1}{2}x^2 \qquad (6.4)$$

Now, since $n^{-\frac{1}{2}}x \to 0$ as $n \to \infty$, we must have

$$\left| x\sqrt{\frac{q}{np}} \right| \leq \frac{1}{2} \quad \text{and} \quad \left| x\sqrt{\frac{p}{nq}} \right| \leq \frac{1}{2}$$

for n sufficiently large, say $n \geq n_0$. For such n, we can expand the logarithmic terms in Taylor series about 0 to obtain

$$\log\left(1 + x\sqrt{\frac{q}{np}}\right) = x\sqrt{\frac{q}{np}} - \frac{1}{2}x^2\frac{q}{np} + R_n$$

$$\log\left(1 - x\sqrt{\frac{p}{nq}}\right) = -x\sqrt{\frac{p}{nq}} - \frac{1}{2}x^2\frac{p}{nq} + R'_n$$

(6.5)

where

$$R_n = \frac{1}{3}\left(\frac{1}{1+\theta}\right)^3\left(x\sqrt{\frac{q}{np}}\right)^3$$

$$R'_n = \frac{1}{3}\left(\frac{1}{1+\theta'}\right)^3\left(-x\sqrt{\frac{p}{nq}}\right)^3$$

with $|\theta| \leq \frac{1}{2}$ and $|\theta'| \leq \frac{1}{2}$. If we next substitute (6.5) into (6.4), we find that

$$\log B_n = -(np + x\sqrt{npq})\left(x\sqrt{\frac{q}{np}} - \frac{1}{2}x^2\frac{q}{np} + R_n\right)$$

$$-(nq - x\sqrt{npq})\left(-x\sqrt{\frac{p}{nq}} - \frac{1}{2}x^2\frac{p}{nq} + R'_n\right) + \frac{1}{2}x^2$$

which simplifies to

$$\log B_n = -(np + x\sqrt{npq})R_n$$

$$- (nq - x\sqrt{npq})R'_n + \frac{1}{2}x^3\left\{q\sqrt{\frac{q}{np}} - p\sqrt{\frac{p}{nq}}\right\}$$

Finally,

$$|np + x\sqrt{npq}||R_n| \leq k_n|R_n|$$

$$\leq n\left(\frac{1}{3}\right)\left(\frac{1}{1+\theta}\right)^3\left|x\sqrt{\frac{q}{np}}\right|^3$$

$$\leq \frac{8}{3}\left(\frac{q}{p}\right)^{3/2}n^{-\frac{1}{2}}|x^3| \to 0$$

as $n \to \infty$ since $n^{-1/6}x \to 0$ as $n \to \infty$, by assumption. Similarly,

$$\lim |nq - x\sqrt{npq}||R'_n| = 0 \quad \text{and} \quad \lim x^3\left\{q\sqrt{\frac{q}{np}} - p\sqrt{\frac{p}{nq}}\right\} = 0$$

as $n \to \infty$, so that $\lim \log B_n = 0$ as $n \to \infty$. That is, $\lim B_n = 1$ as $n \to \infty$, as required. ////

We have shown that the ratio of $\sqrt{npq}\, b(k;n,p)$ to $\phi(x_{nk})$ is close to 1 provided that k is not too distant from np in the sense that $n^{-1/6}x_{nk} \to 0$ as $n \to \infty$. We shall now show that the difference is small for all k. In effect, we show that both $b(k;n,p)$ and $\phi(x_{nk})$ are small if k is distant from np.

Theorem 4.6.2   *For $k = 0, \ldots, n$, define x_{nk} and r_{nk} by*

$$x_{nk} = \frac{k - np}{\sqrt{npq}}$$

and
$$r_{nk} = \sqrt{npq}\, b(k;n,p) - \phi(x_{nk})$$

Then $\lim (\max_k |r_{nk}|) = 0$ *as $n \to \infty$.*

    PROOF   Select integers $i_n < (n + 1)p$ and $j_n > (n + 1)p$ for which $x_{ni_n} \to -\infty$, $x_{nj_n} \to \infty$, and $n^{-1/6}(x_{nj_n} - x_{ni_n}) \to 0$ as $n \to \infty$. Then

$$\sqrt{npq}\, b(i_n;n,p) \sim \phi(x_{ni_n})$$

by Theorem 4.6.1 and $\phi(x_{ni_n}) \to 0$ since $x_{ni_n} \to -\infty$. Now since $b(k;n,p)$ is an increasing function of k for $k < (n + 1)p$ by (1.3), and since $\phi(x)$ is an increasing function of x for $x < 0$, we must have

$$\max_{k \le i_n} |r_{nk}| \le \sqrt{npq}\, b(i_n;n,p) + \phi(x_{ni_n})$$

which tends to zero as $n \to \infty$. Similarly, $\max_{k \ge j_n} |r_{nk}| \to 0$ as $n \to \infty$. Finally, we may select a k_n for which $i_n \le k_n \le j_n$ and

$$|\sqrt{npq}\, b(k_n;n,p) - \phi(x_{nk_n})| = \max_{i_n \le k \le j_n} |r_{nk}| \qquad (6.6)$$

and the left side of (6.6) tends to zero as $n \to \infty$ by Theorem 4.6.1.   ////

We shall now turn our attention to the proof of (5.5). For simplicity, we shall consider only the case that α and β remain bounded as $n \to \infty$, although (5.5) is true without this restriction.

Theorem 4.6.3   *For each n let j_n and k_n be positive integers for which $0 \le j_n < k_n \le n$, and let*

$$\alpha_n = \frac{j_n - np - \tfrac{1}{2}}{\sqrt{npq}} \quad \text{and} \quad \beta_n = \frac{k_n - np + \tfrac{1}{2}}{\sqrt{npq}}$$

If there is a constant c for which $-c \leq \alpha_n < \beta_n \leq c$ *for all* $n = 1, 2, \ldots,$ *then*

$$\sum_{i=j_n}^{k} b(i;n,p) = \Phi(\beta_n) - \Phi(\alpha_n) + r'_n$$

where $r'_n \to 0$ *as* $n \to \infty$.

PROOF We have

$$\sum_{i=j_n}^{k_n} b(i;n,p) = \frac{1}{\sqrt{npq}} \sum_{i=j_n}^{k_n} \phi(x_{ni})$$

$$+\frac{1}{\sqrt{npq}} \sum_{i=j_n}^{k_n} \left[\sqrt{npq}\, b(i;n,p) - \phi(x_{ni})\right] = I_n + R_n, \text{ say}$$

Now, since $x_{ni} = x_{n(i-1)} = 1/\sqrt{npq}$, I_n is a Riemann sum approximating

$$\int_{\alpha_n}^{\beta_n} \phi(x)\, dx = \Phi(\beta_n) - \Phi(\alpha_n)$$

and the remainder term R_n is bounded by

$$R_n \leq \frac{1}{\sqrt{npq}} (k_n - j_n) \max_k |r_{nk}| \leq (\beta_n - \alpha_n) \max_k |r_{nk}|$$

which tends to zero as $n \to \infty$. The theorem follows.   ////

PROBLEMS

4.1 If a bridge player plays 6 hands of bridge during an evening, what is the probability that he will get:
(a) Exactly 2 aces on exactly 2 of the hands?
(b) At least 2 aces on at least 2 of the hands?

4.2 If two balanced dice are rolled 4 times, what is the probability that at least 2 of the rolls will produce at least 9 total spots?

4.3 Two chess players, A and B say, play a series of 10 games. Suppose that the outcomes of the 10 games are independent and that each player has probability 0.5 of winning each game. What is the probability that one of the players will win more games than the other?

4.4 In the previous problem suppose that A and B play 9 games and that A has probability $p = 0.6$ of winning each game. What is the probability that A will win more games than B?

4.5 Mandrake, a magician, claims to have extrasensory perception. In order to test this claim, a balanced coin is tossed 8 times, and he is asked to predict the outcome of each toss. Assuming that Mandrake is actually guessing, what is the probability that he will correctly guess at least 6 of the 8 outcomes?

4.6   In Problem 4.5 suppose that Mandrake actually does have extrasensory perception. Suppose that he can correctly call the toss of a coin with probability $\frac{3}{4}$. What is the probability that he will correctly call at least 6 of the 8 tosses?

4.7   Suppose that items on an assembly line must undergo 10 operations in order to become finished products. Suppose also that each operation malfunctions with probability $p = 0.01$. If 10 items pass through the line, what is the probability that none of the operations will malfunction on (a) exactly 8 of the items; (b) on at least 8 of the items? Assume all 100 operations to be independent.

4.8   Consider a multiple-choice examination with 10 questions, each of which has 4 possible answers. If a student knows the correct answer with probability 0.8 and guesses with probability 0.2, what is the probability that he will correctly answer (a) exactly 8 of the 10 questions; (b) at least 7 of the 10 questions? Assume his answers to the 10 questions to be independent.

4.9   Two fair coins are tossed n times. Given that there were exactly k heads in the $2n$ tosses, what is the conditional probability that there were exactly j heads in the n tosses of the first coin?

4.10   A balanced die is rolled 4 times. Given that no aces and no sixes appear, what is the conditional probability that every other face appears exactly once?

4.11   Let two balanced dice be tossed 6 times. What is the probability that exactly 2 of the tosses produce a total number of spots less than 7, exactly 2 produce a total number of spots equal to 7, and exactly 2 produce a total sum of spots greater than 7?

4.12   *Banach's match-box problem*   A smoker starts out the morning with two boxes, each of which contains n matches. Each time he needs a match, he selects one of the two boxes at random and takes a match from it. What is the probability that the $(n + k)$th match will empty one of the boxes? *Hint:* Let A_i be the event that the ith match is taken from box I, and suppose that the A_i are independent with common probability $\frac{1}{2}$.

> NOTE: Problems 4.13 to 4.16 introduce an application of the binomial and multinomial probabilities to genetics. Heritable characteristics are determined by carriers called *genes*, which appear in pairs. In the simplest case, the genes may assume only two forms a and A, so that there are three possible *genotypes* (pairs) aa, Aa, and AA. There is no distinction between Aa and aA. In sexual reproduction, the genotype of an offspring is determined as follows: one gene is selected at random from each parent, and the selections are independent. Also, the selection of genotypes for different offspring are independent.

4.13   If both parents are of type Aa, what is the probability that an offspring will be of type aa; Aa; AA?

4.14   If two parents of type Aa have 6 offspring, what is the probability that exactly 3 of the offspring will be of type Aa?

4.15   If two parents of type Aa have 6 offspring, what is the probability that exactly 2 of the offspring will be of each genotype?

4.16 Answer Problems 4.13 to 4.15 when the genotypes of the parents are:

    (a) *Aa* and *aa*   (b) *Aa* and *AA*   (c) *AA* and *aa*

4.17 A coin with probability $p > 0$ of turning up heads is tossed until 3 heads have appeared. What is the probability that an even number of tosses will be required?

4.18 A fair coin is tossed until 2 heads have appeared. Let α_k be the probability that at least k tosses will be required and find the smallest integer k for which $\alpha_k \leq \frac{1}{2}$.

4.19 A fair coin is tossed until 2 heads have appeared. Given that more than 3 tosses are required, what is the conditional probability that more than 6 tosses will be required?

4.20 Use the generalized binomial theorem to prove that $\sum_{k=r}^{\infty} a(k;r,p) = 1$ for $p > 0$ and $r \geq 1$.

4.21 A coin is tossed until 2 heads have appeared. Given that exactly k tosses were required, what is the conditional probability that the first toss resulted in heads?

4.22 A coin is tossed until r heads have appeared. Given that exactly k tosses were required, what is the conditional probability that the jth toss resulted in heads, $j = 1, \ldots, k - 1$?

4.23 In $n = 1000$ tosses of a coin which has probability $p = 0.005$ of turning up heads on each toss, estimate the probability that:

    (a) Exactly 5 heads will appear.

    (b) At least 5 heads will appear.

    (c) At most 5 heads will appear.

4.24 The bottle-capping machine at the XYZ Beer Company malfunctions with probability $p = 0.001$ on each bottle that it attempts to cap. If it attempts to cap 2500 bottles in a day, what is the probability that it will malfunction on more than 10 bottles?

4.25 The ABC Cookie Company puts n chocolate chips into a vat of dough from which it makes m cookies and finds that the resulting cookies contain exactly k chocolate chips with probability $p(k;\beta)$, where $\beta = n/m$. If it wishes to make $m = 10,000$ cookies from a particular vat, how many chocolate chips should it put into the vat for 95 percent of the resulting cookies to contain at least 5 chocolate chips?

4.26 A radioactive substance emits α particles with intensity $\beta = 0.1$ per microsecond. What is the probability that there will be more than 2 emissions during the first 10 microseconds?

4.27 In Problem 4.26 find that number t for which the probability of at least 1 emission during the first t microseconds is 0.5.

> NOTE: Problems 4.28 to 4.32 refer to n independent tosses of a coin which has probability p of turning up heads on each toss. X denotes the number of heads.

4.28 If $n = 10$ and $p = \frac{1}{2}$, find the exact and approximate values of the probability that X is less than or equal to k for $k = 1, \ldots, 5$.

4.29 If $n = 100$ and $p = \frac{1}{3}$, estimate the probability that (a) X is greater than 35? (b) that X lies between 25 and 35 inclusive.

4.30 Let $F = X/n$. If $p = \frac{1}{2}$, how large should n be in order that (approximately) the probability that $|F - \frac{1}{2}| \leq 0.1$ is at least 0.95.

4.31 Let $F = X/n$. Find an n for which (approximately) the probability that $|F - p| \le 0.05$ is at least 0.95 for all p, $0 < p < 1$.

4.32 Let $F = X/n$. If $n = 100,000$ and $p = \frac{1}{2}$, estimate the probability that $|F - \frac{1}{2}| \le 0.01$.

4.33 A balanced die is tossed 12,000 times. Estimate the probability that the number of aces lies between 1800 and 2200 inclusive.

4.34 If 12,000 tosses of a die produced a total of 2500 aces, would it be reasonable to conclude that the die is not balanced?

4.35 In Example 4.5.4, how large should n be in order that approximately

$$\Pr\left(|F_n - p| \le 0.01\right) \ge 0.95$$

for all p?

4.36 To estimate the probability p with which a particular treatment will cure a given disease, the treatment is administered independently to n experimental animals. Let X_n denote the number of animals which are cured, and let $F_n = (1/n)X_n$. How large should n be in order that (a) approximately $\Pr\left(|F_n - p| \le 0.02\right) \ge 0.95$ for all p; (b) approximately $\Pr\left(|F_n - p| \le 0.01\right) \ge 0.99$ for all p?

> NOTE: Problems 4.37 and 4.38 develop an approximation to the negative binomial probabilities. Problems 4.39 and 4.40 develop an approximation to the Poisson probabilities.

4.37 Consider the negative binomial probabilities $a(k;r,p)$ as $p \to 0$ and $k \to \infty$ in such a manner that $kp \to x > 0$. Show that

$$\left(\frac{1}{p}\right) a(k;r,p) \to \frac{x^{r-1}e^{-x}}{(r-1)!}$$

4.38 In the notation of Problem 4.37 show that

$$\sum_{a \le kp \le b} a(k;r,p) \to \int_a^b \frac{x^{r-1}e^{-x}}{(r-1)!}\,dx$$

4.39 Let

$$p(k;\beta) = \frac{1}{k!}\,\beta^k e^{-\beta} \quad \text{and} \quad x = \frac{k - \beta}{\sqrt{\beta}}$$

Use Stirling's formula to show that if $k = k_\beta$ depends on β in such a manner that x remains bounded as $\beta \to \infty$, then $\sqrt{\beta}\, p(k;\beta) \sim \phi(x)$ as $\beta \to \infty$.

4.40 In the notation of Problem 4.39 show that

$$\sum_{a\sqrt{\beta} \le k-\beta \le b\sqrt{\beta}} p(k;\beta) \to \Phi(b) - \Phi(a)$$

as $\beta \to \infty$.

4.41 Estimate $\sum_{k=90}^{110} p(k;100)$.

5

RANDOM VARIABLES

5.1 RANDOM VARIABLES

In many problems, we are not concerned with all aspects of the outcome of an experiment but only in a particular numerical characteristic of the outcome, such as the number of red balls in a sample, the height of a randomly selected man, or the income of a randomly selected family. We can abstract the notion of an interesting numerical characteristic as follows. Consider a probability space $(S,\mathscr{S},P)$, and let X denote a real-valued function which is defined on the sample space S. Thus, S may be thought of as the set of possible outcomes of some game or experiment and X as a rule which assigns to each possible outcome $s \in S$ a uniquely defined real number $X(s)$. We shall call X a *random variable* if for every interval I of real numbers the subset of S

$$\{s \in S: X(s) \in I\} \qquad (1.1)$$

is an event, that is, belongs to $\mathscr{S}$. In this case we shall refer to (1.1) as *the event that X belongs to I* and write

$$\Pr (X \in I) = P(\{s \in S: X(s) \in I\}) \qquad (1.2)$$

The requirement that (1.1) be an event guarantees that the right side of (1.2) is well defined. Since many interesting events can be written in the form (1.1), with an appropriate choice of X and I, we shall find the notation (1.2) very useful. More generally, *we shall use the notation* Pr $(\cdot)$ *to denote the probability of the event described inside the parentheses.* For example, Pr $(a < X < b)$ means $P(\{s \in S: a < X(s) < b\})$, Pr $(X = a)$ means $P(\{s \in S: X(s) = a\})$, Pr $(X \leq a)$ means $P(\{s \in S: X(s) \leq a\})$, etc.

We have already considered one random variable and used the notation (1.2) and its variations in the previous chapter, when we considered the number of heads which result from n independent tosses of a coin. Example 5.1.2a gives the details.

EXAMPLE 5.1.1

a Let $(S, \mathscr{S}, P)$ be any probability space, and let A be any event. Then the function X defined by

$$X(s) = \begin{cases} 1 & \text{if } s \in A \\ 0 & \text{if } s \notin A \end{cases}$$

is known as the *indicator of A*. We shall denote X by I_A. Thus, $I_A = 1$ if A occurs, and $I_A = 0$ if A does not occur. We then have Pr $(I_A = 1) = P(\{s: I_A(s) = 1\}) = P(A)$, and Pr $(I_A = 0) = P(A') = 1 - P(A)$.

b Let $A_1, \ldots, A_n$ be any events. Then the function X defined by

$$X(s) = I_{A_1}(s) + \cdots + I_{A_n}(s)$$

for $s \in S$ counts the number of $A_1, \ldots, A_n$ which occur.

c If $A_1, \ldots, A_n$ are mutually exclusive and exhaustive events, then the function X defined by

$$X(s) = \sum_{k=1}^{n} k I_{A_k}(s) \qquad (1.3)$$

computes the index of that A_i which occurs. (All but one of the terms in the sum are 0.) In this case, Pr $(X = k) = P(\{s: X(s) = k\}) = P(\{s: I_{A_k}(s) = 1\}) = P(A_k)$ for $k = 1, \ldots, n$.

Any random variable which assumes only the values $1, \ldots, n$ can be represented in the form (1.3) by simply letting A_k be the event that $X = k$ for $k = 1, \ldots, n$. ////

EXAMPLE 5.1.2

a If a coin is tossed independently n times, the number of heads which appear can be represented as a random variable, as in Example 5.1.1b, by letting A_i be the event that heads appear on the ith toss. From Theorem 4.1.1, we then have

$$\text{Pr } (X = k) = \binom{n}{k} p^k q^{n-k}$$

for $k = 0, \ldots, n$, where p denotes the probability of heads on a single toss and $q = 1 - p$.

b   Similarly, if an unordered random sample of size k is drawn from an urn which contains m red balls and $n - m$ white balls, we can represent the number of red balls in the sample as a random variable, as in Example 5.1.1*c*, and we find that

$$\Pr(X = r) = \frac{\binom{m}{r}\binom{n - m}{k - r}}{\binom{n}{k}}$$

for $r = 0, \ldots, k$ by Theorem 1.4.1.   ////

EXAMPLE 5.1.3   Let $(S, \mathcal{S}, P)$ be any probability space for which S is an interval and $\mathcal{S}$ is the class of Borel subsets of S. Then the function X defined by $X(s) = s$ for $s \in S$ is a random variable. In fact, $\{s: X(s) \in I\} = IS$ is an interval for any interval $I \subset R$. If S is thought of as the set of possible outcomes of the experiment, then X effectively computes the actual outcome of the experiment. Accordingly, we shall refer to X as *the outcome of the experiment*. Observe that if I is a subinterval of S, then $\Pr(X \in I) = P(IS) = P(I)$.

In particular, if $S = [0,1]$ is the unit interval and $P(I)$ is the length of I for every subinterval $I \subset S$, as in Example 2.2.5, then $\Pr(\{X \in I\}) = $ length of IS for every interval $I \subset R$.   ////

EXAMPLE 5.1.4   Consider an experiment in which a point is chosen at random from the unit interval $S = [0,1)$ in such a manner that the probability that the point lies in an interval $I \subset S$ is $P(I) = $ length of I. We can then define many interesting random variables. For example,

$$X(s) = s^2 \quad \text{and} \quad Y(s) = \tan 2\pi s$$

for $s \in S$.

Let us compute $\Pr(X \in I)$. Suppose, for example, that $I = (a,b]$ with $0 < a < b < 1$. Then

$$\Pr(a < X \le b) = P(\{s \in S: s^2 \in (a,b]\})$$

$$= P((\sqrt{a}, \sqrt{b}]) = \sqrt{b} - \sqrt{a}$$

and similar results can be obtained for other intervals.

The computation of $\Pr(Z \in I)$ is more complicated. Examination of Figure 5 shows that for $0 < a < b < \infty$, the event that $a < Z \le b$ is simply

$$\{s \in S: a < Z(s) \le b\} = (c_1, d_1] \cup (c_2, d_2]$$

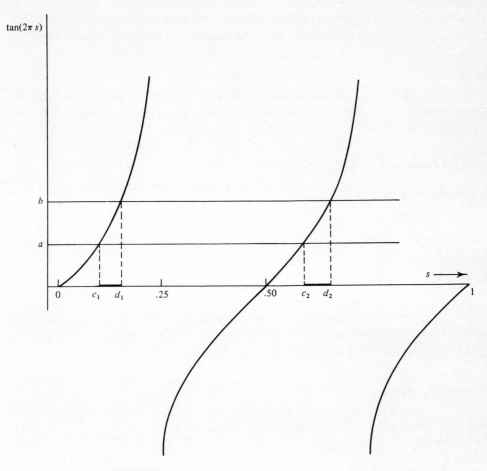

FIGURE 5
The tangent function.

where by definition,

$$c_1 = \frac{1}{2\pi} \arctan a \qquad d_1 = \frac{1}{2\pi} \arctan b$$

$c_2 = c_1 + \frac{1}{2}$, and $d_2 = c_2 + \frac{1}{2}$.[1] It follows that

$$\Pr(a < Z \le b) = P((c_1, d_1]) + P((c_2, d_2])$$
$$= (d_1 - c_1) + (d_2 - c_2)$$
$$= \frac{1}{\pi}(\arctan b - \arctan a)$$

[1] By arctan we mean the principle branch of the arctangent. That is, arctan y is the unique x for which $-\frac{1}{2}\pi < x < \frac{1}{2}\pi$ and tan $x = y$.

and the same result is obtained for any other interval with end points a and b $(a < b)$, for example, (a,b). Y, incidentally, gives the tangent of a randomly selected angle.

////

Many more examples of random variables will be given in the next two sections. We conclude this section by proving that (1.2) defines a probability function, so that the results of Sections 2.3, 2.4, and 2.5 are applicable to it. First, we need a lemma.

Lemma 5.1.1   *Let X be a function from a set S into a set T, and define*

$$X^{-1}(B) = \{s \in S: X(s) \in B\}$$

for all subsets $B \subset T$. Then, for $B, B_1, B_2, \ldots \subset T$, we have

$$X^{-1}(\cup B_i) = \cup X^{-1}(B_i) \qquad (1.4a)$$

$$X^{-1}(\cap B_i) = \cap X^{-1}(B_i) \qquad (1.4b)$$

$$X^{-1}(B') = X^{-1}(B)' \qquad (1.4c)$$

PROOF   The lemma is an easy consequence of the fact that, by definition, $s \in X^{-1}(B)$ if and only if $X(s) \in B$. To prove (1.4a), for example, simply observe that the following statements are equivalent:

1 $s \in X^{-1}(\cup B_i)$
2 $X(s) \in \cup B_i$
3 $X(s) \in B_i$ for some i
4 $s \in X^{-1}(B_i)$ for some i
5 $s \in \cup X^{-1}(B_i)$

This establishes (1.4a), and the proofs of (1.4b) and (1.4c) are similar.   ////

Now, let X be any random variable defined on a probability space $(S, \mathscr{S}, P)$, and let $\mathscr{B}$ be the class of all subsets $B \subset R$ (the set of all real numbers) for which $X^{-1}(B)$ is an event, that is, belongs to $\mathscr{S}$. Condition (1.1) then requires that $\mathscr{B}$ contain all intervals, and we can extend the notation (1.2) by writing $\Pr(X \in B) = P(X^{-1}(B))$ for all $B \in \mathscr{B}$. We now show that $\Pr(X \in B)$ defines a probability function.

Theorem 5.1.1   *Let X be any random variable, and define a function Q by $Q(B) = \Pr(X \in B)$ for $B \in \mathscr{B}$. Then $(R, \mathscr{B}, Q)$ is a probability space.*

↖ Borel set

PROOF   The proof that $\mathscr{B}$ is a σ algebra is left as an exercise (Problem 5.7). We show that Q satisfies axioms (2.1), (2.2), and (2.3) of Chapter 2. Clearly, $0 \leq Q(B) = P(X^{-1}(B)) \leq 1$ since P is a probability measure, and moreover,

$Q(R) = P(X^{-1}(S)) = 1$ for the same reason. To establish (2.2), let A and B be disjoint elements of $\mathcal{B}$. Then, $X^{-1}(A) \cap X^{-1}(B) = X^{-1}(AB) = X^{-1}(\varnothing) = \varnothing$, so that $X^{-1}(A)$ and $X^{-1}(B)$ are mutually exclusive events. Thus,

$$
\begin{aligned}
Q(A \cup B) &= P(X^{-1}(A \cup B)) \\
&= P(X^{-1}(A) \cup X^{-1}(B)) \\
&= P(X^{-1}(A)) + P(X^{-1}(B)) \\
&= Q(A) + Q(B)
\end{aligned}
$$

which is (2.2). Axiom (2.3) can be similarly verified to complete the proof. ////

It follows that the results of Sections 2.3, 2.4, and 2.5 are applicable to Q as well as P, since these results are valid in any probability space. For example, if $A \subset B$, then $\Pr(X \in B - A) = \Pr(X \in B) - \Pr(X \in A)$, and $\Pr(X \in A \cup B) = \Pr(X \in A) + \Pr(X \in B) - \Pr(X \in AB)$ for any A and B in $\mathcal{B}$.

We shall refer to Q as *the distribution of the random variable* X. Thus, the distribution of X specifies the probability that X belongs to B for every set B for which the latter probability is defined and so contains all the information we might ever want to know about probabilities associated with X. Of course, Q is quite complicated, but we shall see in the next few sections how Q can be determined implicitly by much simpler functions.

5.2 DISCRETE DISTRIBUTIONS

We define a (univariate) *mass function* to be a real-valued function f which is defined on $R = (-\infty, \infty)$ and has the following properties:

$$f(x) \geq 0 \qquad \text{for all } x \in R \qquad (2.1)$$

Moreover, there is a finite or countably infinite[1] set C, say $C = \{x_1, x_2, \ldots\}$, for which $f(x) = 0$ for $x \notin C$ and

$$\sum_C f(x) = 1 \qquad (2.2)$$

The term *discrete density* will also be used for a function f which satisfies (2.1) and (2.2). Of course, if (2.2) is satisfied for any choice of C, then it is also satisfied with $C = \{x \in R : f(x) > 0\}$. We shall see that in many cases, the distribution of a random variable can be determined implicitly by a mass function.

---

[1] A set C is countably infinite if there is a one-to-one correspondence between C and the set $Z = \{1, 2, \ldots\}$ of positive integers. $\sum_C f(x)$ denotes the sum of the numbers $f(x)$ for $x \in C$.

We shall say that a random variable X is *discrete* if and only if there is a finite or countably infinite set $C = \{x_1, x_2, \ldots\} \subset R$ for which

$$\text{Pr}\,(X \in C) = 1$$

In particular, this will be the case if the only possible values of X are $x_1, x_2, \ldots,$ and in most cases the x_i will be nonnegative integers.

We shall now show that any discrete random variable X determines a mass function f which in turn determines the distribution of X.

Theorem 5.2.1 *Let X be any discrete random variable. Then the function f defined by*

$$f(x) = \text{Pr}\,(X = x) \qquad (2.3)$$

for $x \in R$ is a mass function. Moreover, if C is any finite or countably infinite set for which $\text{Pr}\,(X \in C) = 1$, *then*

$$\text{Pr}\,(X \in B) = \sum_{BC} f(x) \qquad (2.4)$$

for all B for which the left side of (2.5) is defined.

PROOF We have $f(x) \geq 0$ for all $x \in R$ because probabilities are non-negative. Let C be as in the statement of the theorem. Then for $A \subset C'$, we have $\text{Pr}\,(X \in A) \leq \text{Pr}\,(X \in C') = 1 - \text{Pr}\,(X \in C) = 1 - 1 = 0$ by Theorem 5.1.1. In particular, if $x \notin C$, then $f(x) = \text{Pr}\,(X = x) = \text{Pr}\,(X \in \{x\}) \leq \text{Pr}\,(X \in C') = 0$.

To complete the proof, we must show that f satisfies (2.2) and demonstrate (2.4). Let us first demonstrate (2.4). Let B be as in the statement of the theorem. Then, since $B = BC \cup BC'$ and BC and BC' are mutually exclusive, we have $\text{Pr}\,(X \in B) = \text{Pr}\,(X \in BC) + \text{Pr}\,(X \in BC')$. Moreover, since $BC' \subset C'$, we have $\text{Pr}\,(X \in BC') = 0$, so that $\text{Pr}\,(X \in B) = \text{Pr}\,(X \in BC)$. Since C is finite or countably infinite, the same must be true of BC, and so we can write $BC = \{y_1, y_2, \ldots\}$ with distinct y's. Let $B_j = \{y_j\}$. Then the B_j are mutually exclusive, and their union is BC, so that

$$\text{Pr}\,(X \in BC) = \sum_j \text{Pr}\,(X \in B_j) = \sum_j f(y_j) = \sum_{BC} f(x)$$

This demonstrates (2.4). Equation (2.2) now follows easily. Indeed, taking $B = C$, we have

$$1 = \text{Pr}\,(X \in C) = \sum_C f(x)$$

by (2.4). ////

If X is a discrete random variable, we shall refer to the function f of Equation (2.3) as *the mass function of X*. By (2.4) the mass function of a discrete random variable

X uniquely determines the distribution of X. We have already encountered several discrete random variables, although we did not refer to them as such. We shall now rephrase some of our earlier results in the terminology of random variables.

EXAMPLE 5.2.1 Imagine an n-sided balanced die with k spots on the kth side for $k = 1, \ldots, n$. If the die is rolled once, we can represent the number of spots which appear as a random variable X by letting A_k be the event that exactly k spots appear and

$$X = \sum_{k=1}^{n} k I_{A_k}$$

as in Example 5.1.1c. X is discrete since it may assume only the values $1, \ldots, n$, and its mass function is given by

$$f(k) = \Pr(X = k) = P(A_k) = \frac{1}{n} \qquad (2.5)$$

for $k = 1, \ldots, n$ and $f(x) = 0$ for other values of x. ////

The function of (2.5) is known as the *discrete uniform mass function*. We observe that (2.5) defines not just one mass function but an entire family of mass functions— one for each integer $n = 1, 2, \ldots$. Accordingly, we shall refer to the function f of (2.5) as the *discrete uniform mass function with parameter n*, and we shall say that X has the *discrete uniform distribution with parameter n*.

We shall encounter similar situations below. That is, we shall encounter mass functions f which depend not only on their arguments but also on other free variables, or *parameters*, as we shall call them. The parameters are usually descriptive of the experimental conditions and therefore quite easily interpreted. For example, in Example 5.2.1 the parameter simply describes the number of sides on the die.

EXAMPLE 5.2.2 Consider an urn which contains m red balls and $n - m$ white balls, where m and $n - m$ are nonnegative integers with $n \geq 1$. If a random sample of size $k \leq n$ is drawn from the urn without replacement, then the number X of red balls in the sample is a random variable as in Example 5.1.2.b. X is discrete since it may assume only the values $0, \ldots, k$, and its mass function is given by

$$f(r) = \Pr(X = r) = \frac{\binom{m}{r}\binom{n-m}{k-r}}{\binom{n}{k}} \qquad (2.6)$$

for $r = 0, \ldots, k$ and $f(x) = 0$ for other values of x by Theorem 1.4.1. Equation (2.6) defines the *hypergeometric mass function with parameters m, n, and k* ($0 \leq m \leq n$ and $1 \leq k \leq n$).   ////

EXAMPLE 5.2.3   Consider a coin which has probability p of coming up heads when tossed. If n independent tosses of the coin are made, then the number of heads X which appear is a random variable as in Example 5.1.2a. X is discrete since it can assume only the values $0, \ldots, n$, and its mass function is given by

$$f(k) = \Pr (X = k) = \binom{n}{k} p^k q^{n-k} \qquad (2.7)$$

for $k = 0, \ldots, n$ and $f(x) = 0$ for other values of x. We shall refer to (2.7) as the *binomial mass function with parameters n and p* ($n \geq 1, 0 \leq p \leq 1$).   ////

EXAMPLE 5.2.4   If the coin of Example 5.2.3 is tossed repeatedly, the probability that the first head appears on the kth toss is

$$f(k) = pq^{k-1} \qquad (2.8)$$

for $k = 1, 2, \ldots$ by Theorem 4.2.1. Let $f(x) = 0$ if x is not a positive integer. Then f is a mass function, which we shall refer to as the *geometric mass function with parameter p* ($0 < p < 1$).

We recall from Section 4.2 that the geometric mass function has the lack-of-memory property. With our new terminology, Equation (2.4) of Chapter 4 may be stated as follows: if X has the geometric distribution (mass function), then for all positive integers k and j, the conditional probability that $X > k + j$ given that $X > j$ is

$$\Pr (X > k + j \mid X > j) = \Pr (X > k)$$

the same as the probability that $X > k$. In fact, this property is characteristic of the geometric mass function (see Problem 5.12).   ////

EXAMPLE 5.2.5   If the coin of Example 5.2.3 is tossed repeatedly, the probability that the rth head appears on the kth toss is

$$f(k) = \binom{k-1}{r-1} p^r q^{k-r} \qquad (2.9)$$

for $k = r, r + 1, \ldots$. Equation (2.9) defines the *negative binomial mass function with parameters r and p* ($r \geq 1$ and $0 < p < 1$). The geometric is a special case with $r = 1$. That (2.9) does define a mass function, that is, that condition (2.2) is satisfied, was shown in Section 4.2.   ////

EXAMPLE 5.2.6 A random variable X is said to have the *Poisson distribution* with parameter $\beta > 0$ if and only if X has mass function

$$f(k) = \Pr(X = k) = \frac{\beta^k}{k!}e^{-\beta} \qquad (2.10)$$

for $k = 0, 1, \ldots$ and $f(x) = 0$ for other values of x. That f is a mass function was shown in Section 4.3.

It was also shown in Section 4.3 that (2.10) provides an approximation to the binomial mass function when n is large, p is small, and $\beta = np$ is moderate. Another application of the Poisson distribution is the following. If a radioactive substance is observed for t units of time, where t is small compared with the half-life of the substance, and if the number X of radioactive emissions is recorded, then X may be regarded as a random variable which has the Poisson distribution with parameter βt, where $\beta > 0$ is characteristic of the radioactive substance. β is called the *intensity* of the radiation. We indicated a derivation of this result in Example 4.3.3, and we shall give another derivation of this assertion in Section 7.6. For the present, we accept it as an empirical fact. ////

5.3 ABSOLUTELY CONTINUOUS DISTRIBUTIONS

We define a (univariate) *density function* to be a real-valued function f which is defined on $R = (-\infty, \infty)$ and satisfies

$$f(x) \geq 0 \qquad \text{for } -\infty < x < \infty \qquad (3.1)$$

$$\int_{-\infty}^{\infty} f(x)\,dx = 1 \qquad (3.2)$$

Further, we shall say that a random variable X is *absolutely continuous* if and only if there is a density function f for which

$$\Pr(a < X \leq b) = \int_{a}^{b} f(x)\,dx \qquad (3.3)$$

whenever $a < b$. In this case we shall call f a *density for* X and say that X *has density* f. Since a function can be changed at any finite number of points without affecting its integral, a random variable may have more than one density function.

An interesting property of absolutely continuous random variables is the following. If X is any absolutely continuous random variable and $a \in R$ is any real number, then

$$\Pr(X = a) = 0 \qquad (3.4)$$

To see this observe that for any $\varepsilon > 0$ we have

$$\Pr(X = a) \le \Pr(a - \varepsilon < X \le a) = \int_{a-\varepsilon}^{a} f(x)\, dx = I(\varepsilon), \text{ say}$$

Now, the integrability of f implies that $\lim I(\varepsilon) = 0$ as $\varepsilon \to 0$,[1] so that $I(\varepsilon)$ can be made arbitrarily small by taking $\varepsilon > 0$ sufficiently small. Since $\Pr(X = a) \le I(\varepsilon)$ for all $\varepsilon > 0$, it follows that $\Pr(X = a) = 0$.

It follows from Equation (3.4) that if X is absolutely continuous, then in Equation (3.3) we can replace $a < X \le b$ by any of $a \le X \le b$, $a \le X < b$, or $a < X < b$, since the end points a and b contribute to neither the probability nor the integral. For example,

$$\Pr(a \le X \le b) = \Pr(X = a) + \Pr(a < X \le b)$$
$$= \Pr(a < X \le b) = \int_{a}^{b} f(x)\, dx$$

Equation (3.4) may appear to be somewhat nonintuitive, but in reality it is not. In particular, it does not assert that the events that $X = a$ for $a \in R$ are impossible. From the frequentistic point of view, it simply means that if the experiment to which X refers is repeated n times, the relative frequency with which $X = a$ will tend to zero as $n \to \infty$. From the subjective point of view, it means that for any fixed $a \in R$, the event that $X = a$ is regarded as extremely less probable than the event that $X \in R - \{a\}$. To clarify the latter point, imagine the following game. You are asked to guess a friend's weight exactly—not just to the nearest pound, or tenth of a pound, or millionth of a pound, but exactly. If you succeed, then you win c dollars, and if you fail, then you lose 1 dollar. Suppose also that it is possible to measure your friend's weight, X say, to an arbitrary degree of precision. Is there any value of c for which you would regard the game as fair? If not, then your subjective probability that $X = a$ is zero for every a.

In view of Equation (3.4), density functions are harder to interpret than mass functions (which give probabilities of particular events). However, if a density f is continuous at a point $a \in R$, then $f(a)$ can be interpreted as an approximate ratio of probability to length. To see this let X be absolutely continuous with density f, let $a \in R$, and suppose that f is continuous at a. Then

$$\frac{1}{2h}\Pr(a - h < X \le a + h) = \frac{1}{2h}\int_{a-h}^{a+h} f(x)\, dx$$

---

[1] If f is bounded, say $f(x) \le b$ for all x, then $I(\varepsilon) \le b\varepsilon$, which tends to zero as $\varepsilon \to 0$. For possibly unbounded f, see Problem 5.34.

which converges to $f(a)$ as $h \to 0$ by the fundamental theorem of calculus. That is, $\Pr (a - h < X \le a + h)$ is approximately $2hf(a)$ for small h.

We shall now consider several examples.

EXAMPLE 5.3.1   In Example 5.1.3 we found that if a point X is selected at random from the interval $S = [0,1)$, then $\Pr (X \in I) = $ length of IS for every interval $I \subset R$. This can be written in the form (3.3) with

$$f(x) = \begin{cases} 1 & 0 \le x < 1 \\ 0 & \text{otherwise} \end{cases} \tag{3.5}$$

We shall refer to (3.5) as the *uniform* density on the interval $[0,1)$. More generally, if J is any interval of positive and finite length, we shall refer to the function g defined by

$$g(x) = \begin{cases} \dfrac{1}{|J|} & x \in J \\ 0 & \text{otherwise} \end{cases} \tag{3.6}$$

where $|J|$ denotes the length of J, as the *uniform density on J*, and we shall say that a random variable Y having density g is *uniformly distributed* over J.   ////

EXAMPLE 5.3.2   In Example 5.1.4 we showed that if X denotes the tangent of an angle which is uniformly distributed over the interval $[0,2\pi)$, then

$$\Pr (a < X \le b) = \frac{1}{\pi} (\arctan b - \arctan a)$$

for $a < b$. This can be written in the form (3.3) with

$$f(x) = \frac{1}{\pi(1 + x^2)} \qquad -\infty < x < \infty \tag{3.7}$$

the derivative of $\pi^{-1} \arctan x$. Thus, X is absolutely continuous with density given by (3.7). We shall refer to (3.7) as the *Cauchy density*.   ////

EXAMPLE 5.3.3   A random variable X is said to have the *standard normal distribution* if and only if X has density

$$f(x) = \frac{e^{-\frac{1}{2}x^2}}{\sqrt{\pi 2}} \qquad -\infty < x < \infty \tag{3.8}$$

and we shall refer to (3.8) as the *standard normal density*. The proof that (3.8) does define a density, that is that condition (3.2) is satisfied, was given in Section 4.4

along with a graph of the function. We also showed in Section 4.6 that if Y has the binomial distribution with parameters n and p, $0 < p < 1$, then as $n \to \infty$,

$$\lim \text{Pr}\left(a < \frac{Y - np}{\sqrt{npq}} \le b\right) = \int_a^b f(x)\,dx$$

where f is defined by (3.8). Thus we may think of the standard normal distribution as an approximate distribution for $(Y - np)/\sqrt{npq}$. In fact, the standard normal distribution has a much wider applicability, as we shall see in Section 9.4.   ////

EXAMPLE 5.3.4   For any $\beta > 0$, the function f defined by

$$f(x) = \begin{cases} \beta e^{-\beta x} & x > 0 \\ 0 & x \le 0 \end{cases} \tag{3.9}$$

is a density, because

$$\int_{-\infty}^{\infty} f(x)\,dx = \int_0^{\infty} \beta e^{-\beta x}\,dx = -e^{-\beta x}\big|_0^{\infty} = 1$$

We shall refer the (3.9) as the *exponential* density with parameter β.

The exponential density shares with the geometric mass function the property of lack of memory, as we show in Problem 5.28 and Section 7.6. A derivation of the exponential density will be given in Example 5.5.4.   ////

We conclude this section with an analog of Theorem 5.2.1.

Theorem 5.3.1   *If X is absolutely continuous with density f, then*

$$\text{Pr}\,(X \in B) = \int_B f(x)\,dx \tag{3.10}$$

for every subset $B \subset R$ for which both sides of (3.10) are defined. Moreover, if X is absolutely continuous with density f, then f uniquely determines the distribution of X.

PROOF   Equations (3.3) and (3.4) assert that (3.10) holds whenever B is an interval. Therefore, if

$$B = \bigcup_{k=1}^{n} I_k$$

is the union of a finite number of disjoint intervals, then (3.10) holds since

$$\text{Pr}\,(X \in B) = \sum_{k=1}^{n} \text{Pr}\,(X \in I_k)$$

$$= \sum_{k=1}^{n} \int_{I_k} f(x)\,dx = \int_B f(x)\,dx$$

Thus, we should expect (3.10) to hold for all B which can be approximated by a finite union of disjoint intervals. This is, in fact, true, and the latter class of subsets B contains all sets for which both sides of (3.10) are defined. The details of this approximation are a bit complicated, however, and we omit them.   ////

5.4 THE GAMMA AND BETA DISTRIBUTIONS

In this section we shall introduce two new families of densities. Since both involve the *gamma function* in their definitions, we begin with a discussion of that function. The gamma function is defined on the interval $(0,\infty)$ by

$$\Gamma(\alpha) = \int_0^\infty x^{\alpha-1} e^{-x} \, dx \qquad \alpha > 0 \qquad (4.1)$$

This function has several interesting properties, the most striking of which will now be given.

Lemma 5.4.1 *For $\alpha > 1$, $\Gamma(\alpha) = (\alpha - 1)\Gamma(\alpha - 1)$.*

PROOF   Let $u(x) = x^{\alpha-1}$ and $v(x) = e^{-x}$ for $x > 0$. Then since $\alpha > 1$, $u(x)v(x) \to 0$ as $x \to 0$ or $x \to \infty$, and so we can integrate by parts to obtain

$$\Gamma(\alpha) = -\int_0^\infty u(x)v'(x) \, dx = -uv\big|_0^\infty + \int_0^\infty u'(x)v(x) \, dx$$

$$= (\alpha - 1) \int_0^\infty x^{\alpha-2} e^{-x} \, dx = (\alpha - 1)\Gamma(\alpha - 1)$$

as asserted.   ////

We also have $\Gamma(1) = 1$ by direct computation. In fact

$$\Gamma(1) = \int_0^\infty e^{-x} \, dx = -e^{-x}\big|_0^\infty = 1$$

It now follows that if n is a positive integer, then $\Gamma(n) = (n - 1)\Gamma(n - 1) = (n - 1)(n - 2)\Gamma(n - 2) = \cdots = (n - 1)(n - 2) \cdots 2 \times 1\Gamma(1)$. That is,

$$\Gamma(n) = (n - 1)! \qquad (4.2)$$

$\Gamma(\tfrac{1}{2})$ can also be evaluated.

Lemma 5.4.2 $\Gamma(\tfrac{1}{2}) = \sqrt{\pi}$.

PROOF   To see this make the change of variables $x = \tfrac{1}{2}y^2$ in the integral which defines $\Gamma(\tfrac{1}{2})$ to obtain

$$\Gamma(\tfrac{1}{2}) = \sqrt{2} \int_0^\infty e^{-\frac{1}{2}y^2} \, dy$$

$$= \frac{1}{\sqrt{2}} \int_{-\infty}^\infty e^{-\frac{1}{2}y^2} \, dy = \sqrt{\pi}$$

where the final step follows from the fact that the standard normal density is a density (Lemma 4.4.1). When combined, Lemmas 5.4.1 and 5.4.2 provide an expression for $\Gamma(\alpha)$ when α is a half integer.   ////

EXAMPLE 5.4.1   For any $\alpha > 0$ and $\beta > 0$, the function f defined by $f(x) = 0$ for $x \le 0$ and

$$f(x) = \frac{\beta^\alpha x^{\alpha - 1}}{\Gamma(\alpha)} e^{-\beta x} \qquad (4.3)$$

for $x > 0$ is a density. Indeed, the change of variables $y = \beta x$ yields

$$\int_0^\infty f(x) \, dx = \int_0^\infty \frac{y^{\alpha - 1} e^{-y}}{\Gamma(\alpha)} \, dy = \frac{\Gamma(\alpha)}{\Gamma(\alpha)} = 1$$

We shall refer to (4.3) as the *gamma density with parameters α and β.*

Observe that when $\alpha = 1$, the gamma density is the exponential density. A derivation of the gamma and exponential densities will be given in Example 5.5.4.   ////

EXAMPLE 5.4.2   When $\beta = \tfrac{1}{2}$ and $\alpha = k/2$, where k is a positive integer, the gamma density is known as the *chi-square density.* In this case the free parameter k is called the *degrees of freedom.* The terminology originates with an application to statistics which we shall consider in Section 7.5.   ////

The gamma density may assume a variety of shapes for different values of the parameter α. Some of these are sketched in Figure 6.

EXAMPLE 5.4.3   For any $\alpha > 0$ and $\beta > 0$, the function f defined by

$$f(x) = \frac{\Gamma(\alpha + \beta)}{\Gamma(\alpha)\Gamma(\beta)} x^{\alpha - 1}(1 - x)^{\beta - 1} \qquad 0 < x < 1 \qquad (4.4)$$

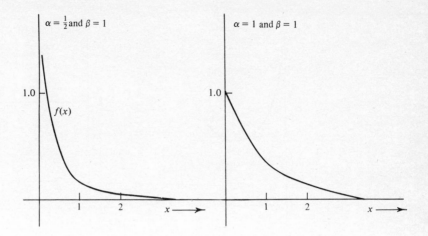

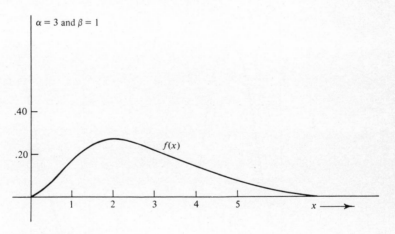

FIGURE 6
Some gamma densities.

and $f(x) = 0$ for $x \notin (0,1)$ is known as the *beta density with parameters α and β*. Since this density will be derived twice in Chapter 7, we defer the proof that (4.4) defines a density.   /////

Like the gamma density, the beta density may assume a wide variety of shapes. Some of these are sketched in Figure 7. Observe that the uniform density on (0,1) is a special case when $\alpha = \beta = 1$.

We conclude this section with an example of a computation with the gamma density.

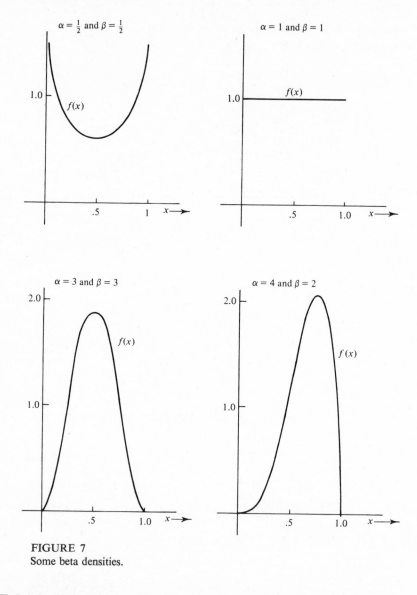

FIGURE 7
Some beta densities.

EXAMPLE 5.4.4   If the length of life in years of male residents of a given state follows the gamma distribution with parameters $\alpha = 2$ and $\beta = 0.02$, what proportion of male residents will live for more than 50 years? We require Pr $(X > 50)$, where X has the gamma distribution with parameters $\alpha = 2$ and $\beta = 0.02$, that is,

$$\text{Pr } (X > 50) = \int_{50}^{\infty} \beta^2 y e^{-\beta y} \, dy$$

The integral can be evaluated by making the change of variables $y = \beta x$ and integrating by parts. The result is

$$\Pr(X > 50) = \int_1^\infty y e^{-y}\, dy = 2e^{-1} = 0.7358 \qquad ////$$

5.4.1 A Proof of Stirling's Formula[1]

In this section we shall give a proof of Stirling's formula (Theorem 1.8.1), which states that

$$n! \sim \sqrt{2\pi}\, n^{n+\frac{1}{2}} e^{-n} \qquad (4.5)$$

as $n \to \infty$. Our starting point is Equation (4.2), which states that $n! = \Gamma(n + 1)$ or equivalently that

$$n! = \int_0^\infty x^n e^{-x}\, dx \qquad (4.6)$$

Let us make the change of variables $x = ny$ in (4.6) to obtain

$$n! = n^{n+1} \int_0^\infty y^n e^{-ny}\, dy$$

$$= n^{n+1} e^{-n} \int_0^\infty y^n e^{-n(y-1)}\, dy$$

$$= n^{n+1} e^{-n} \int_0^\infty e^{n\psi(y)}\, dy$$

where

$$\psi(y) = \log y - (y - 1)$$

for $y > 0$. Stirling's formula (4.5) is thus equivalent to the assertion that

$$I_n = \sqrt{n} \int_0^\infty e^{n\psi(y)}\, dy \to \sqrt{2\pi} \qquad (4.7)$$

as $n \to \infty$.

To establish (4.7) let us study the function ψ. The first two derivatives of ψ are

$$\psi'(y) = \frac{1}{y} - 1 \quad \text{and} \quad \psi''(y) = -\frac{1}{y^2}$$

for $y > 0$. Thus, ψ assumes its maximum value of $\psi(1) = 0$ when $y = 1$, $\psi(y) < 1$

---

[1] This section treats a special topic and may be omitted without loss of continuity.

for $y \neq 1$, and $\psi(y) \to -\infty$ as $y \to \infty$. We now expand ψ in a Taylor series about $y = 1$ to obtain

$$\psi(y) = \tfrac{1}{2}\psi''(y^*)(y - 1)^2 \qquad (4.8)$$

for $y > 0$, where $|y^* - 1| \leq |y - 1|$.

Let $\varepsilon > 0$ be given. Then, since ψ'' is continuous and $\psi''(1) = -1$, there is a $\delta > 0$ for which $-(1 + \varepsilon) \leq \psi''(y) \leq -(1 - \varepsilon)$ for $|y - 1| \leq \delta$. In particular, we must also have

$$-(1 + \varepsilon) \leq \psi''(y^*) \leq -(1 - \varepsilon) \qquad (4.9)$$

in (4.8) for $|y - 1| \leq \delta$.

Let us divide the interval of integration in (4.7) into three subintervals as follows:

$$I_n = \left(\sqrt{n} \int_0^{1-\delta} + \sqrt{n} \int_{1-\delta}^{1+\delta} + \sqrt{n} \int_{1+\delta}^{\infty} \right) e^{n\psi(y)} \, dy$$

$$= I_n' + I_n'' + I_n''', \text{ say}$$

We then have

$$I_n' = \sqrt{n} \int_0^{1-\delta} e^{n\psi(y)} \, dy \leq \sqrt{n} \, e^{n\psi(1-\delta)}$$

which tends to zero as $n \to \infty$ since $\psi(1 - \delta) < \psi(1) = 0$. Similarly, we can show that $I_n''' \to 0$ as $n \to \infty$ (Problem 5.39).

To estimate I_n'' we use (4.8) and (4.9) to deduce that

$$I_n'' \leq \sqrt{n} \int_{1-\delta}^{1+\delta} e^{-\frac{1}{2}n(1-\varepsilon)(y-1)^2} \, dy$$

The change of variables $z = \sqrt{n(1 - \varepsilon)} \, (y - 1)$ then shows that

$$I_n'' \leq \frac{1}{\sqrt{1 - \varepsilon}} \int_{-\delta\sqrt{n(1-\varepsilon)}}^{\delta\sqrt{n(1-\varepsilon)}} e^{-\frac{1}{2}z^2} \, dz$$

which converges to

$$\frac{1}{\sqrt{1 - \varepsilon}} \int_{-\infty}^{\infty} e^{-\frac{1}{2}z^2} \, dz = \sqrt{\frac{2\pi}{1 - \varepsilon}}$$

as $n \to \infty$. Here we used Lemma 4.4.1 to evaluate the integral. Since $I_n' + I_n''' \to 0$ as $n \to \infty$, we must have

$$I_n \leq \sqrt{\frac{2\pi}{1 - \varepsilon}} + \varepsilon \qquad (4.10)$$

for n sufficiently large; and similarly

$$I_n \geq \sqrt{\frac{2\pi}{1 + \varepsilon}} - \varepsilon \qquad (4.11)$$

for n sufficiently large. Since $\varepsilon > 0$ was arbitrary, (4.10) and (4.11) combine to prove (4.7). ////

5.5   DISTRIBUTION FUNCTIONS

If X is a random variable, we define the *distribution function F of X* by

$$F(a) = \Pr(X \le a) \qquad (5.1)$$

for $-\infty < a < \infty$. Thus, if X is discrete with mass function f, then by Theorem 5.2.1

$$F(a) = \sum_{x \le a} f(x) \qquad (5.2)$$

where the summation extends over all $x \le a$ for which $f(x) > 0$; and if X is absolutely continuous with density f, then

$$F(a) = \int_{-\infty}^{a} f(x)\, dx \qquad (5.3)$$

by Theorem 5.3.1. We emphasize, however, that *all random variables have distribution functions*—even those which are neither discrete nor absolutely continuous.

Relations (5.2) and (5.3) can be inverted. Thus, if X is absolutely continuous with density f and distribution function F, we can differentiate (5.3) by the fundamental theorem of calculus to obtain

$$f(a) = F'(a) \qquad (5.3a)$$

for all a at which f is continuous. In particular, (5.3a) holds for all a if F is continuously differentiable. A similar formula holds if X is discrete with mass function f and distribution function F. We shall show later in Theorem 5.6.1 that

$$f(a) = F(a) - F(a-) \qquad (5.2a)$$

where $F(a-)$ denotes the limit of $F(x)$ as $x \to a$ with $x < a$ (see Fig. 8).

We shall call distribution functions of the form (5.2) *discrete*, and we shall refer to f as the *mass function of F*. Similarly, we shall call distribution functions of the form (5.3) *absolutely continuous*, and we shall refer to f as a *density for F*. Further, if F and f are related by either (5.2) or (5.3), we shall call F by the same name (for example, binomial or normal) as f.

EXAMPLE 5.5.1

   a If an n-sided balanced die is rolled once, and if X denotes the number of spots which appear, the probability that $X \le a$ is 0 for $a < 1$, is k/n if $k \le a < k + 1$, where $k = 1, \ldots, n - 1$, and is 1 if $a \ge n$. That is,

$$F(a) = \begin{cases} 0 & \text{for } a < 1 \\ \dfrac{[a]}{n} & \text{for } 1 \le a < n \\ 1 & \text{for } a \ge n \end{cases}$$

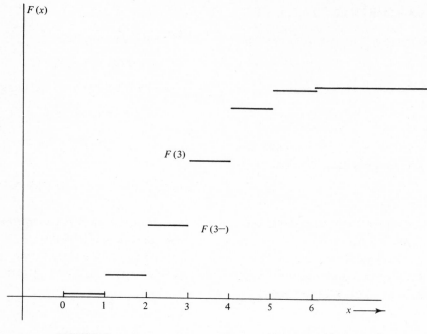

FIGURE 8
The binomial distribution function with $n = 6$ and $p = .5$.

where $[a]$ denotes the greatest integer which is less than or equal to a.

b If X has the geometric distribution (Example 5.2.4), then X has mass function f, where $f(k) = pq^{k-1}$ for $k = 1, 2, \ldots$ and $f(x) = 0$ for other values of x. Let F denote the corresponding distribution function. Then $F(a) = 0$ for $a < 1$, and

$$F(a) = \sum_{k=1}^{[a]} pq^{k-1}$$

for $a \geq 1$ (where $[a]$ denotes the greatest integer which is less than or equal to a). The summation can be evaluated to yield

$$F(a) = \begin{cases} 0 & a < 1 \\ 1 - q^{[a]} & a \geq 1 \end{cases}$$

c If X has the binomial distribution with parameters n and p, then X has distribution function F given by

$$F(a) = \sum_{k=0}^{[a]} \binom{n}{k} p^k q^{n-k}$$

for $0 \leq a < n$, $F(a) = 0$ for $a < 0$, and $F(a) = 1$ for $a \geq n$ (see Figure 8).

$////$

EXAMPLE 5.5.2

a If X has the uniform distribution on the interval $J = (c,d)$ with $c < d$, then X has distribution function F, where

$$F(a) = \begin{cases} 0 & a < c \\ \dfrac{a - c}{d - c} & c \leq a < d \\ 1 & a \geq d \end{cases}$$

This follows from a straightforward integration of the uniform density of Example 5.3.1.

b If X has the exponential distribution with parameter $\beta > 0$ (Example 5.3.4), then X has density $f(x) = 0$ for $x \leq 0$ and $f(x) = \beta e^{-\beta x}$ for $x > 0$. Integration now yields $F(a) = 0$ for $a \leq 0$ and

$$F(a) = \int_0^a \beta e^{-\beta x}\, dx = 1 - e^{-\beta a}$$

for $a > 0$.

c If X has the Cauchy distribution (Example 5.3.2), then X has density $f(x) = 1/\pi(1 + x^2)$ for $-\infty < x < \infty$ by Example 5.3.2. Thus, X has distribution function

$$F(a) = \frac{1}{\pi} \arctan a + \tfrac{1}{2}$$

for $-\infty < a < \infty$ by integration.

d If X has the standard normal distribution, then X has distribution function

$$\Phi(a) = \int_{-\infty}^a \frac{e^{-\frac{1}{2}x^2}}{\sqrt{2\pi}}\, dx$$

for $-\infty < a < \infty$. ////

Relations (5.2a) and (5.3a) may be used to compute the mass function or density of a random variable in some cases. We illustrate by finding the distribution of a linear function of a random variable. If X is a random variable, and if μ and σ are real numbers, we can define a new random variable Y by letting $Y = \sigma X + \mu$. That is, we let $Y(s) = \sigma X(s) + \mu$ for all $s \in S$, the sample space on which X is defined. We may think of Y as X measured in new units. The distributions of X and Y are related in a simple manner, as we shall now show.

Lemma 5.5.1 *Let X be a random variable with distribution function F, and let $Y = \sigma X + \mu$, where $\sigma > 0$. Then the distribution function of Y is given by*

$$G(a) = F\left(\frac{a - \mu}{\sigma}\right) \qquad (5.4a)$$

for $-\infty < a < \infty$. *If* X *is absolutely continuous with a continuous density* f, *then* Y *has density* g, *where*

$$g(a) = \frac{1}{\sigma} f\left(\frac{a - \mu}{\sigma}\right) \qquad (5.4b)$$

for $-\infty < a < \infty$.

  PROOF The set of $s \in S$ for which $Y(s) \leq a$ is the same as the set of $s \in S$ for which $X(s) \leq (a - \mu)/\sigma$. Thus,

$$G(a) = \Pr(Y \leq a)$$
$$= \Pr\left(X \leq \frac{a - \mu}{\sigma}\right) = F\left(\frac{a - \mu}{\sigma}\right)$$

for $-\infty < a < \infty$. This establishes (5.4a). If, in addition, X is absolutely continuous with density f, then (5.4b) follows by differentiation. ////

When dealing with distributions G of the form (5.4), we shall refer to μ and σ as *location and scale parameters*, respectively.

EXAMPLE 5.5.3

a If X has the standard normal distribution, then X has distribution function Φ, as in Example 5.5.2d. Thus, $Y = \sigma X + \mu$ has distribution and density functions given by

$$G(a) = \Phi\left(\frac{a - \mu}{\sigma}\right)$$

$$g(a) = \frac{1}{\sigma\sqrt{2\pi}} \exp\left\{-\frac{1}{2}\left(\frac{x - \mu}{\sigma}\right)^2\right\}$$

for $-\infty < a < \infty$. We shall refer to G and g as the *normal distribution function and density with scale parameter* σ *and location parameter* μ.

b If X has the Cauchy distribution (Examples 5.3.2 and 5.5.2c), then $Y = \sigma X + \mu$ has distribution function and density

$$G(a) = \frac{1}{\pi} \arctan \frac{a - \mu}{\sigma} + \tfrac{1}{2}$$

$$g(a) = \frac{\sigma}{\pi[\sigma^2 + (a - \mu)^2]}$$

for $-\infty < a < \infty$. We shall refer to G and g as the *Cauchy distribution function and density with location parameter* μ *and scale parameter* σ.

c Let g denote the gamma density with parameters α and β. That is, $g(x) = 0$ for $x \leq 0$ and

$$g(x) = \frac{1}{\Gamma(\alpha)} \beta^{\alpha} x^{\alpha - 1} e^{-\beta x}$$

for $x > 0$. Further, let f be g with $\beta = 1$. Then $g(x) = \beta f(\beta x)$ for all x, so that β^{-1} is a scale parameter.   ////

As another application of (5.3a), we shall now give a derivation of the exponential and gamma distributions.

EXAMPLE 5.5.4 Imagine a radioactive substance which emits radioactive particles. If the substance is observed continuously, what is the distribution of the time of the first emission? More generally, what is the distribution of the time of the kth emission, where k is a positive integer? Let X denote the time of the kth emission, let $t > 0$, and let Y denote the number of emissions by time t. Then we may suppose that Y has the Poisson distribution with parameter βt, where $\beta > 0$ is the intensity of the radiation (Example 5.2.6). That is, we suppose that $\Pr(Y = j) = (\beta t)^j e^{-\beta t}/j!$ for $j = 0, 1, 2, \ldots$. Now X is less than or equal to t if and only if there have been at least k emissions by time t. That is, $\Pr(X \leq t) = \Pr(Y \geq k) = 1 - \Pr(Y \leq k - 1)$. Let F denote the distribution function of X. Then we have

$$F(t) = 1 - \sum_{j=0}^{k-1} \frac{1}{j!} (\beta t)^j e^{-\beta t} \qquad (5.5)$$

for $t > 0$. Of course, $F(t) = 0$ for $t \leq 0$ since X is a nonnegative variable.

We can now obtain the density of X by differentiation. If $k = 1$, then $F(t) = 1 - e^{-\beta t}$ for $t > 0$, so that

$$f(t) = \beta e^{-\beta t}$$

for $t > 0$. That is, *the distribution of the time of the first emission is exponential with parameter β*, as claimed in Example 5.3.4.

For $k > 1$, the derivative $f = F'$ can also be computed as

$$f(t) = \sum_{j=0}^{k-1} -\frac{1}{j!} j \beta^j t^{j-1} e^{-\beta t} + \sum_{j=0}^{k-1} \frac{1}{j!} \beta^{j+1} t^j e^{-\beta t}$$

The $(j-1)$st term in the second sum cancels the jth term in the first, leaving

$$f(t) = \frac{1}{(k-1)!} \beta^k t^{k-1} e^{-\beta t} \qquad (5.6)$$

for $t > 0$. Thus, *the time of the kth emission has a gamma distribution with parameters $\alpha = k$ and β.*

As a corollary to our calculations, we see that the distribution function of the gamma density (5.6) is given by (5.5).   /////

5.6 COMPUTATIONS WITH DISTRIBUTION FUNCTIONS

The distribution function F of a random variable X uniquely determines the distribution of X. That is, if X and Y have the same distribution function, then

$$\Pr (X \in B) = \Pr (Y \in B)$$

for all $B \subset R$ for which both symbols are defined. We shall not prove this fact here, since the proof requires some advanced techniques.[1] However, we shall show how $\Pr (X \in I)$ can be computed from the distribution function F of X for any interval I.

If F is a real-valued function on R, we shall say that F is *nondecreasing* if and only if $F(a) \le F(b)$ whenever $a < b$. If F is nondecreasing, then the limit of $F(x)$ as $x \to a$ with $x < a$ exists,[2] and we shall denote this limit by $F(a-)$. Similarly, if F is nondecreasing, then the limit of $F(x)$ as $x \to a$ with $x > a$ exists, and we shall denote this limit by $F(a+)$. Thus, if F is nondecreasing, F is continuous at $a \in R$ if and only if $F(a-) = F(a) = F(a+)$. If F is nondecreasing, we shall say that F is continuous from the right if and only if $F(a) = F(a+)$ for all a and from the left if and only if $F(a) = F(a-)$ for all a.

If F is the distribution function of a random variable X, then F is nondecreasing. Indeed, if $a < b$, the event that $X \le a$ implies that $X \le b$, so that $F(a) = \Pr (X \le a) \le \Pr (X \le b) = F(b)$ by Theorem 2.3.1. Thus, the one-sided limits $F(a-)$ and $F(a+)$ exist for all a. In Section 5.8 we shall show that

$$F(a+) = F(a) = \Pr (X \le a) \qquad (6.1a)$$

$$F(a-) = \Pr (X < a) \qquad (6.1b)$$

for all a, $-\infty < a < \infty$. In particular, F is continuous at a if and only if $F(a) = F(a-)$.

Theorem 5.6.1 *Let X be a random variable, and let F be its distribution function. Then for $a < b$*

$$\Pr (a < X \le b) = F(b) - F(a) \qquad (6.2a)$$

$$\Pr (a \le X \le b) = F(b) - F(a-) \qquad (6.2b)$$

$$\Pr (a < X < b) = F(b-) - F(a) \qquad (6.2c)$$

$$\Pr (a \le X < b) = F(b-) - F(a-) \qquad (6.2d)$$

---

[1] Readers who are familiar with measure theory may consult Neveu (1965), p. 28.
[2] See, for example, Rudin (1964), p. 82.

and for all a

$$\Pr (X = a) = F(a) - F(a-) \qquad (6.3)$$

$$\Pr (X > a) = 1 - F(a) \qquad (6.4a)$$

$$\Pr (X \geq a) = 1 - F(a-) \qquad (6.4b)$$

PROOF Since the proofs of the four equations in (6.2) are all similar, we shall prove only the first. Given $a < b$, let A be the event that $X \leq a$, and let B be the event that $X \leq b$. Then A implies B, and $B - A$ is simply the event that $a < X \leq b$. Therefore, $\Pr (a < X \leq b) = P(B - A) = P(B) - P(A) = \Pr (X \leq b) - \Pr (X \leq a) = F(b) - F(a)$, as asserted. (Here we have used Theorem 2.3.1 to obtain the second equality.) The proof of (6.3) is similar. Let B be the event that $X \leq a$, and let A be the event that $X < a$. Then, $\Pr (X = a) = P(B - A) = P(B) - P(A) = F(a) - F(a-)$, where we have used (6.1) in the final step. Finally, (6.4a) and (6.4b) follow from (6.1) by taking complements. To establish (6.4a), for example, observe that the event that $X > a$ is the complement of the event that $X \leq a$, so that $\Pr (X > a) = 1 - \Pr (X \leq a) = 1 - F(a)$.   ////

Equation (6.3) is of special interest. Since distribution functions are always continuous from the right by (6.1), it can be rephrased as follows. If the distribution function F of the random variable X has a discontinuity of magnitude $\delta = F(a) - F(a-)$ at the point a, then $\Pr (X = a) = \delta$. Conversely, if F is continuous at the point a, then $\Pr (X = a) = 0$. In particular, if F is a continuous function, then $\Pr (X = a) = 0$ for all $a \in R$. Thus, if F is continuous, the four probabilities in (6.2) are all the same, and $\Pr (a < X < b) = \Pr (a < X \leq b) = \Pr (a \leq X < b) = \Pr (a \leq X \leq b) = F(b) - F(a)$.

EXAMPLE 5.6.1

a If X has the standard normal distribution, then $\Pr (a < X < b) = \Pr (a \leq X \leq b) = \Phi(b) - \Phi(a)$ for $a < b$, since Φ is continuous. In particular, we have $\Pr (-1 \leq X \leq 1) = \Phi(1) - \Phi(-1) = 0.683$ and $\Pr (-2 \leq X \leq 2) = 0.954$ from Appendix Table C.3. $\Phi(-1) = [1 - \Phi(1)]$

b More generally, if X has the normal distribution with location parameter μ and scale parameter $\sigma > 0$, then

$$\Pr (a \leq X \leq b) = \Phi\left(\frac{b - \mu}{\sigma}\right) - \Phi\left(\frac{a - \mu}{\sigma}\right) \qquad \text{for } a < b$$

In particular, $\Pr (\mu - \sigma \leq X \leq \mu + \sigma) = \Phi(1) - \Phi(-1) = 0.683$, and $\Pr (\mu - 2\sigma \leq X \leq \mu + 2\sigma) = 0.954$.   ////

EXAMPLE 5.6.2   If X has the gamma distribution with parameters $\alpha = k$, a positive integer, and $\beta > 0$, then

$$\Pr\left(a < X \le b\right) = \sum_{i=0}^{k-1} \frac{1}{i!} \left[(\beta a)^i e^{-\beta a} - (\beta b)^i e^{-\beta b}\right]$$

for $a < b$ by Example 5.5.4. For selected values of a and b, this can be computed from the table of Poisson probabilities in Appendix C.   ////

EXAMPLE 5.6.3

  a   If X has the binomial distribution with parameters $n = 8$ and $p = 0.5$, then $\Pr\left(3 \le X \le 5\right) = F(5) - F(3-) = 0.7109$.   Observe   that $F(5) - F(3) = 0.4922$.

  b   If X has distribution function

$$F(a) = \begin{cases} 0 & a < 1 \\ \dfrac{a}{2} & 1 \le a < 2 \\ 1 & a \ge 2 \end{cases}$$

then $\Pr(X = 1) = F(1) - F(1-) = 0.5$.   However, $\Pr(X = a) = 0$ for $a \ne 1$, since F is continuous at every a except $a = 1$. This random variable is neither discrete nor absolutely continuous.   ////

5.7   MEDIANS AND MODES[1]

If X is any random variable, then any number m for which

$$\Pr\left(X \le m\right) \ge \tfrac{1}{2} \le \Pr\left(X \ge m\right) \qquad (7.1)$$

will be called a *median of X*. If F denotes the distribution function of X, then (7.1) is equivalent to

$$F(m-) \le \tfrac{1}{2} \le F(m) \qquad (7.2)$$

since $\Pr\left(X \ge m\right) = 1 - F(m-)$ by (6.4). Any number m which satisfies (7.2) will be called a *median of F*. If F is continuous, then $F(m-) = F(m)$, and so we must have equality throughout in (7.2) and (7.1). That is, if F is continuous, then X is as likely to lie above as below its median. In this sense, a median may be thought of as a center of the distribution of X.

A random variable may have more than one median. In fact, a random variable

---

[1] This section treats a special topic and may be omitted without loss of continuity.

may have an entire interval of medians. Also, the inequalities in (7.1) and (7.2) may be strict if F is not continuous. Examples will be given below.

EXAMPLE 5.7.1 If X has a density f which is symmetric in the sense that $f(x) = f(-x)$ for all x, then 0 is a median for X. In fact

$$F(0) = \int_{-\infty}^{0} f(x) \, dx = \int_{0}^{\infty} f(x) \, dx = 1 - F(0)$$

so that $F(0) = \frac{1}{2}$. In particular, 0 is a median for the standard normal and standard Cauchy distributions. ////

EXAMPLE 5.7.2 Suppose that the length of time X required for a single radio-active particle to decay has an exponential distribution with parameter $\gamma > 0$. Then the median can be found by solving the equation $Pr(X > m) = 1 - Pr(m \le X) = 1 - (1 - e^{-\gamma m})$

$$\tfrac{1}{2} = \Pr(X > m) = e^{-\gamma m}$$

as $m = (\log 2)/\gamma$. If a radioactive substance contains N such particles, where N is large, and if the particles decay independently, then we expect approximately half of the N particles to have decayed by time m. Accordingly, we call m the *half-life* of the substance. ////

EXAMPLE 5.7.3

a If X has the geometric distribution with parameter $p = 0.5$, then

$$\Pr(X = 1) = \tfrac{1}{2} = \Pr(X \le a)$$

for $1 \le a < 2$. Thus every number m for which $1 \le m \le 2$ is a median for X.
b If X has the geometric distribution with parameter $p = 0.4$, then

$$\Pr(X = 1) = 0.4 \quad \text{and} \quad \Pr(X = 2) = 0.24$$

Thus $F(2-) = 0.4 < 0.64 = F(2)$. That is, 2 is a unique median, and there is strict inequality in (7.1) and (7.2). ////

If X is either a discrete random variable with mass function f or an absolutely continuous random variable with density f, then any number m at which f assumes its maximum is called a *mode* of X or a mode of f. In the discrete case where $f(m) = \Pr(X = m)$, a mode is a most probable value of X or one of several most probable values. A density or mass function may have more than one mode.

EXAMPLE 5.7.4

a The unique mode of the standard normal density

$$f(x) = \frac{1}{\sqrt{2\pi}} e^{-\frac{1}{2}x^2}$$

is $x = 0$.

b The unique mode of the standard Cauchy distribution is also $x = 0$.

EXAMPLE 5.7.5

a Let us find the mode of the gamma density

$$f(x) = \frac{1}{\Gamma(\alpha)} \beta^\alpha x^{\alpha-1} e^{-\beta x}$$

for $x > 0$. Differentiation shows that

$$f'(x) = \frac{1}{\Gamma(\alpha)} \beta^\alpha [(\alpha - 1)x^{\alpha-2} - \beta x^{\alpha-1}] e^{-\beta x}$$

If $\alpha \leq 1$, then $f'(x) < 0$ for all $x > 0$, and so the mode $m = 0$. In fact, if $\alpha < 1$, then $f(x) \to \infty$ as $x \to 0$. If $\alpha > 1$, then $f'(x)$ vanishes when $x = m = (\alpha - 1)\beta^{-1}$.

b Similarly, the mode of the beta density f with parameters $\alpha \geq 1$ and $\beta \geq 1$ and $\alpha + \beta > 2$ is

$$m = \frac{\alpha - 1}{\alpha + \beta - 2}$$

If $\alpha < 1$, then $f(x) \to \infty$ as $x \to 0$, and if $\beta < 1$, then $f(x) \to \infty$ as $x \to 1$.

////

EXAMPLE 5.7.6 Table 9 gives the income x in thousands of dollars of 1000 hypothetical families, where y is the number of families. If a family is selected at random and its income X recorded, what is the mode of X? What is the median of X?

The most probable income is \$9000, which is the mode. At least half of the families have \$11,000 or less, and at least half of the families have \$11,000 or more, so that the median is \$11,000.

////

Table 9

x	8	9	10	11	12	15	20	25	50	100
y	126	186	175	152	121	113	74	42	9	2

5.8 PROPERTIES OF DISTRIBUTION FUNCTIONS[1]

In this section we shall show that distribution functions have certain characteristic properties. As a corollary, we shall see how to construct a probability space on which is defined a random variable having an arbitrary, preassigned distribution function, density, or mass function. We shall use the fact that if $A_1, A_2, \ldots$, is a decreasing sequence of events (that is, if $A_1 \supset A_2 \cdots$), then

$$P\left(\bigcap_{n=1}^{\infty} A_n\right) = \lim_{n \to \infty} P(A_n) \qquad (8.1a)$$

and if $A_1, A_2, \ldots$ is an increasing sequence of events (that is, if $A_1 \subset A_2 \subset \cdots$), then

$$P\left(\bigcup_{n=1}^{\infty} A_n\right) = \lim_{n \to \infty} P(A_n) \qquad (8.1b)$$

See Theorem 2.5.1.

Theorem 5.8.1 *Let X be any random variable, and let F denote its distribution function. Then:*

(i) *F is nondecreasing;*
(ii) *F is continuous from the right; and*
(iii) *F satisfies*

$$\lim F(a) = 0 \qquad as\ a \to -\infty \qquad (8.2a)$$

$$\lim F(a) = 1 \qquad as\ a \to \infty \qquad (8.2b)$$

PROOF That F is nondecreasing has already been shown. It follows that the one-sided limits $F(a+)$ and $F(a-)$ exist for every $-\infty < a < \infty$, and the assertion that F is continuous from the right is equivalent to the assertion that $F(a) = F(a+)$ for every a. To see this, let A_n be the event that $X \leq a + 1/n$ for $n = 1, 2, \ldots$. Then, $A_1 \supset A_2 \supset \cdots$, and the intersection of the A_n is simply the event that $X \leq a$. Therefore

$$F(a+) = \lim_{n \to \infty} F\left(a + \frac{1}{n}\right)$$

$$= \lim_{n \to \infty} P(A_n)$$

$$= P\left(\bigcap_{n=1}^{\infty} A_n\right)$$

$$= \Pr(X \leq a) = F(a)$$

[1] This section may be omitted without loss of continuity.

as asserted, where we have used (8.1) to obtain the crucial middle equality.

We have still to prove (8.2). To establish (8.2a), for example, observe first that as $a \to -\infty$, $\lim F(a)$ exists since F is nondecreasing. For every $n = 1, 2, \ldots$, let A_n be the event that $X \leq -n$. Then, the sequence $A_1, A_2, \ldots$ is decreasing and its intersection is the empty set $\emptyset$. Therefore,

$$\lim_{n \to \infty} F(-n) = \lim P(A_n) = P\left(\bigcap_{n=1}^{\infty} A_n\right) = P(\emptyset) = 0$$

[handwritten: $1 - \lim_{n \to \infty} F(-n)$]

as asserted; (8.2b) can be established similarly. *[handwritten: and $\lim_{n \to \infty} F(n) = 1 ////$]*

[handwritten left margin: other part] *[handwritten: as $a \to$ let A_n be $X \geq n$ then sequ. is dec. event]* *[handwritten: since $Pr(X \geq n) = 1 - Pr(X < n)$]*

Theorem 5.8.1 establishes Equation (6.1a). A similar argument will establish *[handwritten: $1 - F(a)$]* (6.1b), $F(a-) = \Pr(X < a)$. Given a, let A_n be the event that $X \leq a - 1/n$ for $n = 1, 2, \ldots$. Then $A_n \subset A_{n+1}$ for all n, and the union $A = \bigcup_{n=1}^{\infty} A_n$ is the event that $X < a$. Thus, $F(a-) = \lim F(a - 1/n) = \lim P(A_n) = P(A) = \Pr(X < a)$ as $n \to \infty$.

The importance of Theorem 5.8.1 derives in part from the fact that the properties derived there are characteristic of distribution functions. That is, every function F which has the properties asserted by Theorem 5.8.1 is the distribution function of some random variable.

Theorem 5.8.2 *Let F be any nondecreasing, right-continuous function which satisfies (8.2). Then there is a random variable X whose distribution function is F.*

PROOF   We must define a probability space $(S, \mathscr{S}, P)$ and a random variable X and show that the distribution function of X is F. Let $S = (0,1)$, let all subintervals of S be events, that is, elements of $\mathscr{S}$, and let $P(I) = $ length of I if I is a subinterval of S, as in Example 2.2.5. Further, define a function X on S by

$$X(s) = \min \{x \in R : F(x) \geq s\} \qquad (8.3)$$

for each fixed $s \in S$. The set on the right side of (8.3) is nonempty by (8.2), and the minimum is attained because F is continuous from the right.

To see that X is a random variable with distribution function F, observe that $X(s) \leq a$ if and only if $s \leq F(a)$ for each fixed a, $-\infty < a < \infty$. That is, the event that $X \leq a$ is simply $(0, F(a)]$ for every a. Therefore, X is a random variable, and $\Pr(X \leq a) = P((0, F(a)]) = $ length of $(0, F(a)] = F(a)$, as asserted.   ////

Henceforth, we shall use the term "distribution function" to mean any non-decreasing, right-continuous function F which satisfies (8.2). Theorem 5.8.2 guarantees that such functions are distribution functions of random variables.

EXAMPLE 5.8.1

a If *f* is any density function, then its indefinite integral *F* defined by

$$F(a) = \int_{-\infty}^{a} f(y)\, dy$$

is nondecreasing and right-continuous (in fact, continuous) and satisfies (8.2). Therefore, *F* is the distribution of some random variable *X*. By (6.2*a*), we then have

$$\Pr(a < X \le b) = F(b) - F(a) = \int_{a}^{b} f(x)\, dx$$

whenever $a < b$, so that *X* has density *f*. Therefore, we have proved the following corollary to Theorem 5.8.2.   *Given any density function f, there is a random variable X whose density is f.*

b Similarly, *given any mass function f, there is a random variable whose mass function is f.*   ////

EXAMPLE 5.8.2

a The function *F* defined by $F(a) = \sin a$, $-\infty < a < \infty$, is not monotone and does not satisfy (8.2). Therefore, it is not the distribution function of any random variable.

b If *G* and *H* are distribution functions and $0 < \alpha < 1$, then $F = \alpha G + (1 - \alpha)H$ is also a distribution function. Indeed, if *G* and *H* are both non-decreasing and right-continuous and both satisfy (8.2), then *F* will also be non-decreasing and right-continuous and satisfy (8.2). If *G* is absolutely continuous and *H* is discrete, then *F* will be neither absolutely continuous nor discrete.

////

For
Thursday

PROBLEMS

5.1 Let *S* be a set, and for $A \subset S$ let I_A be the indicator of *A*. Verify the following properties:

(a) $I_{AB}(s) = I_A(s)I_B(s)$

(b) $I_{A \cup B}(s) = I_A(s) + I_B(s) - I_{AB}(s)$

(c) $I_{A'}(s) = 1 - I_A(s)$

for all $s \in S$.

5.2 Two balanced dice are thrown. Represent the total number of spots which appear as a random variable *X* on an appropriate sample space. Find $\Pr(X = 6)$.

5.3   Consider the probability space of Example 5.1.4, and let $X(s) = \sin(\pi s/2)$ $s \in S$. Find $\Pr(X \le 0.5)$.

5.4   Consider the probability space of Example 5.1.4 and let $X(s) = \sin 2\pi s$ $s \in S$. Find $\Pr(X \le 0.5)$.

5.5⌐ A point $s = (s_1, s_2)$ is chosen at random from the unit circle S in such a manner that the probability that s belongs to a subregion of S is proportional to the area of the subregion. Let X denote the distance of the point chosen from the origin.
(a) Represent X as a function on an appropriate sample space.
(b) Compute $\Pr(X \le r)$ for $0 < r < 1$.

5.6   Let $(S, \mathscr{S}, P)$ be a probability space, and let X be a real-valued function defined on S. Show that if $\{s \in S: X(s) \le a\}$ is an event for every real number a, then X is a random variable.

5.7   Show that the class $\mathscr{B}$ of Theorem 5.1.1 is a σ algebra. *Hint:* For example, if $B \in \mathscr{B}$, then $X^{-1}(B) \in \mathscr{S}$, so that $X^{-1}(B') = X^{-1}(B)'$ is also in $\mathscr{S}$, because $\mathscr{S}$ is given to be a σ algebra.

5.8   Complete the proof of Lemma 5.1.1.

5.9   Let two balanced dice be rolled, and let X be the total number of spots which appear. Find the mass function of X.

5.10 Cards are drawn sequentially without replacement from a deck until a spade appears. Let X denote the number of draws required. Find the mass function of X.

5.11 Cards are drawn sequentially without replacement until r spades have appeared. Let X denote the number of draws required. Find the mass function of X.

5.12 Let X be any random variable which may assume only the values $1, 2, \ldots$. Show that if $0 < \Pr(X > K) < 1$ and if $\Pr(X > k + 1 \mid X > k) = \Pr(X > 1)$ for all $k = 1, 2, \ldots$, then X has the geometric distribution with parameter $p = \Pr(X = 1)$.

5.13 Let $f(x) = 1/x(x + 1)$ if x is a positive integer, $x = 1, 2, \ldots$, and let $f(x) = 0$ for other values of x. Show that f is a mass function. The distribution determined by this mass function is known as *Zipf's distribution*.

5.14 Let X be a random variable which may assume only positive integer values $1, 2, \ldots$. If $P(X > k + 1 \mid X > k) = (k + 1)/(k + 2)$, what is the mass function of X?

5.15 Suppose that demand on a given product during a given day is a random variable X which has the Poisson distribution with parameter $\beta = 5$:
(a) If a merchant stocks 5 units of the product, what is the probability that the demand will exceed the supply?
(b) How many units should the merchant stock if he wishes the probability that the demand will exceed the supply to be at most 0.01?

5.16 Repeat Problem 5.15 under the assumption that X has the geometric distribution with parameter $p = 0.2$.

5.17 Repeat Problem 5.15 under the assumption that X has the binomial distribution with parameters $n = 10$ and $p = 0.5$.

5.18 If X has the Poisson distribution and $\Pr(X = 0) = \Pr(X = 1)$, find $\Pr(X \le 2)$.

5.19 If X has the geometric distribution and $\Pr(X = 1) = 3 \Pr(X = 2)$, evaluate $\Pr(X \text{ is odd})$.

5.20 If X has the Poisson distribution with parameter β, show that $\Pr{(X \text{ is even})} = e^{-\beta} \cosh \beta$.

5.21 For each $n \geq 1$, let X_n have the binomial distribution with parameters n and p. Further, let $\alpha_n = \Pr{(X_n \text{ is odd})}$. Show that $\alpha_1 = p$ and $\alpha_n = (q - p)\alpha_{n-1} + p$ for $n \geq 2$. Conclude that $\alpha_n = \frac{1}{2}[1 - (q - p)^n]$ for $n = 1, 2, \ldots$.

5.22 Show that the following defines a mass function:

$$f(x) = \frac{2x}{n(n+1)}$$

for $x = 1, 2, \ldots, n$ and $f(x) = 0$ for other values of x.

5.23 For what values of α does there exist a constant c for which

$$f(x) = cx^{-\alpha} \qquad x = 1, 2, \ldots$$

and $f(x) = 0$ for other values of x defines a mass function?

5.24 How should c be chosen for the following functions to define densities?

(a) $f(x) = \begin{cases} cx^2 & 0 \leq x \leq 2 \\ 0 & \text{other values of } x \end{cases}$

(b) $f(x) = \dfrac{c}{(1 + |x|)^2} \qquad -\infty < x < \infty$

5.25 For what values of α does there exist a c such that $f(x) = cx^{\alpha}, x > 0$ and $f(x) = 0$ for $x \leq 0$ defines a density?

5.26 Show that if f and g are densities and $0 < \alpha < 1$, then $h = \alpha f + (1 - \alpha)g$ is also a density.

5.27 Let $f(x) = \frac{1}{2}e^{-|x|}, -\infty < x < \infty$. Show that f is a density. This density is known as the *bilateral exponential density*.

5.28 Show that if X has the exponential distribution with parameter $\beta > 0$, then $\Pr{(X > s + t \mid X > s)} = \Pr{(X > t)}$ for all choices of $s > 0$ and $t > 0$.

5.29 If X has the exponential distribution with parameter $\beta = 2$, find the probability that $1 < X \leq 2$.

5.30 If X has the Cauchy distribution, what is $\Pr{(1 \leq X \leq 3)}$?

5.31 If X has the Cauchy distribution, what is the probability that $1 + X^2 > 3$?

5.32 Professor Smith's class is scheduled to begin at 10 A.M., but he begins his class at a time X which is uniformly distributed over the interval 9:55 to 10:05 A.M. What is the probability that he will begin (a) at least 2 minutes early; (b) at least 2 minutes late?

5.33 Suppose that the time, in hours, required to service a particular kind of sports car is random variable X which has the exponential distribution with parameter $\beta = 1$. What is the probability that more than 2 hours will be required to service the car?

5.34 If f is a density and $a \in R$, show that

$$I(\varepsilon) = \int_{a-\varepsilon}^{a} f(x)\, dx$$

tends to zero as $\varepsilon \to 0$. *Hint:* If f is bounded, by b say, near a, then $I(\varepsilon) \le b\varepsilon$; and if f is unbounded near a, then for $b < a$

$$\int_b^a f(x)\, dx = \lim_{c \to a} \int_b^c f(x)\, dx$$

by definition of the improper Riemann integral.

5.35 Let k be an odd integer. Express $\Gamma(k/2)$ in terms of factorials and powers of 2.

5.36 Show that the beta density is, in fact, a density. That is, show

$$\int_0^1 x^{\alpha-1}(1-x)^{\beta-1}\, dx = \frac{\Gamma(\alpha)\Gamma(\beta)}{\Gamma(\alpha+\beta)}$$

Hint: Write $\Gamma(\alpha)\Gamma(\beta)$ as a double integral and make an appropriate change of variables, as in Lemma 4.4.1.

5.37 For $\alpha > 0$ and $\beta > 0$, let $c = \Gamma(\alpha+\beta)/\Gamma(\alpha)\Gamma(\beta)$, and let $f(x) = cx^{\alpha-1}/(1+x)^{\alpha+\beta}$ for $x > 0$ and $f(x) = 0$ for $x \le 0$. Show that f is a density. This density is known as the *Pareto density* and is sometimes used to describe the distribution of incomes.

5.38 Show that $\int_1^\infty \exp\left[\frac{1}{2}\psi(y)\right] dy$ is finite (see Section 5.4.1).

5.39 Use Problem 5.38 to show that $I_n''' \to 0$ as $n \to \infty$.

5.40 Let X have density $f(x) = \alpha x^{\alpha-1}$ for $0 < x < 1$ and $f(x) = 0$ for other values of x, where $\alpha > 0$. Find the distribution function of X.

5.41 Let X have density $f(x) = \beta/(1+x)^{\beta+1}$ for $x > 0$ and $f(x) = 0$ for $x \le 0$, where $\beta > 0$. Find the distribution function of X.

5.42 Find the density of the random variable X of Example 5.1.4. *Hint:* Find the distribution function and differentiate.

5.43 Find the density of the random variable X of Problem 5.5.

5.44 If X has the gamma distribution with parameters $\alpha = 3$ and $\beta = 1$, find $\Pr(X \le a)$ for $a = 1, 2$.

5.45 Suppose that the lifetime of a light bulb in hours is a random variable X which has the exponential distribution with parameter $\beta = 0.02$. Find the probability that $30 < X < 60$.

5.46 In Problem 5.45 suppose that X has the gamma distribution with parameters $\alpha = 3$ and $\beta = 0.05$. Find the probability that $30 < X < 60$.

5.47 Suppose that family income in a given area in units of $10,000 follows the Pareto distribution with parameters $\alpha = 2$ and $\beta = 1$. That is, suppose that if a family is selected at random from the area and its income X recorded, then X is a random variable which has the Pareto distribution with parameters $\alpha = 2$ and $\beta = 1$. What proportion of the families have incomes between $8000 and $12,000, inclusive?

5.48 Suppose that the demand for electricity, in megawatt-hours, in a given city on a given day is a random variable X which has the normal distribution with parameters $\mu = 500$ and $\sigma = 10$. Find constants a and b for which $\Pr(a < X < b) = 0.95$.

5.49 In Problem 5.48 what is the probability that $X > 530$ megawatt-hours?

5.50 Show that if m is a median (mode) of X, and if $\sigma > 0$, then $\sigma m + \mu$ is a median (mode) of $Y = \sigma X + \mu$.

5.51 Find the mode of the binomial mass function with parameters n and p. *Hint:* Use Equation (1.3) of Chapter 4.

5.52 Find the mode of the negative binomial distribution with parameters r and p.

5.53 Find the mode of the Poisson mass function with parameter β.

5.54 Find the mode of the Pareto density with parameters $\alpha > 1$ and $\beta > 0$.

5.55 Let $F(x) = e^x/(e^x + e^{-x})$ for $-\infty < x < \infty$. Show that F is a distribution function and find its density. F is known as the *logistic* distribution function.

5.56 For $\alpha > 0$, let $F(x) = 1 - e^{-x^\alpha}$ for $x > 0$, and let $F(x) = 0$ for $x \le 0$. Show that F is a distribution function and find a density for F.

5.57 Let $F(x) = \exp(-e^{-x})$ for $-\infty < x < \infty$. Show that F is a distribution function, and find the density function of F. F is known as the *double exponential distribution function*.

5.58 Let $r_1, r_2, \ldots$ be the rational numbers in the interval $[0,1]$, and let X be a random variable for which $\Pr(X = r_n) = 2^{-n}$ for $n = 1, 2, \ldots$. Show that the distribution function of X is discrete but is constant on no subinterval of $[0,1]$.

5.59 Find the medians of (*a*) the logistic distribution function and (*b*) the double exponential distribution function.

6

RANDOM VECTORS

6.1 BIVARIATE DISTRIBUTIONS

Let us now consider the case of two random variables, X and Y say, which are defined on the same probability space $(S, \mathscr{S}, P)$. If I and J are intervals, we might wish to compute either $\Pr(X \in I)$ or $\Pr(Y \in J)$, or both, and we have discussed techniques for computing these probabilities in the previous chapter. We might also wish to compute the probability of the simultaneous occurrence (intersection) of the events that $X \in I$ and $Y \in J$. We shall denote this probability by $\Pr(X \in I, Y \in J)$. Thus

$$\Pr(X \in I, Y \in J) = P(\{s \in S : X(s) \in I \text{ and } Y(s) \in J\}) \qquad (1.1)$$

More generally, if B is a subset of R^2, the set of all ordered pairs of real numbers, then we might wish to compute the probability that the ordered pair (X, Y) belongs to B. We shall denote this probability by $\Pr((X, Y) \in B)$, so that

$$\Pr((X, Y) \in B) = P(\{s \in S : (X(s), Y(s)) \in B\}) \qquad (1.2)$$

As in the previous chapter, we shall employ natural simplifications of the notation (1.1) and (1.2) wherever possible.

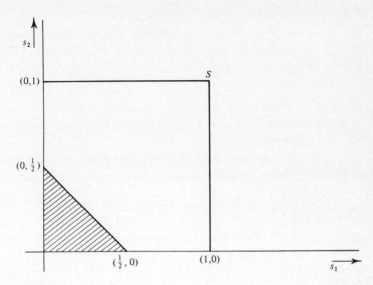

FIGURE 9
The event that $x + y \le \frac{1}{2}$.

EXAMPLE 6.1.1 Consider an urn which contains r red balls, w white balls, and b black balls. If an unordered random sample of size k is drawn, then the number of red balls X and the number of white balls Y in the sample are random variables. Moreover, by Theorem 1.4.2,

$$\Pr\,(X = i,\, Y = j) = \frac{\binom{r}{i}\binom{w}{j}\binom{b}{k - i - j}}{\binom{r + w + b}{k}}$$

whenever i and j are nonnegative integers for which $i + j \le k$. ////

EXAMPLE 6.1.2 Let a point $s = (s_1, s_2)$ be chosen from the unit square

$$S = \{s = (s_1, s_2) : 0 \le s_1 \le 1,\, 0 \le s_2 \le 1\}$$

in such a manner that the probability that s belongs to a subregion $B \subset S$ is the area of B. Then, the coordinate functions $X_1(s_1, s_2) = s_1$ and $X_2(s_1, s_2) = s_2$ define random variables. In this case it is easily verified that $\Pr\,(X_1 + X_2 \le \frac{1}{2}) = \frac{1}{8}$ by drawing an appropriate picture (Figure 9). ////

If X and Y are random variables which are defined on the same probability space, then we shall say that X and Y are *jointly distributed*. Further, we define that *joint distribution* Q of X and Y by

$$Q(B) = \Pr((X, Y) \in B) \qquad (1.3)$$

for all $B \subset R^2$ for which the right side of (1.3) is defined. We shall also refer to Q as the *distribution of the pair* (X, Y). As in the univariate case, it can be shown that Q is a probability measure.

If X and Y are jointly distributed random variables which are each discrete, as in Section 5.2, then we define the *joint mass function* f of X and Y by

$$f(x, y) = \Pr(X = x, Y = y) \qquad (1.4)$$

for $-\infty < x < \infty$ and $-\infty < y < \infty$. We shall also refer to f as the *mass function of the pair* (X, Y). As in the univariate case, it is then easily verified that that f must have the following properties:

$$f(x, y) \geq 0 \qquad (1.5a)$$

for $-\infty < x < \infty$ and $-\infty < y < \infty$; there is a finite or countably infinite set[1] C for which $f(x, y) = 0$ if $(x, y) \notin C$; and

$$\sum_C f(x, y) = 1 \qquad (1.5b)$$

where the summation extends over all $(x, y) \in C$. Also, as in the univariate case, it is easily seen that if X and Y have joint mass function f, then

$$\Pr((X, Y) \in B) = \sum_{BC} f(x, y) \qquad (1.6)$$

for all $B \subset R^2$ for which the left side of (1.6) is defined. In particular the joint mass function f uniquely determines the joint distribution of X and Y.

We shall refer to any function f which vanishes off of a finite or countably infinite set C and satisfies (1.5) as a *bivariate mass function*. We shall see below that if f is any bivariate mass function, then there are discrete random variables X and Y whose joint mass function is f.

Similarly, we define a *bivariate density* to be a real-valued function f which defined on R^2 satisfies

$$f(x, y) \geq 0 \qquad (1.7a)$$

for $-\infty < x < \infty$ and $-\infty < y < \infty$ and

$$\int\int_{-\infty}^{\infty} f(x, y)\, dx\, dy = 1 \qquad (1.7b)$$

Further, if X and Y are jointly distributed random variables, we shall say that the pair

---

[1] If D and E are finite or countably infinite sets for which $\Pr(X \in D) = 1 = \Pr(Y \in E)$, then the cartesian product $C = D \times E$ is finite or countably infinite and $\Pr((X, Y) \notin C) \leq \Pr(X \notin D) + \Pr(Y \notin E) = 0$.

(X, Y) is *absolutely continuous* if and only if there is a bivariate density f for which

$$\Pr (a < X \le b, c < Y \le d) = \int_c^d \int_a^b f(x,y) \, dx \, dy \qquad (1.8)$$

whenever $a < b$ and $c < d$. If (1.8) holds, then we shall call f a *joint density* for X and Y or a *density for the pair* (X, Y). As in the univariate case, it can be shown that if X and Y have joint density f, then

$$\Pr ((X, Y) \in B) = \iint_B f(x,y) \, dx \, dy \qquad (1.9)$$

for all $B \subset R^2$ for which both sides of (1.9) are defined. Moreover, a joint density for two random variables X and Y uniquely determines their joint distribution.

EXAMPLE 6.1.3

a If g and h are univariate densities, then their product f defined by

$$f(x,y) = g(x)h(y)$$

for $-\infty < x < \infty$ and $-\infty < y < \infty$ defines a bivariate density, because f is nonnegative and[1]

$$\int_{-\infty}^{\infty} \!\! \int f(x,y) \, dx \, dy = \left[\int_{-\infty}^{\infty} g(x) \, dx \right] \left[\int_{-\infty}^{\infty} h(y) \, dy \right] = 1^2 = 1$$

This simple remark provides a large class of examples.

b Similarly, if g and h are univariate mass functions, then their product f defined by $f(x,y) = g(x)h(y)$ for $-\infty < x < \infty$ and $-\infty < y < \infty$ defines a bivariate mass function. $////$

EXAMPLE 6.1.4 Let an unordered random sample of size k be drawn from an urn which contains r red balls, b black balls, and w white balls, and let X and Y denote the number of red balls and the number of white balls in the urn, as in Example 6.1.1. Then the joint mass function of X and Y is given by

$$f(x, y) = \frac{\binom{r}{x}\binom{w}{y}\binom{b}{k - x - y}}{\binom{r + w + b}{k}}$$

---

[1] Some rules for manipulating multiple integrals will be given in Section 6.4. Here we anticipate some of the rules.

for nonnegative integers x and y for which $x + y \leq k$ and $f(x,y) = 0$ for other values of x and y.   ////

EXAMPLE 6.1.5   Let B be a region of R^2 with finite positive area $|B|$. Suppose also that the boundary of B has zero area.[1] Then the function f defined by

$$f(x,y) = \begin{cases} \dfrac{1}{|B|} & (x,y) \in B \\ 0 & \text{otherwise} \end{cases}$$

is a density function. We shall refer to f as the *uniform density over B*, and if X and Y have joint density f, we shall say that X and Y are *uniformly distributed over B*.

In the special case that B is the unit circle $B = \{(x,y): x^2 + y^2 \leq 1\}$, we find

$$f(x, y) = \begin{cases} \pi^{-1} & x^2 + y^2 \leq 1 \\ 0 & \text{otherwise} \end{cases} \qquad ////$$

If X and Y are jointly distributed random variables, we define their *joint distribution function F* by

$$F(a,b) = \Pr(X \leq a, Y \leq b) \qquad (1.10)$$

for $-\infty < a < \infty$ and $-\infty < b < \infty$, and we shall also refer to F as the *distribution function of the pair* (X,Y). Thus, if X and Y have joint density f, then

$$F(a,b) = \int_{-\infty}^{b} \int_{-\infty}^{a} f(x,y)\, dx\, dy \qquad (1.11a)$$

for all a and b by (1.8) and

$$f(a,b) = \frac{\partial}{\partial a} \frac{\partial}{\partial b} F(a,b) \qquad (1.11b)$$

at continuity points (a,b) of f.

As in the univariate case, it is possible to express the probability that (X,Y) belongs to any rectangle in terms of its distribution function. For example, we have the following theorem.

Theorem 6.1.1   *Let X and Y have joint distribution function F. If $a < b$ and $c < d$, then* $\Pr(a < X \leq b, c < Y \leq d) = F(b,d) - F(a,d) - F(b,c) + F(a,c)$.

PROOF   Let A (alternatively B, C, and D) be the event that $X \leq a$ (alternatively $X \leq b$, $Y \leq c$, and $Y \leq d$). Then we require the probability

---

[1] The requirement that the area of the boundary be zero ensures that I_B is integrable. See Apostol (1957), p. 258.

of the event $(B - A) \cap (D - C)$. By repeated applications of Theorem 2.3.1, this is

$$P((B - A) \cap (D - C)) = P(B \cap (D - C)) - P(A \cap (D - C))$$
$$= P(BD) - P(BC) - P(AD) + P(AC)$$

which is simply $F(b,d) - F(b,c) - F(a,d) + F(a,c)$, as asserted.   ////

By taking limits, we can now express the probability that (X,Y) belongs to any rectangle in terms of its distribution function (Problems 6.13 to 6.15). However, this expression is much less useful in two dimensions than in one, since bivariate distribution functions are much harder to evaluate and/or tabulate in two dimensions than in one and in two dimensions there are a variety of interesting regions which are not rectangles.

We conclude this section with the two-dimensional analog of Theorems 5.8.1 and 5.8.2. The proof of Theorem 6.1.2, however, is beyond the scope of the book.

Theorem 6.1.2   *Let F be a function which is defined on* R^2. *Then F is the distribution function of a pair* (X,Y) *of random variables if and only if:*

   (i) *F is a nondecreasing and right-continuous in each variable separately;*

   (ii) $\lim F(a,b) = 0$ (1.12a)

   as either $a \to -\infty$ *or* $b \to -\infty$, *and*

                                                              $\lim F(a,b) = 1$ (1.12b)

   as both $a \to \infty$ *and* $b \to \infty$; *and*
   (iii) $F(b,d) - F(a,d) - F(b,c) + F(a,d) \geq 0$ *whenever* $a < b$ *and* $c < d$.

EXAMPLE 6.1.6

   a   If f is any bivariate density, then there are random variables X and Y, whose joint density is f. Indeed, given f, define F by Equation (1.11a). Then it is easily verified that F satisfies conditions (i) to (iii) of Theorem 6.1.2. For example, if $a < b$ and $c < d$, then

$$F(b,d) - F(a,d) - F(b,c) + F(a,c) = \int_c^d \int_a^b f(x,y) \, dx \, dy   \qquad (1.13)$$

which is nonnegative since f is. The verification of (i) and (ii) is left as an exercise (see Example 5.8.1a). It now follows that there is a pair of random variables (X,Y) whose distribution function is F. Finally, it follows from Theorem 6.1.1

that for $a < b$ and $c < d$, $\Pr (a < X \leq b, c < Y \leq d) = F(b,d) - F(a,d) - F(b,c) + F(a,c)$, which is

$$\int_c^d \int_a^b f(x,y) \, dx \, dy$$

by (1.13). Therefore, X and Y have joint density f.

b Similarly, if f is any bivariate mass function, then there are random variables X and Y whose joint mass function is f.   ////

6.2 MARGINAL DISTRIBUTIONS AND INDEPENDENCE

If X and Y have a joint distribution, it is reasonable to hope that the individual distributions of X and Y should be related in some nice way to the joint distribution of X and Y. This is, in fact, the case, and we shall consider this relationship in this section. Let F denote the joint distribution function of X and Y, and let G and H denote the individual distribution functions of X and Y, respectively. Then, since the event that $Y \leq \infty$ is certain to occur, we have formally

$$G(a) = \Pr (X \leq a) = \Pr (X \leq a, Y \leq \infty) = F(a,\infty) \qquad (2.1a)$$

for all a, $-\infty < a < \infty$, and similarly

$$H(b) = F(\infty,b) \qquad (2.1b)$$

for all b, $-\infty < b < \infty$. Relations (2.1a) and (2.1b) are not quite meaningful because the symbols $F(a,\infty)$ and $F(\infty,b)$ have not been defined. However, they become meaningful and correct if we define

$$F(a,\infty) = \lim_{n \to \infty} F(a,n) \qquad \text{and} \qquad F(\infty,b) = \lim_{n \to \infty} F(n,b)$$

See Problem 6.16.

Now, suppose that X and Y have a joint density f. Then the right side of Equation (2.1a) can be written

from above

$$G(a) = F(a,\infty) = \int_{-\infty}^a \int_{-\infty}^\infty f(x,y) \, dy \, dx = \int_{-\infty}^a g(x) \, dx$$

where (by definition)

$$g(x) = \int_{-\infty}^\infty f(x,y) \, dy \qquad (2.2a)$$

for $-\infty < x < \infty$, and it follows that G is absolutely continuous with density g. By symmetry, we find also that H is absolutely continuous with density h, where

$$h(y) = \int_{-\infty}^\infty f(x,y) \, dx \qquad (2.2b)$$

for $-\infty < y < \infty$.

Similar relations can be obtained if X and Y are discrete random variables.

Let f denote the joint mass function of X and Y, and let g and h denote the individual mass functions of X and Y, respectively. Further, let D and E be countable sets for which $\Pr(X \in D) = 1 = \Pr(Y \in E)$. Then, for every x, the event that $X = x$ is the union of the mutually exclusive events $X = x$ and $Y = y$ for $y \in E$ with the event $X = x$ and $Y \notin E$. Since $\Pr(X = x, Y \notin E) \le \Pr(Y \notin E) = 0$, we have

$$\Pr(X = x) = \sum_{y \in E} \Pr(X = x, Y = y)$$

That is, we have

$$g(x) = \sum_{y \in E} f(x,y) \qquad (2.3a)$$

for all x, $-\infty < x < \infty$. Similarly,

$$h(y) = \sum_{x \in D} f(x,y) \qquad (2.3b)$$

for all y, $-\infty < y < \infty$.

In the context of Equations (2.1), (2.2), or (2.3), we shall sometimes refer to G or g as the *marginal distribution function, density, or mass function of X* and to H or h as the *marginal distribution function, density, or mass function of Y*. We summarize our results.

Theorem 6.2.1 *If X and Y are jointly distributed random variables with joint distribution function F, then the distribution functions of X and Y are given by (2.1a) and (2.1b), respectively. If X and Y have joint density f, then X and Y have densities g and h which are given by (2.2a) and (2.2b), respectively, and if X and Y are discrete with joint mass function f, then X and Y have mass functions g and h which are given by (2.3a) and (2.3b), respectively.*

EXAMPLE 6.2.1 If X and Y are discrete random variables which can assume only a finite number of values, then their joint mass function can be given by a table. The marginal mass functions of X and Y can then be obtained by summing across rows and down columns. We illustrate in Table 10.

Table 10

		x			
y	1	2	3	4	$g(x)$
1	0.05	0.05	0.10	0	0.20
2	0.10	0.10	0.05	0.05	0.30
3	0.05	0.05	0	0	0.10
4	0.15	0.05	0.05	0.15	0.40
$h(y)$	0.35	0.25	0.20	0.20	

////

EXAMPLE 6.2.2 Equations (2.1), (2.2), and (2.3) are not the only ways to compute a marginal distribution function, density, or mass function. For example, suppose an unordered random sample of size k is drawn from an urn which contains r red balls, w white balls, and b black balls, and let X and Y denote the number of red balls and the number of white balls in the sample. The joint mass function of X and Y was found in Example 6.1.4. The marginal mass function can be computed from (2.3), but it can also be computed directly. In fact,

$$g(x) = \frac{\binom{r}{x}\binom{w + b}{k - x}}{\binom{r + w + b}{k}}$$

for $x = 0, 1, \ldots, k$ by Theorem 1.4.1.   ////

EXAMPLE 6.2.3 Let X and Y be uniformly distributed over the unit circle $C = \{(x,y): x^2 + y^2 \le 1\}$. Then X and Y have density $f(x,y) = 1/\pi$ if $x^2 + y^2 \le 1$ and $f(x,y) = 0$ otherwise. Therefore, the marginal density of X is

$$g(x) = \int_{-\infty}^{\infty} f(x,y)\, dy = \int_{-\sqrt{1-x^2}}^{\sqrt{1-x^2}} \pi^{-1}\, dy = \frac{2}{\pi}\sqrt{1 - x^2}$$

for $-1 \le x \le 1$. Of course, $g(x) = 0$ if $|x| > 1$, since then $f(x,y) = 0$ for all y.   ////

EXAMPLE 6.2.4 Consider the function f, defined by

$$f(x,y) = C_r \exp\left[-\tfrac{1}{2}Q(x,y)\right]$$

where $-1 < r < 1$,

$$C_r^{-1} = 2\pi\sqrt{1 - r^2} \quad \text{and} \quad Q(x,y) = \frac{x^2 - 2rxy + y^2}{1 - r^2}$$

for $-\infty < x < \infty$ and $-\infty < y < \infty$. We shall show that f is a bivariate density and compute the marginal densities.

By simple algebra we have

$$Q(x,y) = x^2 + z^2 \quad \text{where} \quad z = (y - rx)/\sqrt{1 - r^2}$$

Thus,

$$g(x) = \int_{-\infty}^{\infty} C_r \exp\left[-\tfrac{1}{2}(x^2 + z^2)\right] dy$$

$$= \sqrt{1 - r^2}\, C_r e^{-\frac{1}{2}x^2} \int_{-\infty}^{\infty} e^{-\frac{1}{2}z^2}\, dz$$

$$= \sqrt{2\pi}\sqrt{1 - r^2}\, C_r e^{-\frac{1}{2}x^2} = \frac{1}{\sqrt{2\pi}} e^{-\frac{1}{2}x^2} \tag{2.4}$$

for $-\infty < x < \infty$. Here we have used the fact that the standard normal density has integral 1, and we have used the definition of C_r in the final two steps. It now follows that

$$\int\limits_{-\infty}^{\infty}\!\!\int f(x,y)\, dy\, dx = \int_{-\infty}^{\infty} g(x)\, dx = 1$$

again because the standard normal density has total integral 1, and it follows that f is a bivariate density.

The density f is known as the *standard bivariate normal density with parameter r*. It follows from (2.4) that if X and Y have the standard bivariate normal density with parameter r, then X has the standard (univariate) normal density. By symmetry, Y also has the standard normal density.   ////

We shall say that jointly distributed random variables X and Y are *independent* if and only if

$$\Pr(X \in I, Y \in J) = \Pr(X \in I)\Pr(Y \in J) \qquad (2.5)$$

for all intervals $I \subset R$ and $J \subset R$. That is, X and Y are independent if and only if the events that $X \in I$ and $Y \in J$ are independent for all intervals I and J. Independence can be interpreted as in Chapter 3: *X and Y are independent if and only if the value assumed by X provides no information about Y and conversely.*

We shall now give a criterion for determining whether random variables are independent.

Theorem 6.2.2 *If X and Y have a joint mass function f, then X and Y are independent if and only if*

$$f(x,y) = g(x)h(y) \qquad (2.6a)$$

for all x and y, where g and h denote the marginal mass functions of X and Y, respectively. Similarly, if X and Y are (individually) absolutely continuous with densities g and h, then X and Y are independent if and only if

$$f(x,y) = g(x)h(y) \qquad (2.6b)$$

defines a joint density for the pair (X,Y).

   PROOF   We anticipate the result of Example 6.4.2. Suppose first that X and Y are independent with densities g and h, respectively. If I and J are intervals, then

$$\Pr(X \in I, Y \in J) = \Pr(X \in I)\Pr(Y \in J)$$
$$= \left[\int_I g(x)\, dx\right]\left[\int_J h(y)\, dy\right]$$
$$= \int_I\!\int_J g(x)h(y)\, dy\, dx$$

so that f does define a joint density for X and Y. Conversely, if f defines a joint density for X and Y, then for all intervals I and J [see (1.9)]

$$\Pr(X \in I, Y \in J) = \int_I \int_J g(x)h(y)\,dy\,dx$$

$$= \left[\int_I g(x)\,dx\right]\left[\int_J h(y)\,dy\right]$$

$$= \Pr(X \in I)\Pr(Y \in J)$$

so that X and Y are independent. This establishes the second assertion of the theorem, and the proof of the first is similar.　　////

EXAMPLE 6.2.5　If X and Y have the standard bivariate normal distribution with parameter r, then X and Y are independent if and only if $r = 0$. Indeed, the marginal distributions of X and Y are both standard (univariate) normal, and the product of their marginal densities is therefore

$$\frac{1}{2\pi}\exp\left[-\tfrac{1}{2}(x^2 + y^2)\right] \qquad (2.7)$$

Comparing (2.7) with the joint density of X and Y (Example 6.2.4), we see that (2.7) defines a density for the pair (X,Y) if and only if $r = 0$.　　////

Another criterion for independence may be phrased in terms of distribution functions.

Theorem 6.2.3　*Let X and Y have joint distribution function F. Then X and Y are independent if and only if*

$$F(a,b) = G(a)H(b) \qquad (2.8)$$

for $-\infty < a < \infty$ and $-\infty < b < \infty$, where G and H denote the marginal distribution functions of X and Y, respectively.

PROOF　The "only if" assertion is easy. Indeed, if X and Y are independent, then $\Pr(X \le a, Y \le b) = \Pr(X \le a)\Pr(Y \le b)$ by (2.5), so that $F(a,b) = G(a)H(b)$ for all a and b. To establish the "if" assertion, we argue as follows. Let $a < b$ and $c < d$. Then by Theorem 6.1.1,

$$\Pr(a < X \le b, c < Y \le d) = F(b,d) - F(a,d) - F(b,c) + F(a,c)$$

and if F is of the form (2.8), then $F(b,d) - F(a,d) - F(b,c) + F(a,c) = [G(b) - G(a)][H(d) - H(c)] = \Pr(a < X \leq b)\Pr(c < Y \leq d)$. Thus, (2.8) implies that

$$\Pr(a < X \leq b, c < Y \leq d) = \Pr(a < X \leq b)\Pr(c < Y \leq d) \qquad (2.9)$$

whenever $a < b$ and $c < d$. This establishes (2.5) for intervals of the form $I = (a,b]$ and $J = (c,d]$. We can now show that (2.5) holds for all intervals I and J by simple limiting arguments (Problem 6.17).   ////

6.3 HIGHER DIMENSIONS

The simple notions of the preceding two sections extend without difficulty from two dimensions to several. Thus, consider a probability space $(S, \mathscr{S}, P)$ on which are defined n random variables $X_1, \ldots, X_n$. We shall say that $X_1, \ldots, X_n$ are *jointly distributed*, and we shall refer to $X = (X_1, \ldots, X_n)$ as a *random vector*. If $I_1, \ldots, I_n$ are intervals of real numbers, we shall use the notation $\Pr(X_1 \in I_1, \ldots, X_n \in I_n)$ to denote the probability of the simultaneous occurrence (intersection) of the events $X_j \in I_j, j = 1, \ldots, n$. Thus

$$\Pr(X_1 \in I_1, \ldots, X_n \in I_n) = P(\{s: X_1(s) \in I_1, \ldots, \text{and } X_n(s) \in I_n\}) \qquad (3.1)$$

More generally, if B is a subset of R^n, the set of all ordered n-tuples of real numbers, we shall use the notation $\Pr(X \in B)$ to denote the probability of the set of $s \in S$ for which $X(s) = (X_1(s), \ldots, X_n(s)) \in B$, provided of course, that the latter set is an event. Thus,

$$\Pr(X \in B) = P(\{s: X(s) \in B\}) \qquad (3.2)$$

We define the *joint distribution* Q of $X_1, \ldots, X_n$ by $Q(B) = \Pr(X \in B)$, and we shall also refer to Q as the *distribution of the vector* $X = (X_1, \ldots, X_n)$. As in the univariate and bivariate cases, we can show that Q is a probability measure.

If $X_1, \ldots, X_n$ are all discrete random variables, as defined in Section 5.2, we define the *joint mass function* f of $X_1, \ldots, X_n$ by

$$f(x_1, \ldots, x_n) = \Pr(X_1 = x_1, \ldots, X_n = x_n)$$

for $x = (x_1, \ldots, x_n) \in R^n$. We shall also refer to f as the *mass function of the vector* $X = (X_1, \ldots, X_n)$. It is then easily verified that f has the following properties:

$$f(x) \geq 0 \qquad (3.3a)$$

for all $x = (x_1, \ldots, x_n) \in R^n$; there is a finite or countably infinite subset[1] $C \subset R^n$ for which

$$f(x) = 0 \quad \text{if } x \notin C \qquad (3.3b)$$

and

$$\sum_C f(x) = 1 \qquad (3.3c)$$

where the summation extends over all $x = (x_1, \ldots, x_n) \in C$. Moreover,

$$\Pr (X \in B) = \sum_{BC} f(x) \qquad (3.4)$$

for any $B \subset R^n$ for which the left side of (3.4) is defined. In particular, the joint mass function f uniquely determines the joint distribution.

We shall refer to any function f which satisfies (3.3) as an *n-variate mass function*.

Similarly, we define an *n-variate density* to be a real-valued function f which is defined on R^n and satisfies

$$f(x) \geq 0 \qquad (3.5a)$$

for all $x = (x_1, \ldots, x_n) \in R^n$ and[2]

$$\int_{R^n} f(x) \, dx = 1 \qquad (3.5b)$$

If $X_1, \ldots, X_n$ are jointly distributed random variables, we shall say that $X_1, \ldots, X_n$ are *jointly absolutely continuous with joint density f* if and only if

$$\Pr (X \in B) = \int_B f(x) \, dx \qquad (3.6)$$

for all rectangles $B \subset R^n$. In this case we shall also say that *the vector $X = (X_1, \ldots, X_n)$ is absolutely continuous with density f.* If (3.6) holds for all rectangles $B \subset R^n$, then it can be shown that (3.6) holds for all subsets $B \subset R^n$ for which both sides are defined (compare Theorem 5.3.1). Moreover, a joint density f uniquely determines the joint distribution of $X_1, \ldots, X_n$.

EXAMPLE 6.3.1

a   Consider an urn which contains balls of n different colors, say $r_i \geq 1$ balls of color i, $i = 1, \ldots, n$. If a sample of size $k \leq r = r_1 + \cdots + r_n$ is drawn from the urn without replacement and we let X_i be the number of balls of color i

---

[1] If C_i is a finite or countably infinite set for which $\Pr (X_i \in C_i) = 1$, then $C = C_1 \times \cdots \times C_n$ is a finite or countably infinite set for which $\Pr (X \in C) = 1$.
[2] The integrals appearing in (3.5b) and (3.6) are n-dimensional integrals, and dx denotes the volume element $dx_1 \cdots dx_n$. Rules for evaluating multiple integrals will be given in the next section.

in the sample, $i = 1, \ldots, n$, then $X = (X_1, \ldots, X_n)$ will be a random vector for which

$$\Pr(X_1 = x_1, \ldots, X_n = x_n) = \frac{\binom{r_1}{x_1} \cdots \binom{r_n}{x_n}}{\binom{r}{k}} \qquad (3.7)$$

whenever $x_1, \ldots, x_n$ are nonnegative integers whose sum is k. See Theorem 1.4.2. Equation (3.7) defines the n-variate *hypergeometric* mass function with parameters $r_1, \ldots, r_n$ and k.

b Consider a k-sided die, and let p_i be the probability that the ith face appears on any given roll, $i = 1, \ldots, k$ (the p_i need not be equal). If n independent rolls of the die are made, and if we let X_i be the number of rolls on which the ith face appears, then $X = (X_1, \ldots, X_k)$ will be a random vector for which

$$\Pr(X_1 = x_1, \ldots, X_k = x_k) = \binom{n}{x_1, \ldots, x_k} p_1^{x_1} \cdots p_k^{x_k} \qquad (3.8)$$

whenever $x_1, \ldots, x_k$ are nonnegative integers whose sum is n. See Theorem 4.1.2. We shall refer to (3.8) as the *multinomial* mass function with parameters $n \geq 1$ and $p = (p_1, \ldots, p_k)$. See Problems 6.24 to 6.27 for further properties of the multinomial distribution.   ////

EXAMPLE 6.3.2   Let B be a region of R^n having finite, positive (n-dimensional) volume $|B|$. Suppose also that the boundary of B has zero (n-dimensional) volume. Then, the function f defined by

$$f(x) = \begin{cases} \dfrac{1}{|B|} & x \in B \\ 0 & x \notin B \end{cases} \qquad (3.9)$$

is known as the *uniform* density over B.   ////

As in two dimensions, individual densities or mass functions of random variables $X_1, \ldots, X_n$ can be obtained from a joint density or mass function. We shall give the relevant formulas only for the absolutely continuous case since the formulas for the discrete case can be obtained by replacing integrals by sums. If $X_1, \ldots, X_n$ have a joint density f, then

$$f_1(x) = \int_{-\infty}^{\infty} \cdots \int_{-\infty}^{\infty} f(x, x_2, \ldots, x_n) \, dx_2 \cdots dx_n \qquad (3.10)$$

defines a density for X_1, and densities for X_i, $i \geq 2$, can be obtained from (3.10) and an appropriate relabeling. We shall refer to (3.10) as the marginal density of X_1. More generally, if $k < n$, then

$$g(x_1, \ldots, x_k) = \int_{-\infty}^{\infty} \cdots \int_{-\infty}^{\infty} f(x_1, \ldots, x_n) \, dx_{k+1} \cdots dx_n \qquad (3.11)$$

defines a joint density for $X_1, \ldots, X_k$. We shall sometimes refer to g as the marginal density of $X_1, \ldots, X_k$.

EXAMPLE 6.3.3

a   Let (X_1, X_2, X_3) have the uniform density on the set B of $x = (x_1, x_2, x_3) \in R^3$ for which $0 \le x_1 \le x_2 \le x_3 \le 1$. Thus, $f(x) = 6$ for $x \in B$. Let us find the marginal density of X_1. By Equation (3.10) we have

$$f_1(x) = \int_x^1 \int_{x_2}^1 6 \, dx_3 \, dx_2 = \int_x^1 6(1 - x_2) \, dx_2 = 3(1 - x)^2$$

for $0 \le x \le 1$, and $f_1(x) = 0$ for other values of x.

b   The joint density of (X_1, X_2) is

$$g(x_1, x_2) = \int_{x_2}^1 6 \, dx_3 = 6(1 - x_2)$$

for $0 \le x_1 \le x_2 \le 1$, and $g(x_1, x_2) = 0$ for other values of (x_1, x_2).   ////

EXAMPLE 6.3.4

a   If $X_1, \ldots, X_n$ have the n-variate hypergeometric distribution (Example 6.3.1*a*)

$$f(x_1, \ldots, x_n) = \frac{\binom{r_1}{x_1} \cdots \binom{r_n}{x_n}}{\binom{r}{k}}$$

then X_1 has the univariate hypergeometric distribution

$$f_1(x) = \frac{\binom{r_1}{x} \binom{r_2 + \cdots + r_n}{k - x}}{\binom{r}{k}}$$

$x = 0, \ldots, k$. In fact, X_1 is the number of balls of color 1 in a sample of size k.

b   If $X_1, \ldots, X_k$ have the multinomial distribution

$$f(x_1, \ldots, x_k) = \binom{n}{x_1, \ldots, x_k} p_1^{x_1} \cdots p_k^{x_k}$$

then X_1 has the binomial distribution

$$f_1(x) = \binom{n}{x} p_1{}^x (1 - p_1)^{n-x}$$

$x = 0, \ldots, n$. In fact, X_1 is the number of times an event with probability p_1 occurs in n independent repetitions.   ////

If $X_1, \ldots, X_n$ are jointly distributed random variables, we shall say that $X_1, \ldots,$ X_n are *pairwise independent* if and only if X_i and X_j are independent whenever $i \neq j$ and we shall say that $X_1, \ldots, X_n$ are *mutually independent* if and only if

$$\Pr (X_1 \in I_1, \ldots, X_n \in I_n) = \prod_{j=1}^{n} \Pr (X_j \in I_j) \qquad (3.12)$$

whenever $I_1, \ldots, I_n$ are intervals of real numbers. As in the case of events, the two notions of independence are not equivalent (see Problem 6.29), and it is the latter which is the more interesting. Accordingly, we shall use the unqualified term "independent" to mean mutually independent.

We shall now give a criterion for determining whether random variables are independent.

Theorem 6.3.1 *If $X_1, \ldots, X_n$ are discrete random variables with joint mass function f, then $X_1, \ldots, X_n$ are independent if and only if*

$$f(x_1, \ldots, x_n) = \prod_{i=1}^{n} f_i(x_i)$$

for all $x = (x_1, \ldots, x_n) \in R^n$, where f_i denotes the marginal mass function of X_i, $i = 1, \ldots, n$. Similarly, if $X_1, \ldots, X_n$ are individually absolutely continuous with (marginal) densities $f_1, \ldots, f_n$, then $X_1, \ldots, X_n$ are independent if and only if

$$f(x_1, \ldots, x_n) = \prod_{i=1}^{n} f_i(x_i) \qquad x \in R^n$$

defines a joint density for $X_1, \ldots, X_n$.

The proof of Theorem 6.3.1 is similar to that of Theorem 6.2.1, which it generalizes, and will therefore be omitted.

EXAMPLE 6.3.5

   a   Let $J_1, \ldots, J_n$ be finite intervals. If X_i has the uniform distribution on the interval J_i, $i = 1, \ldots, n$, and if $X_1, \ldots, X_n$ are independent, then $X =$

$(X_1, \ldots, X_n)$ has the uniform density on the rectangle $B = J_1 \times J_2 \times \cdots \times J_n$; for a joint density for $X_1, \ldots, X_n$ is

$$\prod_{i=1}^{n} f(x_i) = \prod_{i=1}^{n} \frac{1}{|J_i|} = \frac{1}{|B|}$$

if $x_i \in J_i$, $i = 1, \ldots, n$ [that is, if $x = (x_1, \ldots, x_n) \in B$], and the product is zero if $x \notin B$. The converse assertion is also true. That is, if X is uniformly distributed over B, then $X_1, \ldots, X_n$ are independent and X_i is uniformly distributed over J_i, $i = 1, \ldots, n$ (see Problem 6.30).

b   If $X_1, \ldots, X_k$ have the multinomial mass function with parameters $n \geq 1$ and $p = (p_1, \ldots, p_k)$, where $0 < p_i < 1$, $i = 1, \ldots, k$, then $X_1, \ldots, X_k$ are not independent, because

$$0 = \Pr(X_1 = n, \ldots, X_k = n) \neq \prod_{i=1}^{k} p_i^n = \prod_{i=1}^{k} \Pr(X_i = n) \qquad ////$$

A final generalization of the notion of independence can now be obtained by letting the X_j of Equation (3.12) themselves be random vectors and interpreting the symbols I_j as rectangles, $j = 1, \ldots, n$. Thus, we shall say that random vectors $X_1, \ldots, X_n$ are *independent* if and only if Equation (3.12) holds whenever $I_1, \ldots, I_n$ are rectangles and I_j is of the same dimension as X_j, $j = 1, \ldots, n$. In this case, Theorem 6.3.1 remains true provided that we interpret the f_j as the mass functions and densities of the random vectors X_j, $j = 1, \ldots, n$.

EXAMPLE 6.3.6   If $X_1, \ldots, X_n$ are independent random variables, and if $1 \leq k < n$, then the random vectors $Y = (X_1, \ldots, X_k)$ and $Z = (X_{k+1}, \ldots, X_n)$ are independent. For if $A = I_1 \times \cdots \times I_k$ and $B = I_{k+1} \times \cdots \times I_n$ are arbitrary rectangles (of dimensions k and $n - k$), then

$$\Pr(Y \in A, Z \in B) = \Pr(X_1 \in I_1, \ldots, X_n \in I_n)$$

$$= \left[\prod_{j=1}^{k} \Pr(X_j \in I_j) \right] \left[\prod_{j=k+1}^{n} \Pr(X_j \in I_j) \right]$$

$$= \Pr(Y \in A) \Pr(Z \in B) \qquad ////$$

6.4   EXAMPLES[1]

Many interesting probabilities are of the form $\Pr(X \in B)$, where X is a random vector and B is a suitable region of R^n. If X is absolutely continuous with density f, then

$$\Pr(X \in B) = \int_B f(x) \, dx \qquad (4.1)$$

---

[1] This section may be omitted without loss of continuity.

and if X is discrete with mass function f, then

$$\Pr (X \in B) = \sum_B f(x)$$

The point is that in order to evaluate these probabilities, we must be able to evaluate multiple sums and integrals, and so we devote this section to a review of techniques for evaluating multiple sums and integrals. For simplicity, we shall restrict our attention to integrals since techniques for handling sums are similar.

The basic idea is the following. Let f be a continuous integrable function on R^2 and let B be a region whose boundary has zero area. Then

$$\iint_B f(x,y) \, dx \, dy = \int_{-\infty}^{\infty} \left[\int_{B_x} f(x,y) \, dy \right] dx$$

$$= \int_{-\infty}^{\infty} \left[\int_{B^y} f(x,y) \, dx \right] dy \qquad (4.2)$$

where for each x, B_x denotes the set of $y \in R$ for which $(x,y) \in B$ and for each y, B^y denotes the set of $x \in R$ for which $(x,y) \in B$ (see Figure 10). Therefore, the evaluation of a double integral can be accomplished by the iterated evaluation of single integrals, that is, by evaluating

$$h(x) = \int_{B_x} f(x,y) \, dy$$

for every x and then computing

$$\int_{-\infty}^{\infty} h(x) \, dx$$

Of course, the roles of X and Y may be reversed; and the assumption that f is continuous may be relaxed.

EXAMPLE 6.4.1 Consider two radioactive particles which decay independently at the same rate, and let X and Y denote their lifetimes. What is the probability that the first particle lasts at least twice as long before decaying as the second? Let X and Y denote the times until the two particles decay. Then we may suppose that X and Y are independent, exponentially distributed random variables with the same parameter β, and we require the probability that $X \geq 2Y$, that is, $\Pr ((X,Y) \in B)$, where B is the set of $(x,y) \in R^2$ for which $x \geq 2y \geq 0$. The joint density of X and Y is

$$f(x, y) = \begin{cases} \beta^2 e^{-\beta(x+y)} & x \geq 0, \, y \geq 0 \\ 0 & \text{otherwise} \end{cases}$$

Thus, we must evaluate (4.1) with the given f and B. Using Equation (4.2), we find that

$$B^y = [2y, \infty)$$

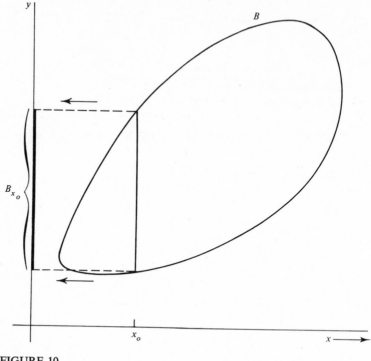

FIGURE 10
Finding B_x.

is the interval from $2y$ to ∞. Therefore, for $y \geq 0$,

$$\int_{B^y} f(x, y) \, dx = \int_{2y}^{\infty} \beta^2 e^{-\beta(x+y)} \, dx = \beta e^{-\beta y} \int_{2y}^{\infty} \beta e^{-\beta x} \, dx$$

$$= \beta e^{-\beta y} e^{-2\beta y} = \beta e^{-3\beta y}$$

so that

$$\int_B f(x,y) \, dx \, dy = \int_0^{\infty} \beta e^{-3\beta y} \, dy = \tfrac{1}{3}$$

By symmetry, the probability that the second particle survives at least twice as long as the first is $\Pr(Y \geq 2X) = \tfrac{1}{3}$. Therefore, the probability that one particle lasts at least twice as long as the other is $\Pr(X \geq 2Y) + \Pr(Y \geq 2X) = \tfrac{2}{3}$. ////

EXAMPLE 6.4.2 Suppose that B is a rectangle, say $B = I \times J$, and that $f(x, y) = g(x)h(y)$, where g and h are integrable functions. Then $B_x = J$ if $x \in I$, and $B_x = \varnothing$, the empty set, if $x \notin I$. Thus,

$$\int_{B_x} f(x,y) \, dy = g(x) \int_J h(y) \, dy \qquad (4.3)$$

if $x \in I$, and the left side of (4.3) is 0 if $x \notin I$. It follows that

$$\iint_{I \times J} f(x,y) \, dx \, dy = \int_I \left[\int_J f(x,y) \, dy \right] dx$$

$$= \left[\int_J h(y) \, dy \right] \left[\int_I g(x) \, dx \right]$$

This fact was used in Example 6.1.3a and in the proof of Theorem 6.2.2. ////

In higher dimensions, it is also possible to reduce multiple integrals to iterated single integrals. For simplicity, we limit ourselves to the case of three variables x_1, x_2, and x_3. The basic formula is

$$\int_B f(x) \, dx = \int_{-\infty}^{\infty} \left[\iint_{B_{x_1}} f(x_1,x_2,x_3) \, dx_3 \, dx_2 \right] dx_1 \qquad (4.4)$$

where B_{x_1} is the set of (x_2,x_3) for which $(x_1,x_2,x_3) \in B$. The inner integral on the right side of (4.4) is now a two-dimensional integral which can be handled by the techniques already discussed.

EXAMPLE 6.4.3 Let $X = (X_1,X_2,X_3)$ be uniformly distributed over the unit cube in R^3. What is the probability that $X_1 \geq 2X_2 \geq 3X_3$? A density for X is

$$f(x_1,x_2,x_3) = \begin{cases} 1 & 0 \leq x_i \leq 1, \ i = 1, 2, 3 \\ 0 & \text{otherwise} \end{cases}$$

and we wish to compute $\Pr(X \in B)$, where B is the set of $x \in R^3$ for which $0 \leq 3x_3 \leq 2x_2 \leq x_1 \leq 1$. For each fixed value of x_1, B_{x_1} is simply the set of (x_2,x_3) for which $0 \leq x_3 \leq 2x_2/3 \leq x_1/3$. Therefore

$$\int_{B_{x_1}} f(x_1,x_2,x_3) \, dx_3 \, dx_2 = \int_0^{(1/2)x_1} \left(\int_0^{(2/3)x_2} dx_3 \right) dx_2$$

$$= \int_0^{x_1/2} \tfrac{2}{3}x_2 \, dx_2 = \tfrac{1}{12}x_1^2$$

Therefore,

$$\int_B f(x) \, dx = \int_0^1 \tfrac{1}{12}x_1^2 \, dx_1 = \tfrac{1}{36} \qquad (4.4a)$$

////

The most general conditions under which (4.2) and (4.4) are valid are somewhat complicated.[1] They are valid, however, if all the integrals appearing in them exist

---

[1] See Thomas (1972), pp. 250–256, for an elementary discussion of (4.2); see Apostol (1957), pp. 260–268, for a more complete discussion of (4.2) and (4.4).

as ordinary (Riemann) integrals or as absolutely convergent, improper integrals. One therefore automatically checks the validity of Equations (4.2) and (4.4) by doing the indicated calculations.

PROBLEMS

6.1 Let an ordered, random sample of size k be drawn without replacement from an urn containing r red balls, w white balls, and b black balls. Let X and Y denote the number of red and white balls in the sample, respectively.
(a) Find the joint mass function of X and Y.
(b) Find the marginal mass functions of X and Y, respectively.
(c) Are X and Y independent?

6.2 In Problem 6.1 let X be the number of the draw on which the first red ball is drawn, and let Y be the number of the draw on which the first white ball is drawn. Answer parts (a) to (c).

6.3 Show that the function f, defined by $f(x,y) = 1/2\pi(1 + x^2 + y^2)^{3/2}$ for $-\infty < x < \infty$ and $-\infty < y < \infty$, is a bivariate density. It is known as the standard *bivariate Cauchy* density.
(a) Show that the marginal densities are both standard univariate Cauchy.
(b) If X and Y have the standard bivariate Cauchy distribution, are X and Y independent?

6.4 Show that the function f, defined by $f(x,y) = Cx^{\alpha-1}y^{\beta-1}(1 - x - y)^{\gamma-1}$ for $x > 0$, $y > 0$, and $x + y \leq 1$, with $C = \Gamma(\alpha + \beta + \gamma)/\Gamma(\alpha)\Gamma(\beta)\Gamma(\gamma)$, is a bivariate density. Here $\alpha > 0$, $\beta > 0$, and $\gamma > 0$. Show that the marginal densities are both beta. Are X and Y independent?

6.5 (a) Show that the function f, defined by $f(x,y) = \frac{1}{2}(1 + x) \exp(-x - y)$, $x > 0$, $y > 0$, is a bivariate density. (b) If X and Y have density f, find the marginal densities of X and Y. (c) Are X and Y independent?

6.6 (a) Show that the function f, defined by $f(x,y) = \frac{1}{2}e^{-x}$ for $x > 0$ and $-x < y < x$ is a bivariate density. (b) If X and Y have density f, find the marginal distributions of X and Y. (c) Are X and Y independent?

6.7 Let X and Y have the density of Problem 6.4. Find the joint density of W and Z, where $W = X/(1 - X)$ and $Z = Y/(1 - Y)$. *Hint:* Use (1.11b).

6.8 Let X and Y have joint density f. If $f(x,y) = g(x)h(y)$ for all x and y, where g and h are not necessarily the densities of X and Y, show that X and Y are independent.

6.9 If X and Y have joint density $f(x,y) = 8xy$ for $0 \leq x \leq y \leq 1$ and $f(x,y) = 0$ otherwise, are X and Y independent?

6.10 Let g be a continuous univariate density, and let $f(x,y) = 2g(x)g(y)$ for $x \leq y$ and $f(x,y) = 0$ if $x > y$. Show that f is a bivariate density. If X and Y have joint density f, find the marginal densities of X and Y. Are X and Y independent?

6.11 If X and Y have joint density, then $\Pr(X = Y) = 0$. Is the result necessarily true only if we assume that X and Y have (univariate) absolutely continuous distributions?

6.12   Let X and Y have a continuous, joint density f. Show that

$$f(x,y) = \frac{-\partial^2}{\partial x\, \partial y} \Pr\,(X \le x,\, Y > y)$$

6.13   Let X and Y have joint distribution function F, and let $a < b$ and $c < d$ be real numbers. Express $F(b-,d) - F(b-,c) - F(a,d) + F(a,c)$ as a probability.

6.14   Let X and Y have joint distribution function F. For $-\infty < a < \infty$ and $-\infty < b < \infty$ express $F(a,b) - F(a-,b) - F(a,b-) + F(a-,b-)$ as a probability.

6.15   Let X and Y have joint distribution function F, and let $a < b$ and $c < d$. Express $\Pr\,(a \le X \le b,\, c < Y \le d)$ in terms of F.

6.16   Let X and Y have joint distribution F, and let G denote the marginal distribution function of X. Show that $G(a) = \lim F(a,n)$, where the limit is taken as $n \to \infty$. *Hint:* Consider the events A_n that $X \le a$ and $Y \le n$ and apply Theorem 2.5.1.

6.17   Complete the proof of Theorem 6.2.3. *Hint:* For example,

$$\Pr\,(a \le X \le b,\, c < Y \le d) = \lim \Pr\,(a - 1/n < X \le b,\, c < Y \le d)$$

as $n \to \infty$.

6.18   Show that the function f, defined by $f(x,y,z) = 1/\pi^2(1 + x^2 + y^2 + z^2)^2$ for $(x,y,z) \in R^3$ is a trivariate density. If X, Y, and Z have joint density f, find the marginal densities of X and of (X,Y). This density is known as the standard *trivariate Cauchy* density.

6.19   Let $\alpha_i > 0$, $i = 1, \ldots, k$, and let $C = \Gamma(\alpha_1 + \cdots + \alpha_k)/\Gamma(\alpha_1) \cdots \Gamma(\alpha_k)$. Show that the function f, defined by

$$f(x_1, \ldots, x_{k-1}) = C \prod_{j=1}^{k-1} x_i^{\alpha_i - 1}(1 - x_1 - \cdots - x_{k-1})^{\alpha_k - 1}$$

for $x_i > 0$, $i = 1, \ldots, k - 1$ and $x_1 + \cdots + x_{k-1} < 1$, is a $(k - 1)$-variate density. A random vector $(X_1, \ldots, X_k)$ is said to have the k-variate *Dirichlet distribution with parameters* $\alpha_1, \ldots, \alpha_k$ if and only if $X_1, \ldots, X_{k-1}$ has density f and $X_k = 1 - X_1 - \cdots - X_{k-1}$.

6.20   Let $X_1, \ldots, X_k$ have the k-variate Dirichlet distribution with parameters $\alpha_1, \ldots, \alpha_k$, and let $j < k$. Show that the marginal distribution of $X_1, \ldots, X_{j-1}$ and $X_j + \cdots + X_k$ is j-variate Dirichlet with parameters $\alpha_i' = \alpha_i$, $i < j$, and $\alpha_j' = \alpha_j + \cdots + \alpha_k$.

6.21   Prove the following theorem: random variables $X_1, \ldots, X_n$ are mutually independent if and only if the events that $X_i \in B_i$, $i = 1, \ldots, n$, are mutually independent for every choice of the intervals $B_1, \ldots, B_n$.

6.22   Let W, X, Y, Z have joint density $f(w,x,y,z) = 24$ for $0 < w < x < y < z < 1$ and $f(w,x,y,z) = 0$ for other values of (w,x,y,z).
(a) Find the marginal densities of the vectors (W,X) and (Y,Z).
(b) Are these vectors independent?

6.23   Let W, X, Y, and Z have joint density $f(w,x,y,z) = 24/(1 + w + x + y + z)^5$ for $w > 0$, $x > 0$, $y > 0$, and $z > 0$ and $f(w,x,y,z) = 0$ for other values of (w,x,y,z). Find the marginal density of W and of (W,X).

(a) Are W, X, Y, and Z independent?

(b) Are (W,X) and (Y,Z) independent?

6.24 Consider a die which is loaded in such a manner that the probability that exactly k spots will appear when the die is rolled is $p_k = k/21$, $k = 1, \ldots, 6$. Let n independent rolls of the die be made, and let X_k be the number of times that exactly k spots appear.

(a) What is the joint distribution of $X_1, \ldots, X_6$?

(b) What is the distribution of $X_1 + X_2 + X_3$?

(c) What is the joint distribution of $(X_1 + X_2 + X_3, X_4, X_5, X_6)$?

6.25 Let $X_1, \ldots, X_k$ have the multinomial distribution with parameters $n \geq 1$ and $p = (p_1, \ldots, p_k)$. If $j < k$, show that the marginal distribution of $X_1, \ldots, X_j$ and $Y = X_{j+1} + \cdots + X_k$ is multinomial with parameters n and $q = (q_1, \ldots, q_{j+1})$, where $q_i = p_i$ for $i \leq j$ and $q_{j+1} = p_{j+1} + \cdots + p_k$. *Hint:* Use induction from Problem 6.24.

6.26 Show that if $X_1, \ldots, X_k$ have the multinomial distribution with parameters n and $p = (p_1, \ldots, p_k)$ and if $j < k$, then $Y = X_1 + \cdots + X_j$ has the binomial distribution with parameters n and $p_1 + \cdots + p_j$.

6.27 Let $X_1, \ldots, X_4$ have the multinomial distribution with parameters $p_1 = 0.1$, $p_2 = 0.2$, $p_3 = 0.3$, and $p_4 = 0.4$. What is the distribution of (a) X_1, (b) $X_1 + X_2$, (c) $(X_1 + X_2, X_3, X_4)$?

6.28 Let an unordered random sample be drawn from an urn which contains 4 red balls, 5 white balls, 6 black balls, and 7 green balls. Further let X_1 denote the number of red balls in the sample, X_2 the number of white, X_3 the number of black, and X_4 the number of green.

(a) What is the joint distribution of $X_1, \ldots, X_4$?

(b) What is the distribution of $X_1 + X_2$?

(c) What is the joint distribution of $(X_1 + X_2, X_3, X_4)$?

6.29 Let $A_1, \ldots, A_n$ be events which are pairwise independent but not mutually independent (see Example 3.3.5). Let $X_i = I_{A_i}$ be the indicator of A_i for $i = 1, \ldots, n$. Show that $X_1, \ldots, X_n$ are pairwise independent but not mutually independent.

6.30 Show that if $X = (X_1, \ldots, X_n)$ is uniformly distributed over the rectangle $B = J_1 \times \cdots \times J_n$, where J_k are finite intervals of positive length, then $X_1, \ldots, X_n$ are independent and X_i is uniformly distributed over J_i, $i = 1, \ldots, n$.

6.31 Let Y and Z be independent random variables which are uniformly distributed over $(0,1)$. Find the probability that the equation $x^2 + 2xY + Z = 0$ has real roots (in x).

6.32 Let X, Y, and Z be independent random variables all of which have the exponential density $f(x) = e^{-x}$, $x > 0$. Find the probability of the simultaneous occurrence of the events that $X \leq 2Y$ and $X \leq 2Z$.

DISTRIBUTION THEORY

7.1 UNIVARIATE DISTRIBUTIONS

We shall often be confronted with the following problem. We are given a random variable or vector X and a function w, and we wish to find the distribution of the random variable or vector $Y = w(X)$. In this chapter we shall consider several techniques for finding the distribution of Y from knowledge of the distribution of X and w.[1] We shall begin with some generalities about functions.

Let w be a function from a set D into another set T. Then D is known as the *domain* of w, and the *range* of w is the set

$$E = \{w(x) \colon x \in D\}$$

Thus, $y \in E$ if and only if there is an $x \in D$ for which $w(x) = y$.

There are two notions of inverse which will be of interest to us. First, we recall from Section 5.1 the notation

$$w^{-1}(B) = \{x \in D \colon w(x) \in B\}$$

for $B \subset T$. Thus, w^{-1} is a function from the class of all subsets of T into the class of

---

[1] We tacitly assume throughout the chapter that $w(X)$ is a random variable or vector, that is, satisfies (1.1) of Chapter V.

all subsets of D, and $x \in w^{-1}(B)$ if and only if $w(x) \in B$. Lemma 5.1.1 asserts that w^{-1} commutes with the operations of union, intersection, and complementation.

w^{-1} is not to be confused with the *inverse function*, which we shall denote by v. We shall say that w is *one to one* if and only if $w(x_1) \neq w(x_2)$ whenever $x_1 \in D$, $x_2 \in D$, and $x_1 \neq x_2$. If w is one to one, then for every $y \in E$, the range, there will be a unique $x \in D$ for which $w(x) = y$, and we can define a function v on E by letting $v(y) = x$ if and only if $w(x) = y$. v is known as the inverse function to w. We emphasize that v can be defined only if w is one to one.

If w is one to one with inverse v, then v is one to one and the inverse of v is w. Moreover, the relations

$$v(w(x)) = x \quad \text{and} \quad w(v(y)) = y \qquad (1.1a)$$

hold for $x \in D$ and $y \in E$, respectively, If D and E are subintervals of R, and if w is continuously differentiable, then relation $(1.1a)$ can be differentiated to yield

$$v'(w(x))w'(x) = 1 \quad \text{and} \quad w'(v(y))v'(y) = 1 \qquad (1.1b)$$

for $x \in D$ and $y \in E$ for which $w'(x) \neq 0 \neq v'(y)$.

EXAMPLE 7.1.1

a Let $D = T = [0,\infty)$, and let $w(x) = x^2$ for $x \in D$. Then w is one to one, and $v(y) = \sqrt{y}$ for $y \in E = [0,\infty)$.

b Let $D = (-\infty,\infty)$, let $T = [0,\infty)$, and let $w(x) = x^2$ for $x \in D$. Then w is not one to one. We have $w^{-1}(\{0\}) = \{0\}$ and $w^{-1}(\{y\}) = \{\sqrt{y}, -\sqrt{y}\}$ for $y \neq 0$.   ////

Let us now return to probability theory. We shall consider the case that X is a random variable and w is a real-valued function defined on a subset $D \subset R$ for which $\Pr(X \in D) = 1$. The case $D = R$ is, of course, not excluded, but in some cases it will be convenient to let D be a proper subset of R. The basic relation between X and $Y = w(X)$ may now be stated.

Lemma 7.1.1 *Let X be a random variable, and let $Y = w(X)$, where w is defined on a subset $D \subset R$ for which $\Pr(X \in D) = 1$. Then*

$$\Pr(Y \in I) = \Pr(X \in w^{-1}(I))$$

for all intervals $I \subset R$.

PROOF   The lemma is a tautology. Indeed, the event that $Y = w(X) \in I$ is the same as the event that $X \in w^{-1}(I)$, and so the probabilities are the same.   ////

If X is a discrete random variable, with mass function f say, then we can let D be the set of possible values of X, that is, a sequence $x_0, x_1, \ldots$ for which $f(x_0) + f(x_1) + \cdots = 1$, in which case the elements of E can also be arranged in a simple sequence, say $E = \{y_0, y_1, \ldots\}$. Thus, if X is discrete, then Y will also be discrete, and the mass function g of Y is given by

$$g(y) = \sum_{x \in w^{-1}(\{y\})} f(x) \qquad (1.2)$$

for $-\infty < y < \infty$. In particular, if w is one to one with inverse function v, then

$$g(y) = f(v(y)) \qquad (1.2a)$$

for $y \in E$ and $g(y) = 0$ for $y \notin E$.

EXAMPLE 7.1.2

a   If X has mass function f, then $Y = X^2$ has mass function g, where $g(0) = f(0)$, $g(y) = f(\sqrt{y}) + f(-\sqrt{y})$ for $y > 0$ and $g(y) = 0$ for $y < 0$. If, in addition, $\Pr(X \geq 0) = 1$, then $g(y) = f(\sqrt{y})$ for $y \geq 0$ and $g(y) = 0$ for $y < 0$.

b   If X has mass function f, then $Y = e^X$ has mass function g, where $g(y) = f(\log y)$ for $y > 0$ and $g(y) = 0$ for $y \leq 0$. ////

Let us now drop the assumption that X is discrete and consider the distribution function of $Y = w(X)$. By Lemma 7.1.1

$$G(y) = \Pr(Y \leq y) = \Pr(X \in w^{-1}((-\infty, y])) \qquad (1.3)$$

for $-\infty < y < \infty$. We illustrate with some examples, in all of which we take $D = R$.

EXAMPLE 7.1.3

a   Let X have distribution function F, and let $Y = aX + b$ be a linear function of X. If $a > 0$, then $Y \leq y$ if and only if $X \leq (y - b)/a$, so that (1.3) requires

$$G(y) = \Pr\left(X \leq \frac{y - b}{a}\right) = F\left(\frac{y - b}{a}\right)$$

for $-\infty < y < \infty$; and if $a < 0$, then we find

$$G(y) = \Pr\left(X \geq \frac{y - b}{a}\right) = 1 - F\left(\frac{y - b}{a} -\right)$$

for $-\infty < y < \infty$.

b   Let X have distribution function F, and let $Y = |X|$. Then, for $y < 0$, the event that $Y \leq y$ is impossible, and for $y \geq 0$, the event that $Y \leq y$ occurs if and only if $-y \leq X \leq y$. Therefore,

$$G(y) = F(y) - F(-y-)$$

for $y \geq 0$ and $G(y) = 0$ for $y < 0$. If F has a continuous density f, then $G(y) = F(y) - F(-y)$ for $y > 0$ and G has density $g = G'$, where $g(y) = f(y) + f(-y)$ for $y > 0$ by differentiation.   ////

We shall say that w is *increasing* if $x < y$ implies $w(x) < w(y)$, and we shall say that w is *decreasing* if $x < y$ implies $w(x) > w(y)$. Further, we shall say that w is *strictly monotone* if w is either increasing or decreasing. Observe that if D is an interval and w has an everywhere positive derivative w', then w will be increasing. In fact, if $x < y$, then $w(y) - w(x) = w'(z)(y - x)$, where $x < z < y$ by the mean-value theorem of differential calculus so $w(y) > w(x)$. Similarly, if D is an interval and w has an everywhere negative derivative, then w will be decreasing.

If w is strictly monotone, then w will be one to one and will have an inverse function v. Moreover, if w is increasing, then we shall have $w(x) \leq y$ if and only if $x \leq v(y)$; and similarly, if w is decreasing, then $w(x) \leq y$ if and only if $x \geq v(y)$ (see Figure 11). Thus, if w is increasing, Equation (1.3) simplifies to

$$G(y) = \Pr (X \leq v(y)) = F(v(y)) \qquad (1.4a)$$

for $y \in E$; and if w is decreasing, then

$$G(y) = 1 - F(v(y)-) \qquad (1.4b)$$

for $y \in E$. Example 7.1.3a treats the special case that w is a linear function.

EXAMPLE 7.1.4   Let X have a distribution function F which is continuous and strictly increasing on D. Then, $Y = F(X)$ has the uniform distribution on the interval $(0,1)$. Indeed, letting H denote the inverse function to F, we find from (1.4a) that $G(y) = F(H(y)) = y$ for $0 < y < 1$.   ////

If F is absolutely continuous, and if the function v is suitably smooth, Equations (1.4) can be differentiated.

Theorem 7.1.1   *Let D be an open interval, and let X have a continuous density f on D. Suppose also that w has continuous derivative w' on D and that $w'(x) \neq 0$ for any $x \in D$. Then, $Y = w(X)$ has density*

$$g(y) = f(v(y))|v'(y)| \qquad y \in E \qquad (1.5)$$

and $g(y) = 0$ for $y \notin E$. Here v denotes the inverse function to w.

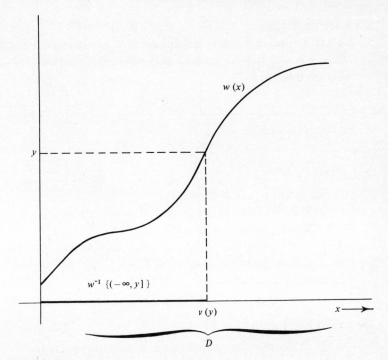

FIGURE 11
An increasing function w.

PROOF   Since w' is continuous and $w'(x) \neq 0$ for any $x \in D$, we must have either $w'(x) > 0$ for all $x \in D$ or $w'(x) < 0$ for all $x \in D$. In either case w is strictly monotone, and E is an interval. Let us consider the case that $w'(x) > 0$ for all $x \in D$. In this case, w is increasing, so that Equation (1.4a) is applicable. Moreover, by (1.1b) v is differentiable with derivative $v'(y) = 1/w'(v(y))$, which is positive. We can therefore differentiate (1.4a) to obtain

$$g(y) = G'(y) = f(v(y))v'(y)$$

which is (1.5) since v' is positive. The case of decreasing w can be handled similarly, and we may take $g(y) = 0$ for $y \notin E$ since $\Pr(Y \in E) = 1$.   ////

Taking $w(x) = ax + b$, $x \in D$, where $a \neq 0$, we find that $w'(x) = a \neq 0$ and that $v(y) = (y - b)/a$ for $y \in E$. Therefore, we have the following corollary, which extends Lemma 5.5.1.

Corollary 7.1.1 *Let D be an open interval, and let X have density f which is continuous on D. Further, let $Y = aX + b$, where $a \neq 0$. Then Y has density*

$$g(y) = |a|^{-1} f\left(\frac{y - b}{a}\right) \qquad y \in E \qquad (1.6)$$

EXAMPLE 7.1.5

a If X has the normal distribution with parameters μ and σ^2, $\sigma^2 > 0$, then $Y = aX + b$ has the normal distribution with parameters $a\mu + b$ and $a^2\sigma^2$. To see this, we let

$$f(x) = \frac{1}{\sigma\sqrt{2\pi}} \exp\left[-\frac{1}{2}\left(\frac{x-\mu}{\sigma}\right)^2\right]$$

in (1.6) and find

$$g(y) = \frac{1}{|a|} f\left(\frac{y-b}{a}\right) = \frac{1}{|a|\sigma\sqrt{2\pi}} \exp\left[-\frac{1}{2}\left(\frac{y-a\mu-b}{a\sigma}\right)^2\right]$$

by simple algebra. In particular, $Z = (X - \mu)/\sigma$ has the standard normal distribution (take $a = \sigma^{-1}$ and $b = -\mu\sigma^{-1}$).

b Similarly, if X has the gamma distribution with parameters α and β (see Section 5.4), and if $c > 0$, then $Y = cX$ has the gamma distribution with parameters α and βc^{-1}. ////

EXAMPLE 7.1.6

a If X has the exponential distribution with parameter $\beta > 0$, then the distribution of $Y = \sqrt{X}$ can be found from Theorem 7.1.1. Indeed, since $\Pr(X > 0) = 1$, we may take $D = (0,\infty)$, and X has density

$$f(x) = \beta e^{-\beta x} \qquad x > 0$$

which is continuous on D. Letting $w(x) = \sqrt{x}$, $x > 0$, we find easily that $E = (0,\infty)$, $v(y) = y^2$, $y > 0$, $v'(y) = 2y$, $y > 0$, and

$$g(y) = 2yf(y^2) = 2\beta ye^{-\beta y^2} \qquad y > 0$$

g is known as the *Rayleigh density with parameter* β.

b If X has the uniform distribution on $(0,1)$, then $Y = -\log X$ has the exponential distribution with parameter $\beta = 1$. Indeed, we may take $D = (0,1)$ and $w(x) = -\log x$, $0 < x < 1$, in which case $E = (0,\infty)$ and $v(y) = e^{-y}$, $y > 0$. Since X has density $f(x) = 1$, $0 < x < 1$, it follows from Theorem 7.1.1 that Y has density

$$g(y) = f(e^{-y})|-e^{-y}| = e^{-y}$$

for $y > 0$. ////

The hypotheses of Theorem 7.1.1 can be relaxed. For example, if w is increasing and f is discontinuous at any finite number of points, then $G(y) = F(v(y))$ will fail to have a derivative at a finite number of points but will still be absolutely continuous with density g, as defined by (1.6) (see Problem 7.18). Similarly, w' may vanish at any finite number of points, provided that it does not change sign, so that w is still monotone (see Problem 7.18).

Theorem 7.1.1 does not apply to functions w which are not monotone, however, and in particular cannot be applied directly to find the density of $Y = X^2$ when X has a density which is positive on an open interval containing zero. We shall now develop a formula for finding the distribution of $Y = w(X)$ when w is symmetric, that is, $w(x) = w(-x)$ for all x. The idea is quite simple. If w is symmetric, then $w(X) = w(Z)$, where $Z = |X|$. We shall apply Theorem 7.1.1 to w and Z. From Example 7.1.3b we know that if X has distribution function F, then Z has distribution function H, where $H(z) = F(z) - F(-z-)$ for $z \geq 0$. Therefore, if X has a continuous density f, then $H(z) = F(z) - F(-z)$ and Z has density $h(z) = H'(z) = f(z) + f(-z)$.

Corollary 7.1.2 *Let X have a continuous density f, and let $C = (0,a)$, $0 < a \leq \infty$, be an interval for which $\Pr(Z \in C) = 1$, where $Z = |X|$. If w is continuously differentiable on C and $w'(x) \neq 0$ for any $x \in C$, then $Y = w(Z)$ has density*

$$g(y) = h(v(y))|v'(y)|$$

for $y \in E$ and $g(y) = 0$ for $y \notin E$. Here v denotes the inverse function to w, and $h(z) = f(z) + f(-z)$, $z \geq 0$.

The corollary follows directly from Theorem 7.1.1. Let us consider an example.

EXAMPLE 7.1.7 If X has a continuous density f, then $Y = X^2$ has density

$$g(y) = \frac{f(\sqrt{y}) + f(-\sqrt{y})}{2\sqrt{y}} \qquad y > 0$$

and $g(y) = 0$ for $y \leq 0$. In fact, we have $w(x) = x^2$, $v(y) = \sqrt{y}$, and $v'(y) = 1/2\sqrt{y}$. In particular, if X has the standard normal distribution with density

$$f(x) = \frac{e^{-\frac{1}{2}x^2}}{\sqrt{2\pi}} \qquad -\infty < x < \infty$$

then Y has the chi-square distribution with one degree of freedom, that is,

$$g(y) = \frac{e^{-\frac{1}{2}y}}{\sqrt{2\pi y}} \qquad y > 0$$

(see Example 5.4.2). ////

7.2 MULTIVARIATE DISTRIBUTIONS

The simple notions of the preceding section extend from one dimension to several. Thus, consider a random vector $X = (X_1, \ldots, X_n)$ with $n \geq 2$; let $w_1, \ldots, w_k$ $(k \geq 1)$ be real-valued functions defined on a region $D \subset R^n$ for which $\Pr(X \in D) = 1$; and let

$$Y_j = w_j(X_1, \ldots, X_n)$$

be random variables, $j = 1, \ldots, k$. We shall consider techniques for finding the joint distribution of $Y_1, \ldots, Y_k$ from the joint distribution of $X_1, \ldots, X_n$ and the functions $w_1, \ldots, w_k$.

As in the previous section, we have the basic relation

$$\Pr(Y_1 \in I_1, \ldots, Y_k \in I_k) = \Pr(X \in w_j^{-1}(I_j), j = 1, \ldots, k) \qquad (2.1)$$

for all intervals $I_j \subset R$, $j = 1, \ldots, k$, since $Y \in I_j$ if and only if $X \in w_j^{-1}(I_j)$, $j = 1, \ldots, k$. Moreover, letting w denote the vector-valued function

$$w(x) = (w_1(x), \ldots, w_k(x))$$

for $x = (x_1, \ldots, x_n) \in D$, and letting Y be the random vector $Y = w(X) = (Y_1, \ldots, Y_k)$, we have the relation

$$\Pr(Y \in B) = \Pr(X \in w^{-1}(B)) \qquad (2.2)$$

for all subsets $B \subset R^k$ for which the right side of (2.2) is meaningful. Let us now consider some examples, in all of which we shall take $D = R^n$.

EXAMPLE 7.2.1

a Let $X_1, \ldots, X_n$ be independent with common distribution function F, and let $Y = \max(X_1, \ldots, X_n)$ (in this case we have $k = 1$). Then the event that $Y \leq y$ occurs if and only if all the events $X_i \leq y$, $i = 1, \ldots, n$ occur. Therefore, the distribution function of Y is given by

$$
\begin{aligned}
G(y) &= \Pr(Y \leq y) \\
&= \Pr(X_1 \leq y, \ldots, X_n \leq y) \\
&= \prod_{i=1}^{n} \Pr(X_i \leq y) = F(y)^n
\end{aligned}
$$

for $-\infty < y < \infty$. If F has a piecewise continuous density f, then G has density

$$g(y) = nF(y)^{n-1}f(y) \qquad -\infty < y < \infty$$

as may be seen by differentiation.

b Let $X_1, \ldots, X_n$ be as above, and let $Y_1 = \min(X_1, \ldots, X_n)$ and $Y_2 =$

$\max (X_1, \ldots, X_n)$. Then, for $y_1 < y_2$, the event that $Y_1 > y_1$ and $Y_2 \leq y_2$ occurs if and only if $y_1 < X_i \leq y_2$, $i = 1, \ldots, n$. Therefore,

$$\Pr (Y_1 > y_1, Y_2 \leq y_2) = \Pr (y_1 < X_i \leq y_2, i = 1, \ldots, n)$$

$$= \prod_{i=1}^{n} \Pr (y_1 < X_i \leq y_2)$$

$$= [F(y_2) - F(y_1)]^n$$

whenever $y_1 < y_2$. Again, if X has a piecewise continuous density f, we can differentiate to obtain a density for $Y = (Y_1, Y_2)$. A density for Y is

$$g(y_1, y_2) = n(n - 1)[F(y_2) - F(y_1)]^{n-2}f(y_1)f(y_2)$$

for $y_1 < y_2$ and $g(y_1, y_2) = 0$ if $y_1 \geq y_2$ (see Problem 6.12). ////

Example 7.2.1a can be generalized as follows. Let $X_1, \ldots, X_n$ be jointly distributed random variables, and let $Y_1, \ldots, Y_n$ be $X_1, \ldots, X_n$ arranged in increasing order. That is, let Y_1 be the minimum of $X_1, \ldots, X_n$, let Y_2 be the second smallest, $\ldots$, and let Y_n be the maximum of $X_1, \ldots, X_n$. $Y_1, \ldots, Y_n$ are known as the *order statistics* of $X_1, \ldots, X_n$. We shall now derive the distribution of Y_k for general k, $1 \leq k \leq n$, in a special case.

Lemma 7.2.1 *Let* $X_1, \ldots, X_n$ *be independent with common (marginal) distribution function* F, *and let* $Y_1, \ldots, Y_n$ *denote their order statistics. Suppose that* F *has a piecewise continuous density* f. *Then for* $1 \leq k \leq n$, Y_k *has density* g_k, *where*

$$g_k(y) = n \binom{n - 1}{k - 1} F(y)^{k-1}[1 - F(y)]^{n-k}f(y)$$

for $-\infty < y < \infty$.

PROOF Let G_k denote the distribution function of Y_k, and let A_i be the event that $X_i \leq y$ for $i = 1, \ldots, n$. Then Y_k, the kth smallest of $X_1, \ldots, X_n$, is less than or equal to y if and only if at least k of the events $A_1, \ldots, A_n$ occur. Moreover, $A_1, \ldots, A_n$ are independent with common probability $P(A_i) = \Pr (X_i \leq y) = F(y)$ for $i = 1, \ldots, n$. Therefore, by Corollary 4.1.1, we have

$$G_k(y) = \sum_{i=k}^{n} \binom{n}{i} F(y)^i[1 - F(y)]^{n-i} \qquad (2.3)$$

We now differentiate, by the product rule, to obtain a density. We have

$$G_k'(y) = \sum_{i=k}^{n} i \binom{n}{i} F(y)^{i-1}[1 - F(y)]^{n-i}f(y)$$

$$- \sum_{i=k}^{n} (n - i) \binom{n}{i} F(y)^i[1 - F(y)]^{n-i-1}f(y)$$

Now, $(i + 1) \binom{n}{i + 1} = n \binom{n - 1}{i} = (n - i) \binom{n}{i}$, so that the $(i + 1)$st term

in the first summation cancels the ith term in the second, leaving

$$G_k'(y) = k \binom{n}{k} F(y)^{k-1}[1 - F(y)]^{n-k}f(y)$$

$$= n \binom{n - 1}{k - 1} F(y)^{k-1}[1 - F(y)]^{n-k}f(y)$$

as asserted.   ////

EXAMPLE 7.2.2   Let $X_1, \ldots, X_n$ be independent random variables which are uniformly distributed over the interval $(0,1)$. We can then picture $X_1, \ldots, X_n$ as points on the line between 0 and 1. The kth point from the left is simply the kth-order statistic Y_k. The density of Y_k can now be obtained by specializing Lemma 7.2.1 to the uniform distribution on $(0,1)$. $F(y) = y$ for $0 \le y \le 1$, $F(y) = 0$ for $y < 0$, and $F(y) = 1$ for $y > 1$. The result is that Y_k has density

$$g_k(y) = n \binom{n - 1}{k - 1} y^{k-1}(1 - y)^{n-k} \qquad (2.4)$$

for $0 < y < 1$ and $g_k(y) = 0$ for other values of y.

We recognize (2.4) as a beta density with parameter $\alpha = k$ and $\beta = n - k + 1$. Thus, the kth (from the left) of n points chosen independently from $(0,1)$ has the beta distribution with parameters $\alpha = k$ and $\beta = n - k + 1$. Moreover, Equation (2.3) now gives an expression for the beta-distribution function, namely

$$G_k(y) = \sum_{i=k}^{n} \binom{n}{i} y^i(1 - y)^{n-i} \qquad (2.5)$$

for $0 \le y \le 1$. Thus, beta-distribution functions can be evaluated from the table of binomial probabilities in Appendix C.   ////

We conclude this section with an extremely useful, if intuitively obvious, result by showing that functions of different independent random variables (or vectors) are themselves independent.

Theorem 7.2.1   *Let $X_1, \ldots, X_n$ be independent random variables, and let $w_1, \ldots, w_n$ be real-valued functions which are defined on R. Further, let $Y_j = w_j(X_j)$, $j = 1, \ldots, n$, be random variables. Then, $Y_1, \ldots, Y_n$ are independent random variables.*

PROOF We shall prove the theorem in the special case that $n = 2$ and the random variables X_1 and X_2 are discrete. Let y_1 and y_2 be arbitrary real numbers, and let $A_i = w_i^{-1}(\{y_i\})$ be the set of $x \in R$ for which $w_i(x) = y_i$, $i = 1, 2$. Then, by (2.1),

$$\Pr(Y_1 = y_1, Y_2 = y_2) = \sum_{(x_1,x_2) \in A_1 \times A_2} \Pr(X_1 = x_1, X_2 = x_2) \qquad (2.6)$$

Now, since X_1 and X_2 are independent, we have $\Pr(X_1 = x_1, X_2 = x_2) = \Pr(X_1 = x_1)\Pr(X_2 = x_2)$, so that the right side of (2.3) is simply

$$\left[\sum_{x_1 \in A_1} \Pr(X_1 = x_1)\right]\left[\sum_{x_2 \in A_2} \Pr(X_2 = x_2)\right]$$

which is $\Pr(Y_1 = y_1)\Pr(Y_2 = y_2)$, by (1.2). Therefore, $\Pr(Y_1 = y_1, Y_2 = y_2) = \Pr(Y_1 = y_1)\Pr(Y_2 = y_2)$ for all y_1 and y_2, so that Y_1 and Y_2 are independent by Theorem 6.2.2. ////

Theorem 7.2.1 is also true if the X_i or Y_j are themselves random vectors, but its proof in this case will be omitted.

EXAMPLE 7.2.3

a If X_1 and X_2 are independent, then so are $Y_1 = X_1^2$ and $Y_2 = X_2^2$.
b If X_1, X_2, X_3, X_4 are independent, then so are $Y_1 = X_1 + X_2$ and $Y_2 = X_3 + X_4$ because the random vectors (X_1, X_2) and (X_3, X_4) are independent by Example 6.3.6. ////

7.3 CONVOLUTIONS

Consider two independent random variables X and Y, and let Z denote their sum $Z = X + Y$. What can be said about the distribution of Z?

If both X and Y are integer-valued random variables, then Z will also be integer-valued. Moreover, for every fixed integer k, the event that $Z = k$ is the union over j of the mutually exclusive events that $X = j$ and $Y = k - j$. Therefore, if f and g denote the mass functions of X and Y, respectively, then the mass function of Z can be computed as follows:

$$h(k) = \Pr(Z = k)$$

$$= \sum_{j=-\infty}^{\infty} \Pr(X = j, Y = k - j)$$

$$= \sum_{j=-\infty}^{\infty} f(j)g(k - j) \qquad (3.1)$$

for $k = 0, \pm 1, \pm 2, \ldots$. The final member of (3.1) defines a function which is known as the *convolution of f and g*. Thus we have shown that the mass function of $Z = X + Y$ is the convolution of f, the mass function of X, and g, the mass function of Y.

There is a corresponding formula for the sum of two independent, absolutely continuous random variables. Namely, if X and Y are independent with densities f and g, respectively, then the sum $Z = X + Y$ has density h, where

$$h(z) = \int_{-\infty}^{\infty} g(z - x)f(x)\, dx \qquad (3.2)$$

for $-\infty < z < \infty$, and h is known as the *convolution of f and g*. To establish (3.2) let H denote the distribution function of Z. Then, $H(z) = \Pr((X,Y) \in B)$, where B is the set of $(x,y) \in R^2$ for which $x + y \leq z$. By Equation (4.2) of Chapter 6, this can also be written

$$H(z) = \iint_{B} f(x)g(y)\, dx\, dy = \int_{-\infty}^{\infty} \left[\int_{-\infty}^{z-x} g(y)f(x)\, dy \right] dx$$

The change of variable $y' = x + y$ in the inner integral now reduces $H(z)$ to

$$H(z) = \int_{-\infty}^{\infty} \left[\int_{-\infty}^{z} g(y' - x)f(x)\, dy' \right] dx$$

$$= \int_{-\infty}^{z} \left[\int_{-\infty}^{\infty} g(y - x)f(x)\, dx \right] dy$$

$$= \int_{-\infty}^{z} h(y)\, dy$$

where h is defined by (3.2). Thus, h is a density for Z, as asserted.

Theorem 7.3.1 *If X and Y are independent, integer-valued random variables with mass functions f and g, then the mass function of $Z = X + Y$ is given by the convolution (3.1). Similarly, if X and Y are independent, absolutely continuous random variables with densities f and g, then a density for their sum $Z = X + Y$ is given by the convolution (3.2).*

EXAMPLE 7.3.1 If X and Y have binomial distributions with the same p, $0 < p < 1$, say

$$\Pr(X = k) = \binom{m}{k} p^k q^{m-k} \qquad k = 0, \ldots, m$$

$$\Pr(Y = k) = \binom{n}{k} p^k q^{n-k} \qquad k = 0, \ldots, n$$

where $q = 1 - p$, then $Z = X + Y$ has the binomial distribution with parameters $m + n$ and p. This is intuitively obvious since X may be thought of as the number of heads which result from m tosses of a coin which has probability p of turning up heads on each toss, Y may be thought of as the number of heads in n tosses of the same coin, and therefore Z may be thought of as the number of heads in $n + m$ tosses. We can also verify the distribution of Z from (3.1). Indeed, we have

$$h(k) = \sum_{j=-\infty}^{\infty} \Pr(X = j) \Pr(Y = k - j)$$

$$= \sum_{j=0}^{m} \binom{n}{k - j} p^{k-j} q^{n-k+j} \binom{m}{j} p^j q^{m-j}$$

$$= p^k q^{m+n-k} \sum_{j=0}^{m} \binom{n}{k - j} \binom{m}{j}$$

$$= \binom{m + n}{k} p^k q^{m+n-k}$$

for $k = 0, \ldots, m + n$. Here we have used the identity of Example 2.3.6a in the final step. We should also remark that the second equality above follows from the fact that $\Pr(X = j) = 0$ for $j < 0$ and $\Pr(Y = k - j) = 0$ for $j > k$. Reductions of this type are quite common in the evaluation of convolutions.   ////

EXAMPLE 7.3.2   Let us now consider two independent Poisson random variables X and Y, say

$$f(k) = \frac{e^{-\alpha} \alpha^k}{k!} \quad \text{and} \quad g(k) = \frac{e^{-\beta} \beta^k}{k!}$$

for $k = 0, 1, \ldots$, where α and β are positive. Then the mass function of $Z = X + Y$ is

$$h(k) = \sum_{j=0}^{k} \left(\frac{1}{j!} \alpha^j e^{-\alpha} \right) \left[\frac{1}{(k - j)!} \beta^{k-j} e^{-\beta} \right]$$

$$= \frac{e^{-(\alpha+\beta)}}{k!} \sum_{j=0}^{k} \binom{k}{j} \alpha^j \beta^{k-j}$$

$$= \frac{1}{k!} (\alpha + \beta)^k e^{-(\alpha+\beta)}$$

for $k = 0, 1, \ldots$. (We used the binomial theorem in the final step.) Thus, the sum $Z = X + Y$ has the Poisson distribution with parameter $\alpha + \beta$.   ////

More examples of discrete convolutions will be found in the problems at the end of this chapter. Let us now consider the absolutely continuous case.

EXAMPLE 7.3.3

a   Let X and Y be independent, exponentially distributed random variables, say

$$f(x) = g(x) = \beta e^{-\beta x} \qquad x > 0$$

and $f(x) = g(x) = 0$ for $x \le 0$. Then, $Z = X + Y$ has density

$$h(z) = \int_{-\infty}^{\infty} g(z - x)f(x)\,dx$$

$$= \int_{0}^{z} \beta e^{-(z-x)}\beta e^{-x}\,dx$$

$$= \beta^2 e^{-\beta z} \int_{0}^{z} dx = \beta^2 z e^{-\beta z}$$

for $z > 0$ and $h(z) = 0$ for $z \le 0$. Thus, the sum of two independent, exponentially distributed random variables (with the same parameter β) has the gamma distribution with parameters $\alpha = 2$ and β (see Section 5.4).

b   More generally, if X and Y are independent, X has the gamma distribution with parameters α_1 and β, and Y has the gamma distribution with parameters α_2 and (the same) β, then $Z = X + Y$ has the gamma distribution with parameters $\alpha = \alpha_1 + \alpha_2$ and β, as can be seen by a computation similar to that given in part *a* (see also Sections 7.4 and 8.4). In particular, if X has the chi-square distribution with j degrees of freedom and Y has the chi-square distribution with k degrees of freedom, then $Z = X + Y$ has the chi-square distribution with $j + k$ degrees of freedom. To see this, simply let $\alpha_1 = j/2$, $\alpha_2 = k/2$, and $\beta = \frac{1}{2}$. ////

EXAMPLE 7.3.4

a   If X and Y are independent, standard normal random variables, then $Z = X + Y$ has the normal distribution with parameters $\mu = 0$ and $\sigma^2 = 2$. Indeed, X and Y have the same density f, where $f(x) = (1/\sqrt{2\pi}) \exp(-\frac{1}{2}x^2)$, $-\infty < x < \infty$. That is, $f = g$. Therefore,

$$g(z - x)f(x) = \frac{1}{2\pi} \exp\left[-\tfrac{1}{2}(z - x)^2 - \tfrac{1}{2}x^2\right]$$

$$= \frac{1}{2\pi} \exp\left[-\tfrac{1}{4}z^2 - \left(x - \frac{z}{2}\right)^2\right]$$

by simple algebra. Therefore,

$$h(z) = \frac{e^{-\frac{1}{4}z^2}}{2\sqrt{\pi}} \int_{-\infty}^{\infty} \frac{e^{-(x-\frac{1}{2}z)^2}}{\sqrt{\pi}} \, dx = \frac{e^{-\frac{1}{4}z^2}}{2\sqrt{\pi}} \qquad (3.3)$$

for $-\infty < z < \infty$, as asserted. [Since the integrand in (3.3) is a normal density with $\mu = z/2$ and $\sigma^2 = \frac{1}{2}$, the integral in (3.3) is 1.]

b More generally, if X and Y are independent, X has the normal distribution with location parameter μ and scale parameter σ, and Y has the normal distribution with location parameter v and scale parameter τ, then $Z = X + Y$ has the normal distribution with parameters $\mu + v$ and $\sqrt{\sigma^2 + \tau^2}$ by a calculation similar to that given in part *a* (see also Section 8.4). ////

EXAMPLE 7.3.5 Let X and Y be independent, standard normal random variables, and let $R = \sqrt{X^2 + Y^2}$ be the distance of (X, Y) from the origin. We shall find the distribution of R by combining previous calculations. X^2 and Y^2 are independent (by Theorem 7.2.1), and both have chi-square distributions with one degree of freedom (by Example 7.1.7). Therefore, $X^2 + Y^2$ has the chi-square distribution with two degrees of freedom (by Example 7.3.3*b*), which is simply the exponential distribution with parameter $\beta = \frac{1}{2}$. Finally, the square root of an exponential random variable has the Rayleigh distribution (by Example 7.1.6*a*), so that R has the Rayleigh distribution with parameter $\beta = \frac{1}{2}$. ////

Examples 7.3.1 to 7.3.4 extend by induction from two summands to many.

Theorem 7.3.2 *Let $X_1, \ldots, X_k$ be independent random variables, and let $S = X_1 + \cdots + X_k$.*

(i) If each X_i has the binomial distribution with parameters n_i and (the same) p, $i = 1, \ldots, k$, then S has the binomial distribution with parameters $n = n_1 + \cdots + n_k$ and p.

(ii) If each X_i has the Poisson distribution with parameter β_i, $i = 1, \ldots, k$, then S has the Poisson distribution with parameter $\beta = \beta_1 + \cdots + \beta_k$.

(iii) If each X_i has the gamma distribution with parameters α_i and (the same) β, $i = 1, \ldots, k$, then S has the gamma distribution with parameters $\alpha = \alpha_1 + \cdots + \alpha_k$ and β.

(iv) If each X_i has the normal distribution with location parameter μ_i and scale parameter σ_i, $i = 1, \ldots, k$, then S has the normal distribution with location parameter $\mu = \mu_1 + \cdots + \mu_k$ and scale parameter σ, where $\sigma^2 = \sigma_1^2 + \cdots + \sigma_k^2$.

PROOF   The theorem follows easily from Examples 7.3.1 to 7.3.4 and mathematical induction. We leave its proof to the reader.   ////

7.4   JACOBIANS[1]

The simple, useful Theorem 7.1.1 generalizes from one dimension to several, and we shall consider its generalization in this section. We begin with a few remarks about transformations of R^n, $n \geq 2$. Consider a region $D \subset R^n$, and let $w_1, \ldots, w_n$ be real-valued functions defined on D. Then, the vector-valued function

$$w(x) = (w_1(x), \ldots, w_n(x)) \qquad (4.1)$$

is defined for $x = (x_1, \ldots, x_n) \in D$. Let E denote the range of the function w. Then we shall refer to w as a *transformation* from D onto E.

If D is an open region and each of the functions w_i is continuously differentiable on D, we shall say that the transformation w is continuously differentiable, and in this case we define the *jacobian* of the transformation w by

$$J_w(x) = \det \left[\frac{\partial}{\partial x_j} w_i(x) \right]$$

for $x \in D$. Thus, $J_w(x)$ is the determinant of the matrix whose (i,j)th entry is the partial derivative of $w_i(x)$ with respect to x_j.

Jacobians play the same role in several dimensions that derivatives played in one.

Theorem 7.4.1   *Let w be a continuously differentiable, one-to-one transformation from an open region $D \subset R^n$ onto another region $E \subset R^n$, and suppose also that $J_w(x) \neq 0$ for any $x \in D$. Further, let $X = (X_1, \ldots, X_n)$ be an absolutely continuous random vector for which $\Pr (X \in D) = 1$, and suppose that X has a density f. If $Y = w(X)$, then Y has density g, where*

$$g(y) = f(v(y))|J_v(y)|$$

for $y \in E$ and $g(y) = 0$ for $y \notin E$, where v denotes the inverse function to w.

PROOF   The theorem follows easily from the change-of-variable formula for multidimensional integrals.[2] Let B be a bounded, closed rectangle, $B \subset E$. Then,

$$\Pr (Y \in B) = \Pr (X \in w^{-1}(B)) = \int_{w^{-1}(B)} f(x) \, dx$$

---

[1] This section treats a special topic and may be omitted without loss of continuity.
[2] See Apostol (1957), p. 271, for the case in which f is continuous on D.

where the latter integral is n-dimensional and dx denotes the volume element in R^n. By the change-of-variable formula for multidimensional integrals, we now have

$$\int_{w^{-1}(B)} f(x)\,dx = \int_B f(v(y))|J_v(y)|\,dy$$

so that

$$\Pr(Y \in B) = \int_B g(y)\,dy$$

for all bounded, closed rectangles $B \subset E$. Since $\Pr(Y \in E) = \Pr(X \in D) = 1$, the theorem follows. ////

There are many conditions placed on the function w in the hypotheses of Theorem 7.4.1. However, we automatically check them by calculating g. Indeed, by computing an inverse transformation v, we show that w is one to one; and since $J_v(y) = 1/J_w(v(y))$ by the chain rule, $J_w(x)$ cannot vanish unless J_v has a singularity at $y = w(x)$. That is, if J_v is continuous, then $J_w(x) \neq 0$ for any $x \in D$. Let us now consider an example.

EXAMPLE 7.4.1 Let X_1 and X_2 have joint density

$$f(x_1, x_2) = 4x_1 x_2$$

for $0 < x_1 < 1$ and $0 < x_2 < 1$ and $f(x_1, x_2) = 0$ for other values of x_1 and x_2, and let

$$Y_1 = X_1 \quad \text{and} \quad Y_2 = X_1 X_2$$

Let us find the joint distribution of Y_1 and Y_2.

In this example we may take D to be the open unit square

$$D = \{(x_1, x_2) : 0 < x_1 < 1, 0 < x_2 < 1\}$$

and the function w to be

$$w_1(x_1, x_2) = x_1 \quad \text{and} \quad w_2(x_1, x_2) = x_1 x_2$$

for $(x_1, x_2) \in D$. The range of w is then easily seen to be

$$E = \{(y_1, y_2) : 0 < y_2 < y_1 < 1\}$$

For $y = (y_1, y_2) \in E$, the equation $y = w(x)$ has a unique solution $x = (x_1, x_2)$, given by

$$x_1 = v_1(y_1, y_2) = y_1 \quad \text{and} \quad x_2 = v_2(y_1, y_2) = \frac{y_2}{y_1}$$

Thus, w is one to one, and v has been found. It remains only to compute $J_v(y)$ and apply Theorem 7.4.1. We have

$$J_v(y) = \det \begin{bmatrix} 1 & 0 \\ -y_2 y_1^{-2} & y_1^{-1} \end{bmatrix} = y_1^{-1}$$

which is positive and continuous. Thus, $J_w(x) \neq 0$ for $x \in D$, and so the condition of the theorem is satisfied. By Theorem 7.4.1, we now obtain a density for $Y = (Y_1, Y_2)$ as

$$g(y_1, y_2) = f(y_1, y_2 y_1^{-1}) y_1^{-1} = 4 y_2 y_1^{-1}$$

for $y = (y_1, y_2) \in E$ and $g(y_1, y_2) = 0$ for other values of y.

Suppose now that instead of the joint distribution of Y_1 and Y_2 we wanted the distribution of Y_2. We might proceed as follows. We define Y_1 and find the joint density of Y_1 and Y_2, as above. Then we compute the marginal density of Y_2 as

$$g_2(y) = \int_{-\infty}^{\infty} g(y_1, y) \, dy_1$$

$$= \int_y^1 4 y y_1^{-1} \, dy_1 = 4y \log y^{-1} \qquad 0 < y < 1 \qquad ////$$

EXAMPLE 7.4.2 Let X_1 and X_2 be independent, and let X_i have the gamma distribution with parameters α_i and (the same) β, $i = 1, 2$. Thus, X_1 and X_2 have joint density

$$f(x_1, x_2) = \frac{1}{\Gamma(\alpha_1 \Gamma)(\alpha_2)} \beta^\alpha x_1^{\alpha_1 - 1} x_2^{\alpha_2 - 1} e^{-\beta(x_1 + x_2)}$$

for $x_1 > 0$ and $x_2 > 0$ and $f(x_1, x_2) = 0$ for other values of x_1 and x_2 where $\alpha = \alpha_1 + \alpha_2$. Let us find the joint density of $Y_1 = X_1/(X_1 + X_2)$ and $Y_2 = X_1 + X_2$.

Let D be the set of (x_1, x_2) for which $x_1 > 0$ and $x_2 > 0$. Then, $\text{Pr}(X \in D) = 1$ and f is continuous on D. Moreover, we can write $Y = w(X)$, where

$$w_1(x_1, x_2) = \frac{x_1}{x_1 + x_2} = y_1 \quad \text{and} \quad w_2(x_1, x_2) = x_1 + x_2 = y_2$$

for $x = (x_1, x_2) \in D$. The range of w is easily found to be the set

$$E = \{(y_1, y_2): 0 < y_1 < 1 \text{ and } y_2 > 0\}$$

and the inverse transformation v [found by solving the equations $w_1(x_1, x_2) = y_1$ and $w_2(x_1, x_2) = y_2$ for $x = (x_1, x_2)$] is

$$v_1(y_1, y_2) = y_1 y_2 \quad \text{and} \quad v_2(y_1, y_2) = (1 - y_1) y_2$$

$$= x_1 \qquad \qquad = \left(1 \cdot (x_1 + x_2) - x_1\right)$$

$$= x_2 \qquad \qquad \text{as desired}$$

for $y = (y_1, y_2) \in E$. Therefore,

$$J_v(y_1, y_2) = \det \begin{bmatrix} y_2 & y_1 \\ -y_2 & 1 - y_1 \end{bmatrix} = y_2 \qquad y \in E$$

which is continuous. It now follows that Y_1 and Y_2 have joint density g, where

$$g(y_1, y_2) = f(y_1 y_2, (1 - y_1) y_2) y_2$$

$$= \frac{1}{\Gamma(\alpha_1)\Gamma(\alpha_2)} \beta^\alpha y_1^{\alpha_1 - 1}(1 - y_1)^{\alpha_2 - 1} y_2^{\alpha - 1} e^{-\beta y_2}$$

for $y = (y_1, y_2) \in E$, where $\alpha = \alpha_1 + \alpha_2$.

This concludes the routine application of Theorem 7.4.1 to find the density of Y_1 and Y_2, but some aspects of this particular example merit consideration, which we now give. Having found a joint density for Y_1 and Y_2, it is now a simple matter to find the marginal densities of Y_1 and Y_2. For example, letting $c^{-1} = \Gamma(\alpha_1)\Gamma(\alpha_2)$, we find that the marginal density of Y_1 is

$$g_1(y) = cy^{\alpha_1 - 1}(1 - y)^{\alpha_2 - 1} \int_0^\infty \beta^\alpha y_2^{\alpha - 1} e^{-\beta y_2} \, dy_2$$

$$= \frac{\Gamma(\alpha_1 + \alpha_2)}{\Gamma(\alpha_1)\Gamma(\alpha_2)} y^{\alpha_1 - 1}(1 - y)^{\alpha_2 - 1}$$

for $0 < y < 1$. (To evaluate the integral, we made the change of variables $u = \beta y_2$ and used the definition of the gamma function.) Thus, we see that Y_1 has the beta distribution with parameters α_1 and α_2. Similarly, we may find the marginal density of Y_2,

$$g_2(y) = \frac{1}{\Gamma(\alpha)} \beta^\alpha y^{\alpha - 1} e^{-\beta y}$$

for $y > 0$, so that Y_2 has the gamma distribution with parameters $\alpha = \alpha_1 + \alpha_2$ and β. Finally, we remark that $g(y_1, y_2) = g_1(y_1)g_2(y_2)$ for all y_1 and y_2, so that Y_1 and Y_2 are independent random variables. Since both Y_1 and Y_2 depend on both X_1 and X_2, the latter remark is somewhat surprising. It depends on the fact that X_1 and X_2 had gamma distributions and would not necessarily be true if they had some other distribution. ////

We shall now consider linear transformations, that is, transformations of the form

$$w(x) = xA + b$$

for $x \in R^n$, where $A = (a_{ij})$ is an n by n matrix and $b = (b_1, \ldots, b_n)$ is a constant vector. If A is nonsingular, then the transformation w is invertible with inverse $v(y) =$

$(y - b)A^{-1}$, where A^{-1} denotes the matrix inverse of A. The jacobian J_v is the constant $J_v(y) = \det A^{-1} = 1/(\det A)$. Thus, we have the following corollary to Theorem 7.4.1.

Corollary 7.4.1 *Let* $X = (X_1, \ldots, X_n)$ *be a random vector with density* f, *and let* $Y = XA + b$, *where* A *is a nonsingular n by n matrix and* $b \in R^n$. *Then* Y *has density* g, *where*

$$g(y) = \left| \frac{1}{\det A} \right| f((y - b)A^{-1})$$

for $y \in R^n$.

EXAMPLE 7.4.3 An n by n matrix A is called *orthogonal* if and only if $AA' = I = A'A$, where the prime denotes transpose and I denotes the n by n identity matrix. If A is orthogonal, then $A^{-1} = A'$, so that A is nonsingular; moreover, $\det A^2 = \det AA' = \det I = 1$, so that $\det A = \pm 1$.

We now claim that if $X_1, \ldots, X_n$ are independent standard normal random variables and if $Y_1, \ldots, Y_n$ are defined by $Y = XA$, where A is orthogonal, $X = (X_1, \ldots, X_n)$, and $Y = (Y_1, \ldots, Y_n)$, then $Y_1, \ldots, Y_n$ are again independent standard normal random variables.

To see this, observe that a density for X is given by

$$f(x_1, \ldots, x_n) = \prod_{i=1}^{n} \frac{1}{\sqrt{2\pi}} e^{-\frac{1}{2}x_i^2} = \left(\frac{1}{\sqrt{2\pi}} \right)^n e^{-\frac{1}{2}(x_1^2 + \cdots + x_n^2)}$$

for $x = (x_1, \ldots, x_n) \in R^n$ by independence (Theorem 6.3.1). Since $xx' = x_1^2 + \cdots + x_n^2$, f can be written in the form

$$f(x) = \left(\frac{1}{\sqrt{2\pi}} \right)^n e^{-\frac{1}{2}xx'}$$

for $x \in R^n$. Now let $Y = XA$, where A is orthogonal. Then, by Corollary 7.4.1, Y has density

$$g(y) = \left| \frac{1}{\det A} \right| f(yA^{-1}) = f(yA')$$

$$= \left(\frac{1}{\sqrt{2\pi}} \right)^n e^{-\frac{1}{2}yA'Ay'} = \left(\frac{1}{\sqrt{2\pi}} \right)^n e^{-\frac{1}{2}yy'}$$

for $y \in R^n$. That is, Y has the same distribution as X, as asserted.   ////

We conclude this section with general formulas for the densities of sums, differences, ratios, and products of two jointly absolutely continuous random variables.

Corollary 7.4.2 *Let X_1 and X_2 have a joint density f. Then $X_1 + X_2$ and $X_1 - X_2$ have densities*

$$g_+(y) = \int_{-\infty}^{\infty} f(y - x, x)\, dx$$

$$g_-(y) = \int_{-\infty}^{\infty} f(x + y, x)\, dx$$

for $-\infty < y < \infty$, respectively. Moreover, if $\Pr(X_2 > 0) = 1$, then X_1/X_2 and $X_1 X_2$ have densities

$$h_1(y) = \int_0^{\infty} x f(xy, x)\, dx$$

$$h_2(y) = \int_0^{\infty} x^{-1} f(yx^{-1}, x)\, dx$$

PROOF We shall prove only the first assertion of the corollary since the proofs of the other three are similar. Let $Y_1 = X_1 + X_2$ and $Y_2 = X_2$. Then the inverse transformation is $X_2 = Y_2$ and $X_1 = Y_1 - Y_2$, the jacobian of which is $J_v(y) = 1$. Thus, Y_1 and Y_2 have joint density $g(y_1, y_2) = f(y_1 - y_2, y_2)$ for $-\infty < y_1, y_2 < \infty$. Thus Y_1 has marginal density

$$g_1(y) = \int_{-\infty}^{\infty} f(y - y_2, y_2)\, dy_2$$

for $-\infty < y < \infty$, as asserted.   ////

EXAMPLE 7.4.4 Let X_1 and X_2 be independent, exponentially distributed random variables with the same parameter β, so that X_1 and X_2 have joint density

$$f(x_1, x_2) = \beta^2 e^{-\beta(x_1 + x_2)}$$

for $x_1 > 0$ and $x_2 > 0$ and $f(x_1, x_2) = 0$ for other values of x_1 and x_2. We shall find the distributions of $Y = X_1 - X_2$ and $Z = X_1/X_2$.

a Let us first find the distribution of Y. For $y > 0$, it is

$$g_-(y) = \beta^2 \int_0^{\infty} e^{-\beta(x+y)} e^{-\beta x}\, dx$$

$$= \beta^2 e^{-\beta y} \int_0^{\infty} e^{-2\beta x}\, dx = \frac{\beta}{2} e^{-\beta y}$$

and by symmetry (Y has the same distribution as $-Y$), we must have $g_-(y) = g_-(-y)$. Therefore,

$$g_-(y) = \frac{\beta}{2} e^{-\beta|y|}$$

for $-\infty < y < \infty$. g_- is known as the *bilateral exponential density with parameter β*.

b Similarly, $Z = X_1/X_2$ has density

$$h_1(z) = \int_0^\infty x\beta^2 e^{-\beta x z} e^{-\beta x}\, dx$$

$$= \int_0^\infty x\beta^2 e^{-\beta(1+z)x}\, dx = \frac{1}{(1+z)^2}$$

for $z > 0$.   ////

EXAMPLE 7.4.5   Let $X_1,\ldots,X_n$ be independent random variables which are uniformly distributed over the interval $(0,1)$, let $Y_1 = \min(X_1,\ldots,X_n)$, and let $Y_2 = \max(X_1,\ldots,X_n)$. Let us find the distribution of the *range* $R = Y_2 - Y_1$. By Example 7.2.1*b*, Y_1 and Y_2 have joint density

$$f(y_1,y_2) = n(n-1)(y_2 - y_1)^{n-2}$$

for $0 \le y_1 \le y_2 \le 1$ and $f(y_1,y_2) = 0$ for other values of y_1 and y_2. Therefore, R has density

$$g_-(r) = \int_0^{1-r} n(n-1)r^{n-2}\, dy = n(n-1)(1-r)r^{n-2}$$

for $0 \le r \le 1$ and $g_-(r) = 0$ for other values of r.   ////

7.5   SAMPLING FROM A NORMAL DISTRIBUTION[1]

As an application of the results of the previous four sections, we now consider an important practical problem, the analysis of measurement errors. Suppose that repeated measurements are made to determine some unknown quantity μ where each measurement involves an error. More precisely, let $X_1,\ldots,X_n$ denote the measurements, and suppose that they can be written in the form

$$X_i = \mu + \sigma Z_i \qquad (5.1)$$

where $Z_1,\ldots,Z_n$ are independent standard normal random variables. Here, the terms σZ_i denote the measurement errors, and $\sigma > 0$ represents the accuracy of the measuring device, with large values of σ corresponding to inaccurate measurements. The parameter σ may or may not be known.

---

[1] This section treats a special topic and may be omitted without loss of continuity.

EXAMPLE 7.5.1

a In order to determine the average nicotine content μ in a particular brand of cigarettes $n = 400$ cigarettes are smoked. Then we let X_i denote the amount of nicotine found in the ith cigarette, $i = 1, \ldots, n$.

b In order to determine the weight gain μ which can be expected from a new diet, n experimental animals are fed the diet. In this example, we let X_i denote the weight gain of the ith animal.　　　　　　　　　　　　　////

In order to estimate the quantity μ of Equation (5.1), it seems natural to take the average of the observations

$$\bar{X} = \frac{1}{n}(X_1 + \cdots + X_n)$$

Therefore, the question which confronts us is: How close can we expect $\bar{X}$ to come to μ? If the parameter σ of Equation (5.1) is known, then the answer to our question is easy. Indeed, we have

$$\frac{\sqrt{n}}{\sigma}(\bar{X} - \mu) = \sqrt{n}\,\bar{Z}$$

where $\bar{Z} = (Z_1 + \cdots + Z_n)/n$. Now, by Theorem 7.3.2, $Z_1 + \cdots + Z_n$ has the normal distribution with location parameter 0 and scale parameter $\sqrt{n}$, so that $\sqrt{n}\,\bar{Z} = (Z_1 + \cdots + Z_n)/\sqrt{n}$ has the standard normal distribution by Example 7.1.5*a*. It follows that

$$\Pr\left(-\frac{a\sigma}{\sqrt{n}} < \bar{X} - \mu < \frac{a\sigma}{\sqrt{n}}\right) = \Pr\left(-a < \frac{\sqrt{n}}{\sigma}(\bar{X} - \mu) < a\right)$$
$$= \Pr\left(-a < \sqrt{n}\,\bar{Z} < a\right)$$
$$= \Phi(a) - \Phi(-a) = 2\Phi(a) - 1$$

where Φ denotes the standard normal distribution function. Thus, we can specify an interval about $\bar{X}$,

$$I = \left(\bar{X} - \frac{a\sigma}{\sqrt{n}},\ \bar{X} + \frac{a\sigma}{\sqrt{n}}\right) \qquad (5.2)$$

which will contain the unknown quantity μ with probability $2\Phi(a) - 1$. The interval I of Equation (5.2) is known to statisticians as a *confidence interval for* μ, and the coverage probability $\gamma = 2\Phi(a) - 1$ is known as its *confidence coefficient*. Typically, a is selected to give γ a desired value, such as 0.95 or 0.99.

EXAMPLE 7.5.2   If in Example 7.5.1a it is known that $\sigma = 1$, and if we find that $\bar{X} = 9.32$ milligrams of nicotine per cigarette, then we may be 95 percent confident that $9.22 < \mu < 9.42$ in the sense that $(9.22, 9.42)$ is a confidence interval for μ with confidence coefficient 0.95. Indeed, taking $a = 1.96$ yields $\gamma = 0.95$ and $a\sigma/\sqrt{n} = 0.098$.   ////

Unfortunately, the parameter σ is usually unknown, so that the end points of the interval (5.2) cannot be computed. In this case σ^2 must also be estimated from the X_i, and it is usually estimated[1] by

$$S^2 = \frac{1}{n-1} \sum_{i=1}^{n} (X_i - \bar{X})^2 \qquad (5.3)$$

We now find the distribution of the random variable

$$T = \frac{\sqrt{n}\,(\bar{X} - \mu)}{S}$$

which can then be used in the same manner that we used the distribution of $(\sqrt{n}/\sigma)(\bar{X} - \mu)$ to place bounds on the error $\bar{X} - \mu$.

Theorem 7.5.1   *Let $k = n - 1$. Then T has density*

$$g_k(t) = \frac{\Gamma\left(\dfrac{k+1}{2}\right)}{\sqrt{k\pi}\,\Gamma\left(\dfrac{k}{2}\right)\left(1 + \dfrac{t^2}{k}\right)^{\frac{1}{2}(k+1)}}$$

for $-\infty < t < \infty$. In particular, $\Pr(|T| > t) = H_k(t)$ for $t > 0$, where

$$H_k(t) = 2 \int_t^{\infty} g_k(s)\,ds$$

We prove Theorem 7.5.1 below, but first let us indicate some applications. It follows from Theorem 7.5.1 that

$$\Pr\left(-\frac{aS}{\sqrt{n}} < \bar{X} - \mu < \frac{aS}{\sqrt{n}}\right) = \Pr(|T| < a) = 1 - H_k(a)$$

for every $a > 0$. Thus, the interval $I = (\bar{X} - aS/\sqrt{n},\ \bar{X} + aS/\sqrt{n})$ is a confidence interval for μ with confidence coefficient $\gamma = 1 - H_k(a)$. Table 11 gives the values of a for which $H_k(a) = 1 - \gamma$ for selected values of k and γ.

---

[1] The summation is divided by $n - 1$ instead of n because the numbers $X_i - \bar{X}$, $i = 1, \ldots, n$ satisfy a linear constraint, $\sum_{i=1}^{n} (X_i - \bar{X}) = 0$.

EXAMPLE 7.5.3   If in Example 7.5.1b we feed $n = 26$ experimental animals and observe an average weight gain of $\overline{X} = 62.5$ grams with an estimated σ of $S = 3.16$ grams, then we may be 95 percent confident that $61.2 < \mu < 63.8$. Here we take $a = 2.06$ and find that $aS/\sqrt{n} = 1.3$.   ////

We shall now prove Theorem 7.5.1. We begin with a preliminary result which is interesting in its own right.

Theorem 7.5.2 *Let $Z_1, \ldots, Z_n$ be independent, standard normal random variables, and define R by*

$$R = \sum_{i=1}^{n} (Z_i - \overline{Z})^2 \qquad (5.4)$$

Then, R has the chi-square distribution with $n - 1$ degrees of freedom and is independent of $\overline{Z}$.

PROOF   Consider the matrix $A = (a_{ij})$, where

$$a_{in} = \frac{1}{\sqrt{n}} \qquad i = 1, \ldots, n$$

$$a_{ij} = \frac{1}{\sqrt{j(j+1)}} \qquad \begin{matrix} i = 1, \ldots, j \\ j < n \end{matrix}$$

$$a_{j(j+1)} = \frac{-j}{\sqrt{j(j+1)}} \qquad j < n$$

$$a_{ij} = 0 \qquad i > j + 1 \quad j < n$$

Then, it is easily verified that the matrix A is orthogonal, that is, $AA' = I$ (the $n \times n$ identity) $= A'A$. Let the random vector $W = (W_1, \ldots, W_n)$ be defined by $W = ZA$, where $Z = (Z_1, \ldots, Z_n)$. That is, let

$$W_n = \sqrt{n}\, \overline{Z}$$

$$W_j = \frac{\sum_{i=1}^{j} Z_i - jZ_{j+1}}{\sqrt{j(j+1)}} \qquad j < n \qquad (5.5)$$

Table 11

$H_k(a)$	\multicolumn{4}{c}{k}			
	5	10	25	∞
0.01	4.03	3.17	2.79	2.60
0.05	2.57	2.23	2.06	1.96
0.10	2.015	1.81	1.71	1.645

Then, since $Z_1, \ldots, Z_n$ are independent standard normal random variables, and since A is orthogonal, it follows that $W_1, \ldots, W_n$ are also independent standard normal random variables (Example 7.4.3). Moreover, since $WW' = (ZA)(ZA)' = ZAA'Z = ZZ'$, we have

$$\sum_{i=1}^{n} W_i^2 = WW' = ZZ' = \sum_{i=1}^{n} Z_i^2 = \sum_{i=1}^{n} (Z_i - \bar{Z})^2 + n\bar{Z}^2 \qquad (5.6)$$

where the last step follows by simple algebra. Since $W_n^2 = n\bar{Z}^2$ by (5.5), we have

$$R = \sum_{i=1}^{n-1} W_i^2$$

from which the theorem follows easily. Indeed, since each W_i has the standard normal distribution, each W_i^2 has the chi-square distribution with one degree of freedom (Example 7.1.7), and therefore R has the chi-square distribution with $n - 1$ degrees of freedom [Theorem 7.3.2(iii)]. Moreover, since R depends only on $W_1, \ldots, W_{n-1}$ while $\bar{Z}$ depends only on W_n, it follows that R and $\bar{Z}$ are independent (Theorem 7.2.1).   ////

PROOF of Theorem 7.5.1   To apply Theorem 7.5.2 to the proof of Theorem 7.5.1, we write $\bar{X}$ and S in terms of $\bar{Z}$ and R. We have already remarked that $(\sqrt{n}/\sigma)(\bar{X} - \mu) = \sqrt{n}\,\bar{Z}$, and a similar computation yields $(n - 1)S^2 = \sigma^2 R$. Therefore,

$$T = \sqrt{n}\,\frac{\bar{X} - \mu}{S} = \frac{\sqrt{n}\,Z}{\sqrt{R/(n - 1)}} \qquad (5.7)$$

is the ratio of two independent random variables.

The distribution of T can now be found from Theorem 7.4.1. Indeed, by independence, $Y = (\sqrt{n}/\sigma)(\bar{X} - \mu)$ and R have joint density

$$f(y,r) = Cr^{\frac{1}{2}k-1}e^{-\frac{1}{2}(r+y^2)} \qquad r > 0$$

where $k = n - 1$ and $C^{-1} = \sqrt{2\pi}\,2^k\Gamma(k/2)$. Consider the transformation

$$T = Y\sqrt{\frac{k}{R}} \qquad \text{and} \qquad U = R$$

The range of this transformation is the set of $(t,u) \in R^2$ for which $u > 0$, and the inverse transformation is

$$R = U \qquad \text{and} \qquad Y = T\sqrt{\frac{U}{k}}$$

The jacobian of the inverse transformation is $J_v(t,u) = \sqrt{u/k}$, $u > 0$, so that the joint density of T and U is

$$g(t,u) = \frac{C}{\sqrt{k}} u^{\frac{1}{2}(k-1)} \exp\left[-\frac{1}{2}\left(1 + \frac{t^2}{k}\right)u\right] \qquad u > 0$$

The marginal density of T is therefore

$$
\begin{aligned}
g_k(t) &= \frac{C}{\sqrt{k}} \int_0^\infty u^{\frac{1}{2}(k-1)} \exp\left[-\frac{1}{2}\left(1 + \frac{t^2}{k}\right)u\right] du \\
&= \frac{C 2^{\frac{1}{2}(k+1)}}{\sqrt{k}} \left(1 + \frac{t^2}{k}\right)^{-\frac{1}{2}(k+1)} \int_0^\infty v^{\frac{1}{2}(k-1)} e^{-v}\, dv \\
&= \frac{\Gamma\left(\dfrac{k+1}{2}\right)}{\sqrt{k\pi}\, \Gamma\left(\dfrac{k}{2}\right)\left(1 + \dfrac{t^2}{k}\right)^{\frac{1}{2}(k+1)}}
\end{aligned}
$$

for $-\infty < t < \infty$, as asserted. $\qquad\qquad ////$

The density g_k found in Theorem 7.5.1 is known as the *t density with k degrees of freedom*. More extensive tables can be found in Beyer (1966).

7.6 RADIOACTIVE DECAY[1]

In this section we shall present a model for radioactive decay. We shall imagine that a radioactive substance, such as radium or uranium, contains a large number, N say, of unstable atoms. Moreover, we shall suppose that each of the unstable atoms may decay by emitting an α particle, a β particle, or a γ ray, at which time the atom becomes stable and is no longer capable of decaying. Let us imagine the N unstable atoms labeled in some manner, and let X_i denote the time at which the particle labeled with i decays, where time is measured from fixed starting point called time zero. We shall make the following assumptions about the manner in which the N atoms decay:

A_1 $X_1,\dots,X_N$ are independent random variables which have a common distribution function F.

A_2 $F(t) = 0$ for $t < 0$, and $F(s) < 1$ for all $s > 0$.

A_3 For all $s > 0$ and $t > 0$, $\Pr(X > s + t \mid X > s) = \Pr(X > t)$.

---

[1] This section treats a special topic and may be omitted without loss of continuity.

The first assumption is self-explanatory, and the second ensures that the conditional probabilities which enter in A_3 are well defined. The third assumption requires that the process of decay be spontaneous in the sense that the decay of a particular unstable atom does not become more or less likely as time progresses.[1] From these three assumptions, we shall derive an exact description of the observable behavior of the radioactive substance.

Let $G(t) = \Pr(X_1 > t) = 1 - F(t)$, $t \geq 0$. Then A_2 requires that $G(s + t)/G(s) = G(t)$ for all $s > 0$ and $t > 0$ or, equivalently,

$$G(s + t) = G(s)G(t) \qquad (6.1)$$

for $s > 0$ and $t > 0$.

Lemma 7.6.1 *There is a constant $\alpha > 0$ for which $G(t) = e^{-\alpha t}$ for $t > 0$.*

PROOF Equation (6.1) requires that $G(m/n) = G(1/n + \cdots + 1/n) = G(1/n)^m$ for positive integer m and n (by mathematical induction). In particular, we must have $G(1) = G(1/n)^n$, $n = 1, 2, \ldots$. Now, since $G(1) \neq 0$ by A_2, we can define a number α by $e^{-\alpha} = G(1)$, and it follows that for all rational numbers r ($= m/n$, where m and n are positive integers),

$$G(r) = G\left(\frac{1}{n}\right)^m = G(1)^r = e^{-\alpha r}$$

Finally, if $t > 0$ is any real number, then there is a sequence of rational numbers $r_1, r_2, \ldots$ for which $r_n \to t$ as $n \to \infty$ and $r_n > t$ for every $n = 1, 2, \ldots$. Since G is continuous from the right (Theorem 5.8.1) and $e^{-\alpha t}$ is continuous, we now have

$$G(t) = \lim_{n \to \infty} G(r_n) = \lim_{n \to \infty} e^{-\alpha r_n} = e^{-\alpha t}$$

for arbitrary $t > 0$. Finally, $\alpha > 0$ since $G(t) \to 0$ as $t \to \infty$. $\quad ////$

Thus, we have shown that the common distribution function F of $X_1, \ldots, X_N$ is the exponential distribution function

$$F(t) = 1 - e^{-\alpha t} \qquad t \geq 0$$

where $\alpha > 0$ is as in Lemma 7.6.1. The median of F [the solution of $F(m) = \frac{1}{2}$],

$$m = \alpha^{-1} \log 2$$

is known as the *half-life* of the substance, since approximately half the particles will have decayed by time m.

---

[1] See, for example, Blackwood, Osgood, and Ruark (1957), p. 271.

Let Y_1 be the minimum of $X_1, \ldots, X_N$, Y_2 the second smallest, ... Thus, Y_i is the time at which the ith decay occurs, $i = 1, \ldots, N$. $Y_1, \ldots, Y_N$ are the order statistics of $X_1, \ldots, X_N$. The marginal distributions of order statistics from an arbitrary distribution function F were found in Lemma 7.2.1. Applying this result where F is the exponential distribution function with parameter α now yields the distribution of Y_k for $k = 1, \ldots, N$. In fact, by Lemma 7.2.1, Y_k has density

$$h_k(y) = N \binom{N-1}{k-1} F(y)^{k-1}[1 - F(y)]^{N-k}f(y)$$

$$= N \binom{N-1}{k-1} (1 - e^{-\alpha y})^{k-1} e^{-(N-k)\alpha y} \alpha e^{-\alpha y}$$

for $y > 0$ and $h_k(y) = 0$ for $y \leq 0$. In particular, Y_1, the time at which the first decay occurs, has the exponential distribution with parameter $\beta = N\alpha$.

Let us now introduce the counting process N_t, defined for $t > 0$ by

$$N_t = k \quad \text{if and only if} \quad Y_k \leq t < Y_{k+1}$$

where (by convention) $Y_0 = 0$ and $Y_{N+1} = \infty$. Thus, N_t is the number of emissions which have occurred by time t.

Theorem 7.6.1 *For $t \geq 0$, N_t has the binomial distribution with parameters N and $p = 1 - e^{-\alpha t}$. That is,*

$$\Pr(N_t = k) = \binom{N}{k} (1 - e^{-\alpha t})^k e^{-\alpha(N-k)t} \qquad (6.2)$$

for $k = 0, \ldots, N$.

PROOF   Let B_i be the event that $X_i \leq t$. That is, let B_i be the event that the particle labeled i has decayed by time t. Then, assumption A_1 implies that $B_1, \ldots, B_N$ are independent with common probability $P(B_i) = \Pr(X_i \leq t) = F(t) = 1 - e^{-\alpha t}$ for $t \geq 0$. Moreover, $N_t = k$ if and only if exactly k of $B_1, \ldots, B_N$ have occurred. Finally, by Theorem 4.1.1, the probability that exactly k of $B_1, \ldots, B_N$ will occur is given by the right side of (6.2).   ////

Since the number of unstable atoms is usually very large, it seems natural to investigate the distribution of N_t as $N \to \infty$ and $\alpha \to 0$ in such a manner that $\beta = N\alpha$ remains fixed. Let $p_N = F(t)$; then $Np_N = N(1 - e^{-\alpha t}) \approx N\alpha t \to \beta t$, so

$$\lim \Pr(N_t = k) = \frac{1}{k!} (\beta t)^k e^{-\beta t} \qquad k = 0, 1, 2, \ldots$$

by Theorem 4.3.1. Thus, N_t has approximately the Poisson distribution with parameter βt. β is sometimes called the intensity of the radiation.

It is also possible to describe the process of decay in terms of the interarrival times (the times between decays),

$$Z_i = Y_i - Y_{i-1} \qquad i = 1, \ldots, N$$

Theorem 7.6.2 $Z_1, \ldots, Z_N$ *are independent random variables. Moreover,* Z_i *has the exponential distribution with parameter* $\beta_i = (N - i + 1)\alpha$ *for* $i = 1, \ldots, N$.

That is, the waiting time until the first decay occurs is exponentially distributed with parameter $\beta = \beta_1 = N\alpha$. Thereafter, the process starts afresh with $N - 1$ unstable atoms, and the waiting time until the next decay occurs is exponentially distributed with parameter $\beta_2 = (N - 1)\alpha$. Thereafter, the process starts afresh with $N - 2$ unstable atoms, etc.

PROOF of Theorem 7.6.2   By Problem 7.22, $Y_1, \ldots, Y_N$ have joint density

$$h(y_1, \ldots, y_N) = N! \, \alpha^N \exp\left[-\alpha(y_1 + \cdots + y_N)\right]$$

for $0 < y_1 < y_2 < \cdots < y_N < \infty$ and $h(y_1, \ldots, y_N) = 0$ for other values of $y_1, \ldots, y_N$. Let

$$W_i = (N - i + 1)Z_i = (N - i + 1)(Y_i - Y_{i-1})$$

for $i = 1, \ldots, N$. Then

$$Y_i = \sum_{j=1}^{i} \frac{W_j}{N - j + 1}$$

for $i = 1, \ldots, N$, so that the transformation is invertible and the jacobian of the inverse transformation is $1/N!$. Moreover, by simple algebra,

$$\sum_{i=1}^{N} W_i = \sum_{i=1}^{N} Y_i$$

It now follows easily from Theorem 7.4.1 that $W_1, \ldots, W_N$ have joint density

$$g(w_1, \ldots, w_N) = \alpha^N \exp\left[-(w_1 + \cdots + w_N)\right]$$

for $w_i > 0, i = 1, \ldots, N$, and $g(w_1, \ldots, w_N) = 0$ for other values of $w_1, \ldots, w_N$. That is, $W_1, \ldots, W_N$ are independent random variables, and W_i has the exponential distribution with parameter α for $i = 1, \ldots, N$. Thus,

$$Z_i = \frac{W_i}{N - i + 1}$$

are independent, by Theorem 7.2.1, and Z_i has the exponential distribution with parameter $\beta_i = (N - i + 1)\alpha$, by Example 7.1.5b.   ////

REFERENCES

For a more complete development of the ideas of Section 7.5, see Hogg and Craig (1970), chap. 6. For a different development of the Poisson distribution as the appropriate model for radioactive decay, see Feller (1968), chap. 17.

PROBLEMS

7.1 A fair coin is tossed n times. Each time heads appear you win a dollar, and each time tails appear you lose a dollar. Let X denote your net winnings (possibly negative). Find the mass function of X.

7.2 Let X have the geometric distribution with parameter p, $0 < p < 1$. Find the mass function of $Y = \min (X, 10)$.

7.3 Let X be uniformly distributed over $(0,1)$. Find a density for $Y = \sin \frac{1}{2}\pi X$.

7.4 Let X be uniformly distributed over $(0,1)$. Find densities for $Y = \sin 2\pi X$ and $Z = \cos 2\pi X$.

7.5 Let X be uniformly distributed over $(0,1)$. Find a density for $Y = X/(1 - X)$.

7.6 Let X have the uniform distribution on $(0,1)$. Find a density for X^α, where $\alpha > 0$.

7.7 Let X have the normal distribution with location parameter μ and scale parameter σ. Find a density for $Y = e^X$. The distribution of Y is known as the *log normal distribution*.

7.8 Let X have the standard normal distribution function Φ. What is the distribution of $\Phi(X)$?

7.9 Let X have the standard normal distribution function Φ. What is the distribution of $\Phi(X)^2$?

7.10 Let X have the normal distribution with location parameter μ and scale parameter σ. Find a density for $Y = X^2$.

7.11 Let X have the standard Cauchy distribution. Find a density for $Y = X^2$.

7.12 Show that if X has the standard Cauchy distribution with density $f(x) = 1/\pi(1 + x^2)$ for $-\infty < x < \infty$, then $1/X$ also has the standard Cauchy density.

7.13 Let X have density f; let $Y = -X$ if $X < 0$, and let $Y = 2X$ if $X > 0$. Find a density for Y.

7.14 Let X have the exponential distribution with parameter β, and let $Y = [X]$ be the greatest integer which is less than or equal to X. Find the mass function of Y.

7.15 Let X and Y be as in Problem 7.14, and let $Z = X - Y$. Find a density for Z.

7.16 Let Y and Z be as in Problem 7.15. Show that Y and Z are independent.

7.17 Let G be a continuous distribution function which has a continuous derivative G' at all but a finite number of points, say $a_1, \ldots, a_n$, where G' may fail to exist. Show that

$$G(x) = \int_{-\infty}^{x} g(y) \, dy$$

for all x, $-\infty < x < \infty$, where $g(y) = G'(y)$, $y \neq a_i$, and g may be defined arbitrarily at the points $a_1, \ldots, a_n$. *Hint:* For $a_1 < x < a_2$, write $G(x) = G(x) - G(a_1 + \varepsilon) +$

$G(a_1 + \varepsilon) - G(a_1 - \varepsilon) + G(a_1 - \varepsilon)$, use the fundamental theorem of calculus, and let $\varepsilon \to 0$. Then proceed by induction.

7.18   Prove Theorem 7.1.1:

(a) Under the assumption that f is piecewise continuous, that is, continuous at all but a finite number of points.

(b) Under the assumption that $w'(x) = 0$ at a finite number of points and w' does not change sign.

7.19 Let $X_1, \ldots, X_n$ be independent random variables which are uniformly distributed over $(0,1)$. Let $Y_1 = \min(X_1, \ldots, X_n)$, and $Y_2 = \max(X_1, \ldots, X_n)$. Find a joint density for Y_1 and Y_2.

7.20   Find densities for Y_1, Y_2, and (Y_1, Y_2), as defined in Problem 7.19, when $X_1, \ldots, X_n$ are independent random variables which are exponentially distributed with parameter $\beta > 0$.

7.21   Let $X_1, \ldots, X_5$ be independent random variables which are uniformly distributed over $(0,1)$, and let $Y_1, \ldots, Y_5$ denote the order statistics. Find the probability that $0.3 < Y_3 < 0.7$.

7.22   Let $X_1, X_2, \ldots, X_n$ be independent random variables with common density f, and let $Y_1, \ldots, Y_n$ be the order statistics. Show that a joint density for $Y_1, \ldots, Y_n$ is

$$g(y_1, \ldots, y_n) = n! \, f(y_1) \cdots f(y_n)$$

if $-\infty < y_1 < \cdots < y_n < \infty$ and $g(y_1, \ldots, y_n) = 0$ for other values of $y_1, \ldots, y_n$. *Hint:* Suppose that f is continuous, and consider $h^{-n} \Pr(y_1 - h < Y_i \le y_i, i = 1, \ldots, n)$ for small h.

7.23   (a) Let $X_1, \ldots, X_n$ be independent with common density f and distribution function F, and let $Y_1, \ldots, Y_n$ denote the order statistics. Further, let $1 \le k_1 < k_2 < \cdots < k_r \le n$ be integers. Show that $Y_{k_1}, \ldots, Y_{k_r}$ have joint density

$$\frac{n!}{(k_1 - 1)! \, (k_2 - k_1 - 1)! \cdots (n - k_r)!} F(y_1)^{k_1 - 1}$$
$$\times \, [F(y_2) - F(y_1)]^{k_2 - k_1 - 1} \cdots [1 - F(y_r)]^{n - k_r} f(y_1) \cdots f(y_r)$$

for $-\infty < y_1 < \cdots < y_r < \infty$.

(b) Specialize part (a) to the case that F is the uniform distribution on $(0,1)$.

7.24   Let X_1 and X_2 be independent random variables which are uniformly distributed over $(0,1)$. Find a density for $Y = X_1 - X_2$. *Hint:* Draw a picture.

7.25   Let X and Y be independent random variables which have the Poisson distribution with the same parameter β. Find the mass function of $Z = X - Y$.

7.26   Let $X_1, \ldots, X_k$ be uniformly distributed over the unit ball

$$B = \{x \in R^k : x_1^2 + \cdots + x_k^2 \le 1\}$$

Find the distribution function of R, where $R^2 = X_1^2 + \cdots + X_k^2$.

7.27   Let X and Y be independent random variables, and let u and v be increasing functions. Show directly that $u(X)$ and $v(Y)$ are independent.

7.28. Let X and Y be independent random variables which are uniformly distributed over (0,1). Show that $Z = X + Y$ has the triangular density

$$f(z) = \begin{cases} z & 0 \le z \le 1 \\ 1 - z & 1 \le z \le 2 \end{cases}$$

7.29 Let X and Y be independent random variables which are geometrically distributed with the same parameter p. Find the mass function of $X + Y$.

7.30 Let X_1, X_2, and X_3 be independent, standard normal random variables. Find a density for $R = \sqrt{X_1{}^2 + X_2{}^2 + X_3{}^2}$.

7.31. Let X_1 and X_2 be independent random variables which are uniformly distributed over (0,1). Find a density for $X_1 X_2$. *Hint:* Consider logarithms.

7.32. Let $X_1, \ldots, X_n$ be independent, random variables which are uniformly distributed over (0,1). Find the distribution of $Y = X_1 X_2 \cdots X_n$.

7.33 Let X and Y be independent with common density $f(x) = \frac{1}{2} e^{-|x|}$, $-\infty < x < \infty$. Find a density for $X + Y$.

7.34. Show that if X and Y are independent with densities f and g, respectively, then $Z = X/Y$ has density

$$h(z) = \int_{-\infty}^{\infty} |y| f(zy) g(y) \, dy$$

for $-\infty < z < \infty$. *Hint:* Compute the distribution function and differentiate.

7.35. Show that if X and Y are independent, standard normal random variables, then $Z = X/Y$ has the standard Cauchy distribution.

7.36 Let X have distribution function F, let Y be uniformly distributed over $(-a,a)$, $a > 0$, and let X and Y be independent. Show that the distribution function of $Z = X + Y$ is

$$H(z) = \frac{1}{2a} \int_{-a}^{a} F(z - y) \, dy$$

7.37 If X and Y are independent with distribution functions F and G, respectively, show that $Z = X + Y$ has distribution function

$$H(z) = \int_{-\infty}^{\infty} F(z - y) \, dG(y)$$

If F is absolutely continuous, show that H is also absolutely continuous, even if G is discrete.

7.38 Let X_1 and X_2 be independent, standard normal random variables. Show that $X_1 - X_2$ and $X_1 + X_2$ are also independent.

7.39 Let X and Y be independent, standard normal random variables. Find the joint distribution of R and Θ, where $X = R \cos \Theta$ and $Y = R \sin \Theta$.

7.40 Let X, Y, and Z be independent, standard normal random variables. Define R, ϕ, and Θ by $-\pi/2 < \phi \le \pi/2$, $-\pi < \Theta \le \pi$, $R \ge 0$, $X = R \cos \phi \cos \Theta$, $Y = R \cos \phi \sin \Theta$, and $Z = R \sin \phi$.

(a) Show that R, ϕ, and Θ have joint density $f(r,\phi,\theta) = (1/2\pi)^{\frac{3}{2}} \cos \phi \, r^2 \exp \left(-\frac{1}{2}r^2\right)$ for $-\pi/2 < \phi \leq \pi/2$, $-\pi < \theta \leq \pi$, and $r \geq 0$.

(b) Show that (ϕ,Θ) have marginal density $g(\phi,\theta) = (1/4\pi) \cos \phi$ for $-\pi/2 < \phi \leq \pi/2$, $-\pi < \theta \leq \pi$.

(c) What is the distribution of R?

(d) Are R, ϕ, and Θ independent?

7.41 In the notation of Problem 7.40, let $U = \cos \phi \cos \Theta$, $V = \cos \phi \sin \Theta$, and $W = \sin \phi$. Show that (U,V,W) has the uniform distribution on the surface of the unit sphere in R^3 in the sense that $\Pr ((U,V,W) \in B) =$ (surface area of B)/4π for subsets B of the surface of the sphere.

7.42 Let X_1, X_2, and X_3 be independent, exponentially distributed random variables with the same parameter. Find the joint distribution of $Y_1 = X_1/(X_1 + X_2 + X_3)$, $Y_2 = X_2/(X_1 + X_2 + X_3)$, and $Y_3 = X_1 + X_2 + X_3$.

7.43 More generally, let $X_1, \ldots, X_{k+1}$ be independent, and let X_i have the gamma distribution with parameters α_i and β, $i = 1, \ldots, k$. Find the joint distribution of $Y_1, \ldots, Y_{k+1}$, where $Y_{k+1} = X_1 + \cdots + X_{k+1}$ and $Y_j = X_j/Y_{k+1}$, $j = 1, \ldots, k$.

7.44 Show that the marginal (joint) distribution of $Y_1, \ldots, Y_k$ and $(1 - Y_1 - \cdots - Y_k)$ in Problem 7.43 is the Dirichlet distribution with parameters $\alpha_1, \ldots, \alpha_{k+1}$.

7.45 If X and Y have the standard, bivariate normal distribution with parameter r, $-1 < r < 1$, find densities for $X - Y$ and $X + Y$.

7.46 Let X_1 and X_2 be independent random variables, and let X_i have the gamma distribution with parameters α_i and β, $i = 1, 2$. Find the distribution of X_1/X_2.

7.47 Let g_k denote the density of the t distribution with k degrees of freedom. Show that

$$\lim g_k(t) = \frac{1}{\sqrt{2\pi}} e^{-\frac{1}{2}t^2}$$

as $k \to \infty$ for all t, $-\infty < t < \infty$.

7.48 A matrix B is called a *projection matrix* if and only if $B' = B = B^2$. Show that if B is a projection matrix and $X_1, \ldots, X_n$ are independent, standard normal random variables, then XBX' has the chi-square distribution on r degrees of freedom, where r is the rank of B. *Hint:* If B is any symmetric matrix, then there is an orthogonal matrix A such that ABA' is diagonal.

7.49 For $0 < s < t < \infty$, find the distribution of $N(s,t) = N_t - N_s$.

7.50 Show that as $N \to \infty$ in such a manner that $\beta = N\alpha$ remains fixed,

$$\lim \Pr (N(s,t) = k) = \frac{1}{k!} \beta^k (t - s)^k e^{-\beta(t-s)}$$

for each fixed $k = 0, 1, \ldots$.

7.51 Show that if $s_1 < t_1 < s_2 < t_2$, then $N(s_1,t_1)$ and $N(s_2,t_2)$ are asymptotically independent in the following sense. As $N \to \infty$ and $\alpha \to 0$ with $\beta = N\alpha$ fixed

$$\lim \Pr (N(s_1,t_1) = j, N(s_2,t_2) = k) = \frac{\beta^{k+j}}{j! \, k!} (t_1 - s_1)^j (t_2 - s_2)^k e^{-\beta(t_1-s_1)-\beta(t_2-s_2)}$$

for all fixed j and k.

8

EXPECTATION

8.1 EXPECTATION

Let X be a discrete random variable with mass function f, and let $C = \{x_0, x_1, \ldots\}$ be a finite or countably infinite set for which $\Pr(X \in C) = 1$. We define the *expectation*, or *expected value*, of X to be the number

$$E(X) = \sum_C xf(x) \qquad (1.1)$$

provided that the summation on the right side of (1.1) converges absolutely.[1]

We may interpret expectation as follows. Suppose, for simplicity, that $C = \{x_0, x_1, \ldots, x_k\}$ is a finite set, and imagine the experiment to which X refers to be repeated n times, where n is large. Further, let X_j denote the value of X on the jth trial (repetition). Then we can compute the observed empirical average of X over the n repetitions as

$$\bar{X}_n = \frac{1}{n} \sum_{j=1}^n X_j = \sum_{i=0}^k x_i f_n(x_i)$$

[1] Recall that $\sum_C xf(x)$ means that the numbers $xf(x)$, $x \in C$, are to be summed. The sum converges absolutely if and only if $\sum_C |x| f(x)$ is finite. This will always be the case if C is a finite set but may fail if C is infinite.

where $f_n(x_i)$ is the relative frequency with which $X_j = x_i$ during the n repetitions. That is, $f_n(x_i) = 1/n$ (the number of times $X_j = x_i$), as in Section 2.1. Now according to the frequentistic interpretation of probability, $f_n(x_i)$ converges to $\Pr(X = x_i) = f(x_i)$ as $n \to \infty$. This suggests that $\bar{X}_n$ will converge to

$$\sum_{i=0}^{k} x_i f(x_i) = \sum_{C} x f(x) = E(X)$$

as $n \to \infty$, and in Section 9.2 we show that this is, in fact, the case. *That is, $E(X)$ represents an idealized empirical average for X in the same manner that probabilities represent idealized relative frequencies.*

EXAMPLE 8.1.1

a If $\Pr(X = c) = 1$, where c is a constant, then we may take $C = \{c\}$ and we obtain $E(X) = c \times 1 = c$ by Equation (1.1). In particular, $E(0) = 0$ and $E(1) = 1$.

b Consider a gambler who wins a dollar with probability p and loses a dollar with probability $q = 1 - p$. If X denotes his gain (positive or negative), we may take $C = \{-1,1\}$ and we obtain $E(X) = 1 \times p - 1 \times q = p - q$.

c If a balanced n-sided die is rolled once and X denotes the number of spots which appear, then X has the discrete uniform distribution; that is, $\Pr(X = k) = 1/n$, $k = 1, \ldots, n$. Therefore,

$$E(X) = \sum_{k=1}^{n} k\,\frac{1}{n} = \frac{1}{n}\,\frac{n(n+1)}{2} = \frac{n+1}{2}$$

In particular, if $n = 6$, then the expected number of spots is 3.5. Observe that in parts *b* and *c*, $E(X)$ need not be a possible value of X.   ////

Let us now consider absolutely continuous random variables. If X is absolutely continuous with density f, then by analogy with (1.1), we define the *expectation*, or *expected value*, of X to be

$$E(X) = \int_{-\infty}^{\infty} xf(x)\,dx \qquad (1.2)$$

provided that an integral appearing in (1.2) converges absolutely.[1]

---

[1] The integral converges absolutely if and only if $\int_{-\infty}^{\infty} |x| f(x)\,dx$ is finite; see Appendix B.

EXAMPLE 8.1.2

a If X has the uniform distribution on the interval (a,b), $a < b$, then X has density $f(x) = 1/(b - a)$ for $a < x < b$ and $f(x) = 0$ for $x \notin (a,b)$. Therefore,

$$E(X) = \int_a^b \frac{x \, dx}{b - a} = \frac{b^2 - a^2}{2(b - a)} = \frac{a + b}{2}$$

the midpoint of the interval.

b If X has a symmetric density f, that is, $f(x) = f(-x)$ for all x, and if $E(X)$ is defined, then $E(X) = 0$; for if f is symmetric, then

$$\int_{-\infty}^0 xf(x) \, dx = - \int_0^\infty xf(x) \, dx$$

In particular, if X has the standard normal distribution, then $E(X) = 0$.

c If X denotes the waiting time until the first emission from a radioactive substance, then X has the exponential distribution with parameter β, the intensity of the radiation (see Example 5.5.4). Therefore,

$$E(X) = \int_0^\infty x\beta e^{-\beta x} \, dx = \frac{1}{\beta} \int_0^\infty ye^{-y} \, dy = \frac{1}{\beta} \Gamma(2) = \frac{1}{\beta}$$

d If X has the Cauchy distribution [with density $f(x) = 1/\pi(1 + x^2)$, $-\infty < x < \infty$], then

$$\int_{-a}^b xf(x) \, dx = \frac{1}{2\pi} \left[\log (1 + b^2) - \log (1 + a^2) \right]$$

which does not approach any limit as $a \to \infty$ and $b \to \infty$ independently. Therefore, $E(X)$ is not defined. ////

Equations (1.1) and (1.2) can be combined into one equation by using the Riemann-Stieltjes integral.[1] Indeed, if X is either discrete or absolutely continuous, with distribution function F, then

$$E(X) = \int_{-\infty}^\infty x \, dF(x) \qquad (1.3)$$

provided that the integral appearing in (1.3) converges absolutely. Moreover, the integral appearing in (1.3) will exist for many distribution functions which are neither discrete nor absolutely continuous. We can therefore extend the definition of expectation as follows. If X is any random variable with distribution function F, we define

---

[1] The Riemann-Stieltjes integral is discussed in Appendix B. Readers who wish to do so may regard $\int_{-\infty}^\infty x \, dF(x)$ as notation which means $\sum_c xf(x)$ if F is discrete with mass function f and means $\int_{-\infty}^\infty xf(x) \, dx$ if F is absolutely continuous with density f.

the expectation, or expected value, of X by (1.3) provided only that the integral in (1.3) converges absolutely.

If X is a nonnegative random variable, that is, if $\Pr(X \geq 0) = 1$, then $F(x) = 0$ for $x < 0$, so that (1.3) reduces to

$$E(X) = \int_0^\infty x \, dF(x) \qquad (1.4)$$

which is meaningful even if the integral diverges (necessarily to ∞). We can therefore make a final extension of the definition of expectation by allowing nonnegative random variables to have infinite expectation. That is, if X is nonnegative, we define $E(X)$ by (1.4), even if the integral diverges.

EXAMPLE 8.1.3

a Let X have the uniform distribution on $(0,2)$, and let $Y = \max(1,X)$. Then the distribution function of Y is

max

$P_r(1 \leq y) \cdot P_r(x \leq y) = P_r(Y \leq y) = $

$G(y) = \begin{cases} 0 & y < 1 \\ \dfrac{y}{2} & 1 \leq y < 2 \\ 1 & y \geq 2 \end{cases}$

Therefore,

$$E(Y) = 1[G(1) - G(1-)] + \frac{1}{2}\int_1^2 x \, dx = \tfrac{1}{2} + \tfrac{3}{4} = \tfrac{5}{4}$$

b If X and Y are independent exponentially distributed random variables (with the same parameter β), then the ratio $Z = X/Y$ has density $f(z) = 1/(1 + z)^2$ for $z > 0$ (see Example 7.4.4*b*). Now, for $a > 0$,

$$\int_0^a zf(z) \, dz \geq \frac{1}{2}\int_1^a (z + 1)^{-1} \, dz = \tfrac{1}{2}[\log(a + 1) - \log z]$$

which diverges to ∞ as $a \to \infty$. Therefore, $E(Z) = \infty$. Observe that $E(Z)$ does not equal $E(X)/E(Y) = 1$. ////

We conclude[1] this section with an interesting and useful geometrical description of expectation.

---

[1] The remainder of this section treats a special topic and may be omitted.

$$E(X) = A_1 - A_2$$

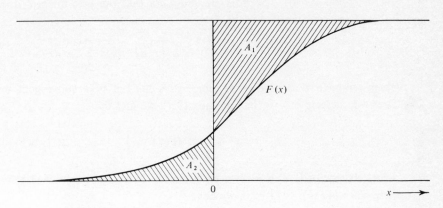

FIGURE 12
A geometrical interpretation of $E(X)$.

Theorem 8.1.1 *Let X be any random variable, and let F denote its distribution function. Then*

$$E(X) = \int_0^\infty [1 - F(x)]\, dx - \int_{-\infty}^0 F(x)\, dx \qquad (1.5)$$

in the following sense: if either side of (1.5) is finite, then so is the other and they are equal.

Theorem 8.1.1 admits the following geometrical interpretation. In Figure 12 $E(X)$ is the area between the graph of F and the line of height 1 for $x \geq 0$ minus the area between the graph of F and the line of height 0 for $x \leq 0$.

PROOF To prove the theorem, we shall integrate by parts the two integrals which appear in the following expression for $E(X)$:

$$E(X) = \int_0^\infty x\, dF(x) + \int_{-\infty}^0 x\, dF(x) \qquad (1.6)$$

The first of these is the limit as $b \to \infty$ of

$$\int_0^b x\, dF(x) = bF(b) - \int_0^b F(x)\, dx$$

$$= b[F(b) - 1] + \int_0^b [1 - F(x)]\, dx \qquad (1.7)$$

(see Theorem B.5 of Appendix B). Suppose that the first integral in (1.6) is finite. Then we have

$$0 \le b[1 - F(b)] = b \int_b^\infty dF(x) \le \int_b^\infty x \, dF(x)$$

which tends to zero as $b \to \infty$, since it is the tail of a convergent integral. Therefore, letting $b \to \infty$ in Equation (1.7), we find that

$$\int_0^\infty x \, dF(x) = \int_0^\infty [1 - F(x)] \, dx \qquad (1.8)$$

if the left side of (1.8) is known to be finite. If the right side of (1.8) is finite, then, since $b[F(b) - 1] \le 0$, we have

$$\int_0^\infty x \, dF(x) = \lim_{b \to \infty} \int_0^b x \, dF(x)$$

$$\le \lim_{b \to \infty} \int_0^b [1 - F(x)] \, dx$$

$$= \int_0^\infty [1 - F(x)] \, dx < \infty$$

so that (1.8) also holds. Thus we have shown that one side of (1.8) is finite if and only if the other is, in which case they are equal. A similar argument will show that

$$\int_{-\infty}^0 x \, dF(x) = - \int_{-\infty}^0 F(x) \, dx$$

in the same sense. The theorem follows.   ////

The following corollaries to the theorem are useful.

Corollary 8.1.1   *If X is a nonnegative random variable with distribution F, then*

$$E(X) = \int_0^\infty [1 - F(x)] \, dx$$

finite or infinite.

Corollary 8.1.2   *If X is a nonnegative integer-valued random variable, then*

$$E(X) = \sum_{k=1}^\infty \Pr(X \ge k)$$

finite or infinite.

PROOF   The first corollary follows directly from the theorem, and the second then follows from the fact that if X is integer-valued, then $\Pr(X > x) = \Pr(X \geq k)$ for $k - 1 \leq x < k$.   ////

EXAMPLE 8.1.4   If X has the geometric distribution with parameter p, then $\Pr(X \geq k) = q^{k-1}, k = 1, 2, \ldots$, so that $E(X) = 1 + q + q^2 + \cdots = 1/(1 - q) = 1/p$.   ////

8.2   PROPERTIES OF EXPECTATION

In this section we shall develop several general properties of expectation, just as we developed general properties of probability in Sections 2.3, 2.4, and 2.5. We begin with an important theorem.

Theorem 8.2.1   *Let X be a random variable with distribution function F, and let D be a subset of R for which $\Pr(X \in D) = 1$. Further, let w be a real-valued function on D, and let the random variable Y be defined by $Y = w(X)$. Then*

$$E(x) = \sum_{j} x_j \, f(x_j)$$

$$E(Y) = \int_D w(x) \, dF(x) \qquad (2.1)$$

provided that the integral appearing on the right side of (2.1) converges absolutely.[1]

PROOF   We shall prove the theorem only in the case that X has a discrete distribution, although it is true in the generality stated. Let f denote the mass function of X, and, for simplicity, let D be the set of $x \in R$ for which $f(x) > 0$. Then

$$\int_D w(x) \, dF(x) = \sum_{x \in D} w(x) f(x) \qquad (2.2)$$

is an absolutely convergent series (by assumption). Let E denote the range of Y, and for each $y \in E$ let $B_y = w^{-1}(\{y\})$ be the set of $x \in D$ for which $w(x) = y$. Then since the terms of an absolutely convergent series can be summed in any

---

[1] Readers who are unfamiliar with the Riemann-Stieltjes integral may regard $\int_D w(x) \, dF(x)$ as notation which means $\sum_D w(x) f(x)$ if F is discrete with mass function f and means $\int_D w(x) f(x) \, dx$ if F is absolutely continuous with density f.

order to the same value,[1] we write

$$\sum_{x \in D} w(x)f(x) = \sum_{y \in E} \left[\sum_{x \in B_y} w(x)f(x) \right]$$

$$= \sum_{y \in E} y \left[\sum_{x \in B_y} f(x) \right]$$

$$= \sum_{y \in E} y \operatorname{Pr}(Y = y) = E(Y) \qquad (2.3)$$

where we have used Equation (1.2) of Chapter 7 in the third equality and the definition of $E(Y)$ in the final one. Equations (2.2) and (2.3) clearly combine to prove the theorem in the discrete case. ////

In the absolutely continuous case (when F has density f), Equation (2.1) assumes the form

$$E(Y) = \int_D w(x)f(x)\,dx \qquad (2.4)$$

and a proof of (2.4) is sketched in Problem 8.14. Let us now consider some examples.

EXAMPLE 8.2.1

a Let X have the uniform distribution on the interval $(0,2\pi]$, and let $Y = \sin X$. Then, the distribution function of Y is tedious to compute, but the expectation of Y can be computed quite easily, since

$$E(Y) = \int_0^{2\pi} \frac{\sin x}{2\pi}\,dx = \frac{1}{2\pi}(\cos 0 - \cos 2\pi) = 0$$

b If X is any random variable and $Y = X^2$, then

$$E(Y) = \int_{-\infty}^{\infty} x^2\,dF(x)$$

In particular, if X has the exponential distribution with parameter β, then

$$E(X^2) = \int_0^{\infty} \beta x^2 e^{-\beta x}\,dx = \Gamma(3)\beta^{-2} = 2\beta^{-2} \qquad ////$$

Equations (2.3) and (2.4) also remain valid if $X = (X_1, \ldots, X_n)$ is a random vector, f denotes the mass function or density of X, and D denotes the set of $x \in R^n$ for which $f(x) > 0$ [and dx is interpreted as the volume element in R^n in Equation (2.4)].

---

[1] See, for example, Rudin (1964), p. 66.

EXAMPLE 8.2.2

a Let X_1 and X_2 be independent, exponentially distributed random variables with common parameter $\beta = 1$, and let $Y = X_1 X_2$. Then

$$E(Y) = \int\int_0^\infty x_1 x_2 e^{-(x_1 + x_2)} \, dx_1 \, dx_2 = \left(\int_0^\infty x e^{-x} \, dx \right)^2 = \Gamma(2)^2 = 1$$

b Let (X, Y) have joint mass function f, and let D be a finite or countably infinite set for which $f(x, y) = 0$ if $(x, y) \notin D$. If $E(X)$ is finite, then

$$E(X) = \sum_D x f(x, y)$$

where the summation extends over all $(x, y) \in D$. This follows from Theorem 8.2.1 by taking $w(x, y) = x$ for $(x, y) \in D$. ////

We now turn to some basic properties of expectation—its *linearity* and *monotonicity* as an operator on random variables.

Theorem 8.2.2 *Let X and Y be any two jointly distributed random variables with finite expectations, and let α and β be any real numbers. Then*

$$E(\alpha X + \beta Y) = \alpha E(X) + \beta E(Y)$$

Moreover, if $\Pr(X \geq Y) = 1$, *then* $E(X) \geq E(Y)$.

PROOF   Again, we shall prove (2.5) in the discrete case only, although it is true in the generality stated. Let f denote the joint mass function of X and Y, and let $D \subset R^2$ denote a finite or countably infinite set for which $f(x, y) = 0$ if $(x, y) \notin D$. Then, by Theorem 8.2.1,

$$E(\alpha X + \beta Y) = \sum_D (\alpha x + \beta y) f(x, y)$$

$$= \alpha \sum_D x f(x, y) + \beta \sum_D y f(x, y) = \alpha E(X) + \beta E(Y)$$

where all summations extend over all $(x, y) \in D$. This establishes the first assertion of the theorem.

The second assertion of the theorem is trivial if $Y = 0$, for then $\Pr(X \leq x) = 0$ for $x < 0$ [see Equation (1.4)]. In general, we may let $X' = X - Y$. Then, $E(X') \geq 0$ and $E(X') = E(X) - E(Y)$. ////

Corollary 8.2.1 *If X has finite expectation and α and β are real numbers, then* $E(\alpha X + \beta) = \alpha E(X) + \beta$.

Corollary 8.2.2 *Let* $X_1, \ldots, X_n$ *be jointly distributed random variables with finite expectation, and let* $S = X_1 + \cdots + X_n$. *Then*

$$E(S) = E(X_1) + \cdots + E(X_n)$$

PROOF   Corollary 8.2.1 follows by taking $Y = 1$ in Theorem 8.2.2, and Corollary 8.2.2 can be established by induction.   ////

EXAMPLE 8.2.3

a   Let X denote the number of heads which result from n independent tosses of a coin which has probability p of turning up heads on each toss. Then we write $X = X_1 + \cdots + X_n$, where X_i is 1 if the ith toss results in heads and 0 if it results in tails, $i = 1, \ldots, n$. Since $E(X_i) = 1 \times p + 0 \times q = p$, $i = 1, \ldots, n$, it now follows from Corollary 8.2.2 that $E(X) = E(X_1) + \cdots + E(X_n) = p + \cdots + p = np$. Observe that X has the binomial distribution with parameters n and p.

b   The independence of $X_1, \ldots, X_n$ is *not* required by Corollary 8.2.2. Thus, if a sample of size k is drawn without replacement from an urn which contains m red balls and $n - m$ white balls, the expected number of red balls is $E(X) = km/n$; for we may write $X = X_1 + \cdots + X_k$, where $X_i = 1$ if the ith ball drawn is red and $X_i = 0$ if it is white, and clearly, $E(X_i) = m/n$, $i = 1, \ldots, k$.

c   Let X have the normal distribution with parameters μ and σ^2. Then, by Example 7.1.5*a*, $Z = (X - \mu)/\sigma$ has the standard normal distribution, so that $E(Z) = 0$ by Example 8.1.2*b*. Since $X = \sigma Z + \mu$, it now follows that $E(X) = \sigma \times 0 + \mu = \mu$. *Therefore, the parameter* μ *of a normal distribution is the expectation.*   ////

The expectation of the product of two *independent* random variables is the product of their expectations.

Theorem 8.2.3   *Let* X *and* Y *be independent random variables with finite expectations. Then* $E(XY) = E(X)E(Y)$.

PROOF   Again, we shall prove Theorem 8.2.3 only in the case that X and Y are discrete random variables, although it is true in the generality stated. Let X and Y be independent discrete random variables with finite expectations. Let g and h denote the (marginal) mass functions of X and Y, and let D and E be countable sets for which $\Pr(X \in D) = 1 = \Pr(Y \in E)$. Then the joint mass function of X and Y is f, where $f(x,y) = g(x)h(y)$, and clearly $f(x,y) = 0$ unless $(x,y) \in C = D \times E$. Therefore,

$$E(XY) = \sum_C xyf(x,y)$$

$$= \left[\sum_D xg(x)\right]\left[\sum_E yh(y)\right] = E(X)E(Y)$$

as asserted. Here the summations extend over all $(x,y) \in C$, $x \in D$, and $y \in E$, respectively. ////

Theorem 8.2.3 extends by induction from two random variables to several.

Corollary 8.2.4 *Let $X_1,\ldots,X_n$ be independent random variables with finite expectation, and let $Y = \prod_{i=1}^n X_i$. Then*

$$E(Y) = \prod_{i=1}^n E(X_i)$$

8.3 THE MEAN AND VARIANCE

The expectation of a random variable X is also known as the *mean* of X, denoted by μ. Thus,

$$\mu = E(X) = \int_{-\infty}^\infty x\, dF(x) \qquad (3.1)$$

where F denotes the distribution function of X. Similarly, the expectation of the random variable $Y = (X - \mu)^2$ is known as the *variance* of X and denoted by σ^2 or $D(X)$. Thus, $\sigma^2 = D(X) = E[(X - \mu)^2]$, where $\mu = E(X)$. By Theorem 8.2.1, we have

$$\sigma^2 = D(X) = \int_{-\infty}^\infty (x - \mu)^2\, dF(x) \qquad (3.2)$$

where F denotes the distribution function of X. The positive square root σ of the variance is known as the *standard deviation* of X.

Of course, the Riemann-Stieltjes integrals appearing in (3.1) and (3.2) simplify to sums if F is discrete and to ordinary integrals if F is absolutely continuous.

The mean and variance of a random variable provide important information about its distribution. As we indicated in Section 8.1, the mean $\mu = E(X)$ may be thought of as a long-term average of X over many trials of the experiment to which X refers. Similarly, the variance $\sigma^2 = E[(X - \mu)^2]$ may be thought of as a long-term average of $(X - \mu)^2$. Thus, *the variance σ^2 provides a measure of the tendency*

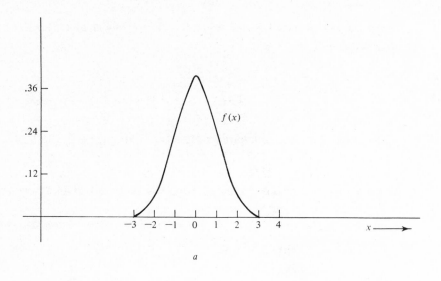

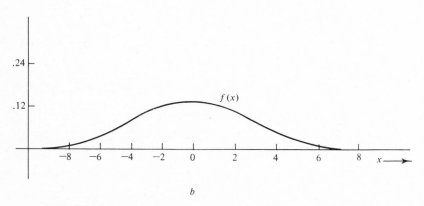

FIGURE 13
Normal distributions with different variances: (a) $\sigma^2 = 1$; (b) $\sigma^2 = 9$.

of X to deviate from its mean. That is, large values of σ^2 indicate a tendency toward appreciable deviations, while small values of σ^2 indicate that the distribution of X concentrates closely about μ. In Section 9.1 we shall, in fact, give an inequality which bounds $\Pr\,(|X - \mu| \geq a)$ by a simple function of a and σ^2.

We illustrate the difference between large and small variances in Figure 13.

Another, complementary way of considering the mean and variance is provided by the following lemma.

Lemma 8.3.1 *Let X be any random variable for which $E(X^2)$ is finite. Then, among all real numbers a, $E[(X - a)^2]$ is minimized by taking $a = \mu = E(X)$, in which case the minimum is $\sigma^2 = E[(X - \mu)^2]$.*

PROOF   For any a we have $E[(X - a)^2] = E(X^2) - 2aE(X) + a^2$ by Theorem 8.2.2. The derivative of this expression with respect to a is $-2E(X) + 2a$, which vanishes if and only if $a = \mu = E(X)$. Moreover, the second derivative is $2 > 0$, and so the extremum is a minimum.   ////

We can rephrase the lemma as follows. Suppose that we had to predict the value of X by a single number a, and suppose that by so doing we incur a loss of $(X - a)^2$. If we wish to minimize our expected loss, then we should predict X by its mean $a = \mu = E(X)$, in which case we incur the loss $\sigma^2 = E[(X - \mu)^2]$. Thus, the mean μ may be thought of as the best constant guess or *predictor* for the random variable X, and σ^2 as a measure of our ability to predict.

For our examples, the following lemmas will be convenient.

Lemma 8.3.2 *Let X be any random variable for which $E(X^2)$ is finite. Then the variance of X is*

$$\sigma^2 = E(X^2) - \mu^2$$

where $\mu = E(X)$. That is, σ^2 is the expectation of the square of X minus the square of the expectation of X.

PROOF   We have

$$\sigma^2 = E[(X - \mu)^2] = E(X^2) - 2\mu E(X) + \mu^2$$
$$= E(X^2) - 2\mu^2 + \mu^2 = E(X^2) - \mu^2$$

as asserted.   ////

Lemma 8.3.3 *Let X be a random variable with mean μ and variance σ^2. Then, the mean and variance of the random variable $Y = aX + b$ are $v = a\mu + b$ and $\tau^2 = a^2\sigma^2$, respectively.*

PROOF $v = E(Y) = E(aX + b) = aE(X) + b = a\mu + b$ by Corollary 8.2.1. Therefore,

$$\tau^2 = E[(Y - v)^2] = E[(aX - a\mu)^2] = a^2E[(X - \mu)^2] = a^2\sigma^2$$

by the same corollary.   ////

EXAMPLE 8.3.1

a Let X have the uniform distribution on $(0,1)$. Then

$$E(X) = \int_0^1 x \, dx = \tfrac{1}{2}$$

$$E(X^2) = \int_0^1 x^2 \, dx = \tfrac{1}{3}$$

so that $\sigma^2 = \tfrac{1}{3} - (\tfrac{1}{2})^2 = \tfrac{1}{12}$.

b If X has the exponential distribution with parameter β, then $E(X) = \beta^{-1}$ by Example 8.1.2c, and, similarly, $E(X^2) = 2\beta^{-2}$ by Example 8.2.1b. Thus, $\sigma^2 = 2\beta^{-2} - \beta^{-2} = \beta^{-2}$.

c More generally, if X has the gamma distribution with parameters α and β, then

$$E(X^k) = \int_0^\infty \frac{x^k \beta^\alpha x^{\alpha-1}}{\Gamma(\alpha)} e^{-\beta x} \, dx$$

$$= \beta^{-k} \int_0^\infty \frac{y^{k+\alpha-1}}{\Gamma(\alpha)} e^{-y} \, dy = \frac{\beta^{-k}\Gamma(\alpha+k)}{\Gamma(\alpha)}$$

for $k = 1, 2, \ldots$. In particular, $\mu = E(X) = \Gamma(\alpha + 1)/\beta\Gamma(\alpha) = \alpha/\beta$, and $E(X^2) = \alpha(\alpha + 1)/\beta^2$, so that $\sigma^2 = \alpha/\beta^2$. In particular, the mean and variance of a chi-square distribution with k degrees of freedom ($\alpha = k/2$ and $\beta = \tfrac{1}{2}$) are $\mu = k$ and $\sigma^2 = 2k$.   ////

EXAMPLE 8.3.2   If Z has the standard normal distribution, then Z has mean $E(Z) = 0$ by Example 8.1.2b and Z^2 has the chi-square distribution with one degree of freedom by Example 7.1.7. That is, Z^2 has the gamma distribution with parameters $\alpha = \tfrac{1}{2}$ and $\beta = \tfrac{1}{2}$. Therefore, the variance of Z is $E(Z^2) = 1$ by Example 8.3.1. More generally, if X has the normal distribution with parameters μ and σ, then $X = \sigma Z + \mu$, where $Z = (X - \mu)/\sigma$ has the standard normal distribution (see Example 7.1.5a). Thus, the mean and variance of X are $E(X) = \mu$ and σ^2, respectively, by Lemma 8.3.3. That is, *the parameters μ and σ of the normal distribution are its mean and standard deviation, respectively.*   ////

For *independent* random variables, not only the mean but also the variance is additive.

Theorem 8.3.1   *Let $X_1, \ldots, X_n$ be independent random variables with variances $\sigma_1^2, \ldots, \sigma_n^2$, respectively. Then the variance of the sum $S = X_1 + \cdots + X_n$ is*

$$\sigma^2 = \sigma_1^2 + \cdots + \sigma_n^2$$

PROOF We shall prove the theorem in the special case that $n = 2$. The general case then follows easily by mathematical induction. Let μ_1 and μ_2 denote the means of X_1 and X_2, respectively, and let $Y_i = X_i - \mu_i$, $i = 1, 2$. Then, $E(Y_i) = 0$, $i = 1, 2$, and the variance of X_i is $E(Y_i^2) = \sigma_i^2$, $i = 1, 2$. Moreover, since the mean of S is $\mu = \mu_1 + \mu_2$ by Theorem 8.2.2, we have $S - \mu = Y_1 + Y_2$. Therefore, the variance of S is

$$\sigma^2 = E[(S - \mu)^2] = E[(Y_1 + Y_2)^2]$$
$$= E(Y_1^2) + 2E(Y_1 Y_2) + E(Y_2^2)$$

Now Y_1 and Y_2 are independent, so that $E(Y_1 Y_2) = E(Y_1)E(Y_2) = 0$ by Theorem 8.2.3. Therefore,

$$\sigma^2 = E(Y_1^2) + E(Y_2^2) = \sigma_1^2 + \sigma_2^2$$

as asserted. $////$

EXAMPLE 8.3.3 Let $X_1, \ldots, X_n$ be independent random variables for which $\Pr(X_i = 1) = p$ and $\Pr(X_i = 0) = q = 1 - p$, $i = 1, \ldots, n$. Then, $S = X_1 + \cdots + X_n$ has the binomial distribution with parameters n and p. Let us compute the variance of S from Theorem 8.3.1. We have $E(X_i) = p$ and $E(X_i^2) = E(X_i) = p$, so that the variance of each X_i is $\sigma_i^2 = p - p^2 = pq$, $i = 1, \ldots, n$. Therefore, the variance of S is $\sigma^2 = npq$ by Theorem 8.3.1. $////$

8.4 THE MOMENT-GENERATING FUNCTION

Let X be any random variable, and let F denote its distribution function. Then the numbers

$$\mu_k = E(X^k) \qquad (4.1)$$

$k = 1, 2, \ldots$, are called the *moments* of X. More precisely, μ_k is called the kth *moment* of X, provided that $E(|X|^k) < \infty$. Thus, the first moment is simply the mean $\mu = E(X)$, and the variance of X is $\sigma^2 = \mu_2 - \mu_1^2$ by Lemma 8.3.2.

Let F denote the distribution function of X. Then, by Theorem 8.2.1, we have

$$\mu_k = \int_{-\infty}^{\infty} x^k \, dF(x) \qquad (4.1a)$$

provided that the integral appearing on the right side of (4.1a) converges absolutely. Of course, the Riemann-Stieltjes integral in (4.1a) simplifies to an ordinary integral if F is absolutely continuous and to a sum if F is discrete.

EXAMPLE 8.4.1

a   Let X have the beta distribution with parameter α and β. Then

$$\mu_k = \frac{\Gamma(\alpha + \beta)}{\Gamma(\alpha)\Gamma(\beta)} \int_0^1 x^k x^{\alpha-1} (1 - x)^{\beta-1}\, dx$$

$$= \frac{\Gamma(\alpha + \beta)}{\Gamma(\alpha)\Gamma(\beta)} \frac{\Gamma(\alpha + k)\Gamma(\beta)}{\Gamma(\alpha + k + \beta)}$$

$$= \frac{\Gamma(\alpha + \beta)\Gamma(\alpha + k)}{\Gamma(\alpha)\Gamma(\alpha + \beta + k)} = \frac{(\alpha + k - 1)_k}{(\alpha + \beta + k - 1)_k}$$

where for real x, $(x)_k = x(x - 1) \cdots (x - k + 1)$. In particular, the mean and variance are $\mu = \mu_1 = \alpha/(\alpha + \beta)$ and $\sigma^2 = \mu_2 - \mu_1^2 = \alpha\beta/(\alpha + \beta)^2 \times (\alpha + \beta + 1)$ by simple algebra.

b   Similarly, if X has the gamma distribution with parameters α and β, then

$$\mu_k = \frac{(\alpha + k - 1)_k}{\beta^k}$$

by Example 8.3.1c.   ////

Again, let X be any random variable. We define the *moment-generating function* of X by

$$M(t) = E(e^{tX}) \tag{4.2}$$

provided that the expectation defining $M(t)$ is finite for all t in some nondegenerate interval (a,b), $a < b$. If F denotes the distribution function of X, then

$$M(t) = \int_{-\infty}^{\infty} e^{tx}\, dF(x) \tag{4.2a}$$

by Theorem 8.2.1, provided that the expectation on the right side of (4.2a) converges absolutely. Again, the Riemann-Stieltjes integral in (4.2a) simplifies to an ordinary integral if F is absolutely continuous and to a sum if F is discrete.

Both the moments and the moment-generating function of a random variable X depend only on the distribution function of X by (4.1) and (4.2). We shall sometimes refer to the moment-generating function of X as the moment-generating function of F. The name moment-generating function derives from the fact that the moments of X can be computed by differentiating $M(t)$ at $t = 0$.

Theorem 8.4.1   *Let X be any random variable with a moment-generating function which is finite on some open interval containing zero, say $M(t) < \infty$ for $-h < t < h$, where $h > 0$. Then X has moments*

$$\mu_k = M^{(k)}(0)$$

the kth derivative of M at $t = 0$, $k = 1, 2, \ldots$.

PROOF   The idea is that since both expectation and differentiation are linear operations, they should commute. That is, we should have

$$\frac{d^k}{dt^k} M(t) = \frac{d^k}{dt^k} E(e^{tX}) = E\left(\frac{d^k}{dt^k} e^{tX}\right) = E(X^k e^{tX})$$

from which the theorem follows on setting $t = 0$. Of course, the crucial (middle) equality does require justification since an interchange of limits is involved. We shall return to this point in the next section.   ////

By taking logarithms we can obtain a simple method for computing the mean and variance of a random variable.

Corollary 8.4.1   *Let X be a random variable with a moment-generating function which is finite on an open interval containing zero. Further, let $\psi(t) = \log M(t)$. Then the mean and variance of X are given by*

$$\mu = \psi'(0) \quad and \quad \sigma^2 = \psi''(0)$$

PROOF   We first remark that $M(0) = E(e^{0X}) = E(1) = 1$. Therefore, by the chain rule, $\psi'(0) = M'(0)/M(0) = \mu_1 = \mu$, and

$$\psi''(0) = \frac{M''(0)M(0) - M'(0)^2}{M(0)^2} = \mu_2 - \mu_1{}^2 = \sigma^2$$

as asserted.   ////

Let us now consider some examples.

EXAMPLE 8.4.2   If X has the binomial distribution with parameters n and p, then

$$M(t) = \sum_{k=0}^{n} e^{kt} \binom{n}{k} p^k q^{n-k} = \sum_{k=0}^{n} \binom{n}{k} (pe^t)^k q^{n-k} = (q + pe^t)^n$$

for $-\infty < t < \infty$, where the final equality follows from the binomial theorem. By differentiation we can rederive the result that $\mu = np$ and $\sigma^2 = npq$.   ////

EXAMPLE 8.4.3   If X has the Poisson distribution with parameter β, then

$$M(t) = \sum_{k=0}^{\infty} e^{kt} \frac{1}{k!} \beta^k e^{-\beta} = e^{-\beta} \sum_{k=0}^{\infty} \frac{1}{k!} (\beta e^t)^k = e^{-\beta} e^{\beta e^t} = e^{\beta(e^t - 1)}$$

for all t, $-\infty < t < \infty$. In this example, $\psi(t) = \log M(t) = \beta(e^t - 1)$ is easy to differentiate, and we find that $\mu = \psi'(0) = \beta$ and $\sigma^2 = \psi''(0) = \beta$. Therefore, *the mean and variance of the Poisson distribution are both β.*   ////

EXAMPLE 8.4.4

a If X has the negative binomial distribution [with mass function $f(k) = \binom{k-1}{r-1} p^r q^{k-r}$, $k = r, r + 1, \ldots$], then

$$M(t) = \sum_{k=r}^{\infty} e^{kt} \binom{k-1}{r-1} p^r q^{k-r}$$

$$= \left(\frac{pe^t}{p_1}\right)^r \sum_{k=r}^{\infty} \binom{k-1}{r-1} p_1^r q_1^{k-r} \qquad (4.3)$$

where we have set $q_1 = qe^t$ and $p_1 = 1 - q_1$. Now, if $q_1 < 1$, then the final sum in (4.3) is the sum of negative binomial probabilities with parameters r and p_1 and is therefore 1. It follows that

$$M(t) = \left(\frac{pe^t}{1 - qe^t}\right)^r$$

for $t < -\log q$, that is, $q_1 < 1$. The sum diverges if $t \geq -\log q$.

The mean and variance of the negative binomial distribution can now be computed by differentiation. Indeed, letting $\psi(t) = \log M(t)$, we find

$$\psi'(t) = r + \frac{rqe^t}{1 - qe^t} \qquad \text{and} \qquad \psi''(t) = \frac{rqe^t}{(1 - qe^t)^2}$$

so that $\mu = r + rq/(1 - q) = rp^{-1}$ and $\sigma^2 = rq/(1 - q)^2 = rqp^{-2}$.

b The geometric distribution is a special case of the negative binomial with $r = 1$. The mean and variance are $\mu = qp^{-1}$ and $\sigma^2 = qp^{-2}$.   ////

EXAMPLE 8.4.5

a If X has the gamma distribution with parameters α and β, then

$$M(t) = \int_0^{\infty} e^{tx} \frac{\beta^\alpha x^{\alpha-1} e^{-\beta x}}{\Gamma(\alpha)} \, dx$$

$$= \int_0^{\infty} \frac{\beta^\alpha x^{\alpha-1}}{\Gamma(\alpha)} e^{-(\beta-t)x} \, dx$$

$$= \left(\frac{\beta}{\beta - t}\right)^\alpha \int_0^{\infty} \frac{y^{\alpha-1} e^{-y}}{\Gamma(\alpha)} \, dy = \left(\frac{\beta}{\beta - t}\right)^\alpha$$

for $t < \beta$, and the integral diverges if $t \geq \beta$. The third equality above follows from the change of variables $y = (\beta - t)x$, and the final one from the definition of the gamma function.

The mean and variance of the gamma distribution were found in Example 8.3.1b to be $\mu = \alpha\beta^{-1}$ and $\sigma^2 = \alpha\beta^{-2}$. This result can be checked by differentiation.

b The exponential distribution is a special case of the gamma with $\alpha = 1$. Thus, the moment-generating function of the exponential distribution is

$$M(t) = \frac{\beta}{\beta - t}$$

for $t < \beta$. The mean and variance are $\mu = \beta^{-1}$ and $\sigma^2 = \beta^{-2}$.   ////

For our next example, it will be convenient to have the following lemma.

Lemma 8.4.1 *Let X be a random variable with moment-generating function M, and let a and b be real numbers. Then, the moment-generating function of $Y = aX + b$ is $N(t) = e^{bt}M(at)$.*

PROOF We have

$$N(t) = E(e^{tY}) = E(e^{atX+bt}) = e^{bt}E(e^{atX}) = e^{bt}M(at)$$

for all t for which $M(at)$ is finite.   ////

EXAMPLE 8.4.6 If Z has the standard normal distribution, then Z has moment-generating function

$$M(t) = e^{\frac{1}{2}t^2} \qquad (4.4)$$

for $-\infty < t < \infty$. Indeed, since $x^2 - 2tx = (x - t)^2 - t^2$, we have

$$M(t) = \int_{-\infty}^{\infty} e^{tx}\frac{e^{-\frac{1}{2}x^2}}{\sqrt{2\pi}}\, dx = \int_{-\infty}^{\infty} \frac{1}{\sqrt{2\pi}}e^{-\frac{1}{2}x^2 + tx}\, dx$$

$$= e^{\frac{1}{2}t^2}\int_{-\infty}^{\infty}\frac{e^{-\frac{1}{2}(x-t)^2}}{\sqrt{2\pi}}\, dx$$

and the latter integral is simply the integral of a normal density with mean $\mu = t$ and variance $\sigma^2 = 1$ and is, therefore, 1. This establishes (4.4).

Now suppose that X has the normal distribution with mean μ and variance σ^2. Then we can write $X = \sigma Z + \mu$, where $Z = (X - \mu)/\sigma$ has the standard normal distribution (see Example 7.1.5a). Therefore, letting N and M denote the moment-generating functions of X and Z, respectively, we have

$$N(t) = e^{\mu t}M(\sigma t) = \exp\left(\mu t + \tfrac{1}{2}\sigma^2 t^2\right) \qquad ////$$

We have seen that the moment-generating function is a useful tool for computing means and variances. It is also useful in finding the distribution of a sum of independent random variables. The technique depends on the following two results.

Theorem 8.4.2 *Let X and Y be random variables with distribution functions F and G, respectively. Also, let X have a moment-generating function M, and let Y have a moment-generating function N. If $M(t) = N(t)$ (finite) for all t in some nondegenerate interval, then $F(x) = G(x)$ for all x, $-\infty < x < \infty$.*

That is, if X and Y have the same moment-generating function, they have the same distribution. *Thus, we can determine the distribution of a random variable by finding its moment-generating function.*

The proof of Theorem 8.4.2 is beyond the scope of the book; we prove a related result in Section 8.4.1, however.

In applications of Theorem 8.4.2, it will be useful to have the following theorem.

Theorem 8.4.3 *Let $X_1, \ldots, X_n$ be independent random variables with moment-generating functions $M_1, \ldots, M_n$, respectively. If $M_1, \ldots, M_n$ are all finite on the same interval (a,b), $a < b$, then the sum $S = X_1 + \cdots + X_n$ has moment-generating function*

$$M(t) = \prod_{i=1}^{n} M_i(t)$$

for $a < t < b$.

PROOF For any t, the random variables

$$e^{tX_1}, \ldots, e^{tX_n}$$

are independent by Theorem 7.2.1. Therefore, by Corollary 8.2.4,

$$E(e^{tS}) = E(e^{t(X_1 + \cdots + X_n)}) = E\left(\prod_{i=1}^{n} e^{tX_i}\right) = \prod_{i=1}^{n} E(e^{tX_i}) = \prod_{i=1}^{n} M_i(t)$$

for $a < t < b$, as asserted. ////

We can now derive again the results of Theorem 7.3.2.

EXAMPLE 8.4.7 If $X_1, \ldots, X_n$ are independent, normally distributed random variables, then the moment-generating function of X_i is

$$M_i(t) = \exp\left(t\mu_i + \tfrac{1}{2}\sigma_i^2 t^2\right)$$

for $-\infty < t < \infty$, where μ_i denotes the mean and σ_i^2 the variance of X_i, $i = 1, \ldots, n$.

It now follows from Theorem 8.4.3 that the moment-generating function of $S = X_1 + \cdots + X_n$ is

$$M(t) = \exp\left(t\mu + \tfrac{1}{2}\sigma^2 t^2\right)$$

for $-\infty < t < \infty$, where $\mu = \mu_1 + \cdots + \mu_n$ and $\sigma^2 = \sigma_1{}^2 + \cdots + \sigma_n{}^2$. Since M is the moment-generating function of a normal distribution with mean μ and variance σ^2, it now follows from Theorem 8.4.2 that S has the normal distribution with mean μ and variance σ^2. ////

In a similar manner one can establish the following assertions.

Theorem 8.4.4 *Let $X_1, \ldots, X_k$ be independent random variables, and let S be the sum $S = X_1 + \cdots + X_k$.*

(i) If each X_i has the binomial distribution with parameters n_i and the same p, then S has the binomial distribution with parameters $n = n_1 + \cdots + n_k$ and p.

(ii) If each X_i has the negative binomial distribution with parameters r_i and the same p, then S has the negative binomial distribution with parameters $r = r_1 + \cdots + r_k$ and p.

(iii) If X_i has the Poisson distribution with parameter β_i, then S has the Poisson distribution with parameter $\beta = \beta_1 + \cdots + \beta_k$.

(iv) If each X_i has the gamma distribution with parameters α_i and the same β, then S has the gamma distribution with parameters $\alpha = \alpha_1 + \cdots + \alpha_k$ and β.

8.4.1 Generating Functions[1]

Let $a_0, a_1, a_2, \ldots$ be a sequence of real numbers. Then we define the *generating function of the sequence* $a_0, a_1, \ldots$ to be

$$P(z) = \sum_{n=0}^{\infty} P(X=n) z^n, \quad |z| \le 1$$

$$A(t) = \sum_{k=0}^{\infty} a_k t^k \qquad (4.5)$$

provided that the series converges for all t in some nondegenerate interval about zero, say for $-h < t < h$. If X is a nonnegative integer-valued random variable, and if

$$a_k = \Pr(X = k)$$

---

[1] In this section we prove a special case of Theorems 8.4.1 and 8.4.2. The results of this section are used only in Sections 10.6 and 12.5, and it may be omitted without loss of continuity.

prob. gen. funct.

for $k = 0, 1, 2, \ldots$, then we shall refer to A as the *generating function of X*. In this case $A(1) = 1$, since $a_0, a_1, a_2, \ldots$ are probabilities and the series (4.5) converges for $-1 \le t \le 1$. Also, if A is the generating function of X, then

$$A(t) = E(t^X) \qquad (4.6)$$

for $-1 \le t \le 1$. Thus, if A denotes the generating function of X, and if M denotes the moment-generating function of X, then

$$M(t) = A(e^t) \qquad (4.7)$$

EXAMPLES 8.4.8

a If $a_k = 1$ for $k = 0, 1, 2, \ldots$, then the generating function of $a_0, a_1, \ldots$ is $A(t) = 1/(1 - t)$ for $-1 < t < 1$ by Example 1.7.3*b*.

b If $a_k = 1/k!$ for $k = 0, 1, 2, \ldots$, then the generating function of $a_0, a_1, \ldots$ is $A(t) = e^t$ for $-\infty < t < \infty$ by Example 1.7.3*a*.

c If X has the binomial distribution with parameters n and p, then X has generating function $A(t) = (q + pt)^n$ for $-\infty < t < \infty$ by (4.7) and Example 8.4.2.

d Similarly, if X has the Poisson distribution with parameter β, then X has generating function $A(t) = e^{\beta(t-1)}$ for $-\infty < t < \infty$.

e If X has the negative binomial distribution with parameters r and p, then X has generating function $A(t) = [pt/(1 - qt)]^r$ for $-q^{-1} < t < q^{-1}$.   ////

Generating functions have properties similar to those of moment-generating functions, but since only discrete random variables are involved, generating functions are simpler than moment-generating functions, and it will be possible to treat them more fully; in particular, we shall prove analogs to Theorems 8.4.1 and 8.4.2.

We use the fact that a power series can be differentiated termwise.[1] Thus if A is the generating function of $a_0, a_1, \ldots$, and if $A(t)$ converges for $-h < t < h$, where $h > 0$, then

$$A'(t) = \sum_{k=1}^{\infty} k a_k t^{k-1} \qquad (4.8)$$

and $A'(t)$ also converges for $-h < t < h$. More generally, the jth derivative of A exists and is given by

$$A^{(j)}(t) = \sum_{k=j}^{\infty} (k)_j a_k t^{k-j} \qquad (4.9)$$

for $-h < t < h$, where $(k)_j = k(k - 1) \ldots (k - j + 1)$. Observe that $A^{(j)}$ is the

---

[1] See, for example, Rudin (1964), p. 158.

generating function of the sequence a_0', a_1', ..., where $a_k' = (k + j)_j a_{k+j}$ for $k = 0, 1, \ldots$.

Theorem 8.4.5 *Let X be any nonnegative integer-valued random variable, and let A denote its generating function. Then*

$$E(X) = A(1-) \qquad (4.10)$$

the limit of $A(t)$ as $t \to 1$ with $t < 1$. Equation (4.10) is valid whether $E(X)$ is finite or infinite.

PROOF We have $a_k = \Pr(X = k) \geq 0$ for $k = 0, 1, 2, \ldots$, and so A' is a nondecreasing function, by (4.8). Therefore, $A'(1-)$ exists. Now

$$A'(t) = \sum_{k=0}^{\infty} k a_k t^{k-1} \leq \sum_{k=0}^{\infty} k a_k = E(X)$$

for all $t < 1$, so that $A(1-) \leq E(X)$. Moreover, for any $n = 1, 2, \ldots$

$$A(1-) = \lim \sum_{k=0}^{\infty} k a_k t^{k-1} \geq \lim \sum_{k=0}^{n} k a_k t^{k-1} = \sum_{k=0}^{n} k a_k$$

where the limit is taken as $t \to 1$ with $t < 1$. Finally

$$\sum_{k=0}^{n} k a_k \to \sum_{k=0}^{\infty} k a_k = E(X)$$

as $n \to \infty$, and so the theorem follows. ////

Theorem 8.4.6 *Let $a_0, a_1, \ldots$ have generating function A, and let $b_0, b_1, \ldots$ have generating function B. If $A(t) = B(t)$ for $-h < t < h$ for some $h > 0$, then $a_k = b_k$ for all $k = 0, 1, 2, \ldots$.*

PROOF If $A(t) = B(t)$ for $-h < t < h$, then by (4.9)

$$a_k = \frac{1}{k!} A^{(k)}(0) = \frac{1}{k!} B^{(k)}(0) = b_k$$

for all $k = 0, 1, 2, \ldots$. ////

If $a_0, a_1, a_2, \ldots$ and $b_0, b_1, b_2, \ldots$ are two sequences of real numbers, then we define their *convolution* to be the sequence $c_0, c_1, \ldots$, where

$$c_n = \sum_{k=0}^{n} a_k b_{n-k} \qquad (4.11)$$

for $n = 0, 1, 2, \ldots$. If $a_k = \Pr(X = k)$ and $b_k = \Pr(Y = k)$ for $k \geq 0$, where X and Y are independent nonnegative integer-valued random variables, then $c_n =$

$\Pr(X + Y = n)$ for $n \geq 0$ by Theorem 7.3.1. However, we do not require the sequences $a_0, a_1, \ldots$ and $b_0, b_1, \ldots$ to be probabilities in the definition (4.11).

Theorem 8.4.7 *Let $a_0, a_1, \ldots$ have generating function A, and let $b_0, b_1, \ldots$ have generating function B. If both $A(t)$ and $B(t)$ converge for $-h < t < h$, then $c_0, c_1, \ldots$ has generating function C, where $C(t) = A(t)B(t)$ for $-h < t < h$.*

PROOF Let us first suppose that a_k and b_k are nonnegative for $k \geq 0$. Then, since nonnegative terms may be summed in any order to the same limit, we have

$$C(t) = \sum_{n=0}^{\infty} c_n t^n = \sum_{n=0}^{\infty} \sum_{k=0}^{n} a_k b_{n-k} t^n$$

$$= \sum_{k=0}^{\infty} a_k t^k \left(\sum_{n=k}^{\infty} b_{n-k} t^{n-k} \right) = B(t) \sum_{k=0}^{\infty} a_k t^k = A(t)B(t) \qquad (4.12)$$

for $-h < t < h$, as asserted.

For the general case, replace a_k and b_k by $|a_k|$ and $|b_k|$ in (4.12), and deduce that the series converges absolutely for $-h < t < h$. The interchange of orders of summation can then be justified by the absolute convergence. ////

EXAMPLE 8.4.9 Let X_1 and X_2 be independent random variables, and let X_i have geometric distribution with parameter p_i, $i = 1, 2$ where $p_1 \neq p_2$. What is the distribution of $Y = X_1 + X_2$? The generating function of X_i is

$$A_i(t) = \frac{p_i t}{1 - q_i t}$$

for $-q_i^{-1} < t < q_i^{-1}$ for $i = 1, 2$. Thus, the generating function of Y is

$$C(t) = \frac{p_1 p_2 t^2}{(1 - q_1 t)(1 - q_2 t)}$$

and C converges for $-1 < t < 1$. Let us expand C as a partial fraction as

$$C(t) = \frac{p_1 p_2 t^2}{q_1 - q_2} \left(\frac{q_1}{1 - q_1 t} - \frac{q_2}{1 - q_2 t} \right)$$

$$= \frac{p_1 p_2}{q_1 - q_2} \sum_{k=0}^{\infty} (q_1^{k+1} - q_2^{k+1}) t^{k+2}$$

The coefficient of t^k in the expansion of $C(t)$ is simply $c_k = \Pr(Y = k)$. Thus

$$\Pr(Y = k) = \frac{p_1 p_2}{q_1 - q_2}(q_1^{k-1} - q_2^{k-1})$$

for $k = 2, 3, \ldots.$ $////$

8.5 COVARIANCE AND CORRELATION[1]

In Section 8.3 we characterized the mean μ of a random variable X as that number (constant a) which minimizes $E[(X - a)^2]$, and we described μ as the best constant predictor of X. Now suppose that we are allowed to predict X by a linear function of some other random variable Y. For example, this problem might arise if X were the unobservable state of some system and $Y = X + Z$, where Z represents observation error. On the basis of observing Y, we would wish to estimate or predict the value of X. If we restrict ourselves to estimates which are linear functions of Y, say $aY + b$, and if we measure the estimation error by mean square error

$$E[(X - aY - b)^2]$$

then the problem can be stated as follows. Find constants a and b which minimize $E[(X - aY - b)^2]$.

Theorem 8.5.1 *Let X and Y be jointly distributed random variables with means μ and v and variances σ^2 and τ^2, respectively. If $\tau^2 > 0$, then $E[(X - aY - b)^2]$ is minimized by taking $b = \mu - av$, where*

$$a = \frac{E[(X - \mu)(Y - v)]}{\tau^2} \qquad (5.1)$$

PROOF By Lemma 8.3.1 we know that for any a, $E[(X - aY - b)^2]$ is minimized by taking $b = E(X - aY) = \mu - av$. With this choice of b, we have

$$E[(X - aY - b)^2] = E[(X_1 - aY_1)^2] \qquad (5.2)$$

where $X_1 = X - \mu$ and $Y_1 = Y - v$. Expanding (5.2), we now find that

$$E[(X_1 - aY_1)^2] = E(X_1{}^2) - 2aE(X_1 Y_1) + a^2 E(Y_1{}^2)$$
$$= \sigma^2 - 2aE(X_1 Y_1) + a^2 \tau^2$$

Differentiating, we now find that the minimum occurs when $a = E(X_1 Y_1)/\tau^2$, as asserted. $////$

---

[1] This section treats a special topic and may be omitted.

If $\sigma^2 > 0$, then we find from Theorem 8.5.1 that the minimum of $E[(X - aY - b)^2]$ is

$$\sigma^2 - 2aE(X_1Y_1) + a^2\tau^2 = \sigma^2 - E(X_1Y_1)\tau^{-2} = \sigma^2(1 - r^2) \qquad (5.3)$$

where (by definition)

$$r = \frac{E[(X - \mu)(Y - \nu)]}{\sigma\tau} \qquad (5.4)$$

Thus if we predict X by a linear function of Y, it is possible to reduce the expected error by a factor of r^2 [from σ^2 to $\sigma^2(1 - r^2)$] from the error which would have been incurred if we had just predicted X by its mean μ. We may therefore think of r as a measure of the *linear dependence* between X and Y. We shall call r the *correlation coefficient* of X and Y. Further, we define the *covariance* of X and Y to be

$$C(X,Y) = E[(X - \mu)(Y - \nu)] \qquad (5.5a)$$

so that

$$r = \frac{C(X,Y)}{\sigma\tau} \qquad (5.5b)$$

Both covariance and correlation are symmetric in X and Y.

Before we consider examples, let us remark that the value of (5.3) must always be nonnegative, since it is simply $E[(X - aY - b)^2]$. Therefore, we must always have $r^2 \le 1$, or equivalently,

$$-1 \le r \le 1 \qquad (5.6a)$$

In terms of the covariance and variances, (5.6a) can also be written

$$|C(X,Y)| \le \sigma\tau \qquad (5.6b)$$

where σ^2 and τ^2 denote the variances of X and Y, respectively. Expression (5.6b) is a special case of *Schwarz's inequality*, which is discussed further in Problem 8.18.

In the computation of covariances and correlations, the identity

$$C(X,Y) = E(XY) - E(X)E(Y) \qquad (5.7)$$

is quite useful. Its proof is similar to that of Lemma 8.3.2 and will be left as an exercise.

Equation (5.7) has an interesting theoretical consequence, namely, *if X and Y are independent random variables, then $C(X,Y) = 0$*, and consequently $r = 0$ also. In fact, if X and Y are independent, then $E(XY) = E(X)E(Y)$ by Theorem 8.2.3. Since we have interpreted r as a measure of dependence between X and Y, we should certainly anticipate this result. *It is possible, however, for dependent random variables to be uncorrelated*, that is, have zero correlation, as we shall see in the following examples.

EXAMPLE 8.5.1 Let X and Y have joint density

$$f(x,y) = \begin{cases} 2 & 0 \leq x \leq y \leq 1 \\ 0 & \text{otherwise} \end{cases}$$

Then, simple computations yield $\mu = \frac{1}{3}$, $v = \frac{2}{3}$, $\sigma^2 = \frac{1}{18} = \tau^2$, and

$$E(XY) = 2 \int_0^1 \int_0^y xy \, dx \, dy = \int_0^1 y^3 \, dy = \frac{1}{4}$$

so that

$$r = \frac{\frac{1}{4} - (\frac{1}{3})(\frac{2}{3})}{18} = \frac{1}{2} \qquad ////$$

EXAMPLE 8.5.2 Let X and Z be independent random variables with means zero and variance σ^2 and θ^2, respectively. If $Y = X + Z$, then the variance of Y is $r^2 = \sigma^2 + \theta^2$ by Theorem 8.3.1, and

$$C(X,Y) = E(XY) = E(X^2) + E(XZ) = \sigma^2$$

Therefore, the correlation is $r = \sigma^2/\sigma\sqrt{\sigma^2 + \theta^2} = \sigma/\sqrt{\sigma^2 + \theta^2}$. The best linear predictor of X is aY, where $a = C(X,Y)/\tau^2 = \sigma^2/(\sigma^2 + \theta^2)$. $////$

EXAMPLE 8.5.3 It is possible for dependent random variables to be uncorrelated. For example, let X have the uniform distribution on $(-1,1)$, and let $Y = X^2$. Then, X and Y are highly dependent since Y is, in fact, a function of X. However,

$$E(X) = \frac{1}{2} \int_{-1}^1 x \, dx = 0$$

$$E(XY) = E(X^3) = \frac{1}{2} \int_{-1}^1 x^3 \, dx = 0$$

so that $C(X,Y) = 0$ and consequently $r = 0$. $////$

For our remaining examples, we shall find the following lemma useful.

Lemma 8.5.1 *Let X and Y be jointly distributed random variables with correlation coefficient r, and let $X' = aX + b$ and $Y' = cY + d$, where $ac \neq 0$. Then, the correlation coefficient of X' and Y' is $r' = acr/|ac|$. In particular, $|r'| = |r|$.*

The proof of Lemma 8.5.1 is similar to that of Lemma 8.3.3 and will therefore be omitted.

EXAMPLE 8.5.4

a Let X and Y have the standard bivariate normal distribution with parameter r, $-1 < r < 1$. That is, let X and Y have joint density

$$f(x,y) = \frac{1}{2\pi\sqrt{1-r^2}} \exp\left(-\frac{1}{2}\frac{x^2 - 2rxy + y^2}{1-r^2}\right)$$

for $-\infty < x, y < \infty$. Then r *is the correlation between X and Y.* To see this observe first that X and Y both have the standard univariate normal distribution by Example 6.2.4, so that $E(X) = E(Y) = 0$ and $D(X) = D(Y) = 1$ by Example 8.3.2. Therefore, the correlation between X and Y is simply $C(X,Y) = E(XY)$. Now, since $x^2 - 2rxy + y^2 = (x - ry)^2 + (1 - r^2)y^2$,

$$E(XY) = \int\int_{-\infty}^{\infty} \frac{1}{2\pi\sqrt{1-r^2}} xy \exp\left(-\frac{1}{2}\frac{x^2 - 2rxy + y^2}{1-r^2}\right) dx\, dy$$

$$= \int_{-\infty}^{\infty} \left\{\int_{-\infty}^{\infty} x \frac{1}{\sqrt{2\pi(1-r^2)}} \exp\left[-\frac{1}{2}\frac{(x-ry)^2}{1-r^2}\right] dx\right\}$$

$$\times \frac{1}{\sqrt{2\pi}} ye^{-\frac{1}{2}y^2} dx$$

Moreover, the inside integral is simply the mean of a normal distribution with mean $\mu = ry$ and variance $\sigma^2 = 1 - r^2$, that is, ry. Thus,

$$E(XY) = \int_{-\infty}^{\infty} ry^2 \frac{1}{\sqrt{2\pi}} e^{-\frac{1}{2}y^2} dy = r$$

as asserted.

b More generally, let W and Z have the standard bivariate normal distribution with parameter r, $-1 < r < 1$, and let

$$X = \sigma W + \mu \qquad \text{and} \qquad Y = \tau Z + v$$

where $-\infty < \mu, v < \infty$, $\sigma > 0$, and $\tau > 0$. Then, by a simple application of Corollary 7.4.1, X and Y have joint density

$$g(x,y) = \frac{1}{2\pi\sigma\tau\sqrt{1-r^2}} \exp\left[-\frac{1}{2}\frac{Q(x,y)}{1-r^2}\right]$$

where $\qquad Q(x,y) = \left(\frac{x-\mu}{\sigma}\right)^2 - 2r\frac{x-\mu}{\sigma}\frac{y-v}{\tau} + \left(\frac{y-v}{\tau}\right)^2$

The density g is known as the general bivariate normal density with parameters μ, v, σ, τ, and r.

The parameters are quite easy to interpret. By Lemma 8.3.3 and Example 8.3.2, μ and σ^2 are the mean and variance of X, and v and τ^2 are the mean and variance of Y. Moreover, by part a and Lemma 8.5.1, the correlation between X and Y is r.   ////

EXAMPLE 8.5.5 Let X be any random variable with a finite positive variance σ^2, and let $Y = aX + b$, where $a \neq 0$. Then the correlation between X and Y is either 1 or -1. Indeed, it is obvious that the correlation between X and X is simply $\sigma^2/\sigma^2 = 1$, so the correlation between X and $Y = a/|a|$ by Lemma 8.5.1. The converse of this statement is also true. If X and Y are random variables with correlation coefficient r which is either 1 or -1, then there are constants a and b for which $\Pr(Y = aX + b) = 1$ (see Problem 9.3).   ////

Theorem 8.5.2 *Let* $X_1, \ldots, X_m$ *and* $Y_1, \ldots, Y_n$ *be jointly distributed random variables with finite variances. Further, let* $S = X_1 + \cdots + X_m$ *and* $T = Y_1 + \cdots + Y_n$. *Then*

$$C(S,T) = \sum_{i=1}^{m} \sum_{j=1}^{n} C(X_i, Y_j) \qquad (5.8a)$$

In particular, if σ_i^2 *denotes the variance of* X_i, $i = 1, \ldots, m$, *then the variance of* S *is*

$$\sigma^2 = \sum_{i=1}^{m} \sigma_i^2 + 2 \sum_{i=2}^{m} \sum_{j=1}^{i-1} C(X_i, X_j) \qquad (5.8b)$$

PROOF   We have

$$E(ST) = E\left(\sum_{i=1}^{m} \sum_{j=1}^{n} X_i Y_j \right) = \sum_{i=1}^{m} \sum_{j=1}^{n} E(X_i Y_j)$$

and

$$E(S)E(T) = \left[\sum_{i=1}^{m} E(X_i) \right] \left[\sum_{j=1}^{n} E(Y_j) \right] = \sum_{i=1}^{m} \sum_{j=1}^{n} E(X_i)E(Y_j)$$

so that

$$C(S,T) = E(ST) - E(S)E(T)$$

$$= \sum_{i=1}^{m} \sum_{j=1}^{n} [E(X_i Y_j) - E(X_i)E(Y_j)] = \sum_{i=1}^{m} \sum_{j=1}^{n} C(X_i, Y_j)$$

This establishes (5.8a), from which (5.8b) follows on taking $m = n$ and $X_i = Y_i$, $i = 1, \ldots, m$.   ////

As a corollary to Theorem 8.5.2, we can rederive the result that the variance of a sum of independent random variables is the sum of their variances. Indeed, if

$X_1, \ldots, X_m$ are independent, then $C(X_i, X_j) = 0$ for $i \neq j$, so that the result follows from (5.8b). As a second application of Theorem 8.5.2, we shall compute the variance of the hypergeometric distribution.

EXAMPLE 8.5.6 Let an ordered random sample of size k be drawn without replacement from an urn which contains m red balls and $n - m$ white balls, where $k \leq n$. Further let $X_i = 1$ if the ith ball drawn is red, and let $X_i = 0$ if the ith ball drawn is white. Then $S = X_1 + \cdots + X_k$ gives the total number of red balls in the sample, so that S has the hypergeometric distribution with parameters m, n, and k (see Example 5.2.2). Now

$$E(X_i) = \frac{m}{n} = E(X_i^2) \quad \text{and} \quad E(X_i X_j) = \frac{m(m-1)}{n(n-1)}$$

for $i = 1, \ldots, k$, $j = 1, \ldots, k$, and $i \neq j$. Thus, the mean and variance of X_i are $\mu = p$ and $\sigma^2 = pq$, where $p = m/n$ and $q = 1 - p$. Moreover, by (5.7) the covariance between X_i and X_j is $C(X_i, X_j) = E(X_i X_j) - E(X_i)E(X_j) = -pq/(n-1)$. Therefore, by Theorem 8.5.2, the mean and variance of S are

$$E(S) = kp \quad \text{and} \quad D(S) = kpq \frac{n-k}{n-1} \tag{5.9}$$

respectively. Since S has the hypergeometric distribution, and since the mean and variance of a random variable depend only on its distribution function, we have shown that the mean and variance of the hypergeometric distribution are given by (5.9).

/////

8.6 EXAMPLES[1]

EXAMPLE 8.6.1 Stratified sampling Consider a population which consists of t strata, $t \geq 2$, as in Example 3.3.6. We wish to conduct an opinion poll to learn, for example, what proportion of the population favors a particular political candidate or issue. Let n_i denote the size of the ith stratum, and let m_i denote the number of people in the ith stratum who favor the candidate or issue in question. Then the total population size is $n = n_1 + \cdots + n_t$, of which $m = m_1 + \cdots + m_t$ favor the candidate or issue. Further, let $\alpha_i = n_i/n$ be the proportion of the population in the ith stratum, and let $p_i = m_i/n_i$ be the proportion of those individuals in the ith

---

[1] This section treats a special topic and may be omitted without loss of continuity.

stratum who favor the candidate or issue. Then the proportion of the population who favor the candidate or issue is

$$p = \alpha_1 p_1 + \alpha_2 p_2 + \cdots + \alpha_t p_t$$

Of course, $\alpha_1 + \cdots + \alpha_t = 1$.

Suppose now that $\alpha_1, \ldots, \alpha_t$ are known to us, that $p_1, \ldots, p_t$ and p are unknown, and that we wish to conduct an opinion poll in order to learn about them. Suppose also that we have enough resources to take a sample of size k from the population.

Two possibilities present themselves. We might take a *simple random sample*. That is, we might take a random sample from the entire population. If we do so, and if we denote by X the number of people in the sample who favor the candidate or issue, then we might estimate p by

$$\hat{p} = \frac{1}{k} X$$

the proportion of the sample who favor the candidate or issue. By Example 8.5.6, the mean and variance of $\hat{p}$ are then

$$E(\hat{p}) = p \quad \text{and} \quad D(\hat{p}) = \frac{1}{k} pq \frac{n-k}{n-1} \tag{6.1}$$

where $q = 1 - p$.

Another possibility is to take a *stratified sample*. That is, we divide the sample size k into groups of sizes $k_1, \ldots, k_t$, where $k_1 + k_2 + \cdots + k_t = k$, and we take a random sample of size k_i from the ith stratum for $i = 1, \ldots, t$. If we let X_i denote the number of people in the sample from the ith stratum who favor the candidate or issue, then we might estimate p_i by $\hat{p}_i = X_i/k_i$ and p by

$$\hat{\hat{p}} = \alpha_1 \hat{p}_1 + \cdots + \alpha_t \hat{p}_t$$

The mean and variance of $\hat{\hat{p}}$ are then $E(\hat{\hat{p}}) = \alpha_1 E(\hat{p}_1) + \cdots + \alpha_t E(\hat{p}_t) = \alpha_1 p_1 + \cdots + \alpha_t p_t = p$ and

$$D(\hat{\hat{p}}) = \sum_{i=1}^{t} \frac{1}{k_i} \alpha_i^2 p_i q_i \frac{n_i - k_i}{n_i - 1} \tag{6.2}$$

by Example 8.5.6, Lemma 8.3.3, and Theorem 8.3.1.

How might we compare these two sampling schemes? Which of the two is better?

The answer to the first of these questions is quite simple. Since both $\hat{p}$ and $\hat{\hat{p}}$ have mean p, and since the variance measures the tendency of a random variable to deviate from its mean, it seems reasonable to compare them on the basis of the variances. That is, *one estimate will be judged better than the other if and only if it has a smaller variance.*

The second question is more subtle, but it can be answered as follows. If the sample sizes $k_1, \ldots, k_t$ are chosen to be proportional to strata sizes $n_1, \ldots, n_t$, then stratified sampling is better than simple random sampling. To see why, we shall make the simplifying assumption that n is large when compared to k and that n_i is large when compared to k_i, so that factors $(n - k)/(n - 1)$ and $(n_i - k_i)/(n_i - 1)$ may be neglected in (6.1) and (6.2). We find that

$$D(\hat{p}) \approx \frac{1}{k} pq \quad \text{and} \quad D(\hat{p}) \approx \sum_{i=1}^{t} \frac{1}{k_i} \alpha_i^2 p_i q_i$$

Let us now select the sample sizes k_i to be proportional to the strata sizes. That is, let $k_i = \alpha_i k$ for $i = 1, \ldots, t$. Then we have

$$D(\hat{p}) \approx \frac{1}{k} \sum_{i=1}^{t} \alpha_i p_i q_i$$

Simple algebra now shows that

$$\frac{1}{k} \sum_{i=1}^{t} \alpha_i p_i q_i = \frac{1}{k} pq - \frac{1}{k} \sum_{i=1}^{t} \alpha_i (p_i - p)^2$$

This is strictly less than $(1/k)pq \approx D(\hat{p})$ unless $p_1 = p_2 = \cdots = p_t$. Thus, we have $D(\hat{p}) < D(\hat{p})$ unless $p_1 = \cdots p_t$, in which case $D(\hat{p}) \approx D(\hat{p})$.   ////

EXAMPLE 8.6.2 An inventory problem A merchant has to stock an amount of goods z in order to meet a random demand X. The merchant buys his goods at a fixed cost c and sells them at a fixed price $p > c$. How much should he order if he wishes to maximize his expected profit?

We shall suppose, for simplicity, that the demand X has an absolutely continuous distribution function F with density f, which is positive on $(0, \infty)$. We also suppose that the demand is nonnegative, so that $F(x) = 0$ for $x \leq 0$. If the merchant orders z units of inventory, his profit is

$$Y = p \min(X, z) - cz$$

Thus, his expected profit is

$$E(Y) = \int_0^z pxf(x)\,dx + pz[1 - F(z)] - cz$$

Here we have used Theorem 8.2.1 to compute the expectation of $\min(X, z)$. If we now differentiate $E(Y)$ with respect to z, we find

$$\frac{d}{dz} E(Y) = p[1 - F(z)] - c$$

which is zero if and only if $F(z) = (p - c)/p$. Moreover, since the second derivative of $E(Y)$ is $-pf(z)$, which is negative, we see that the extremum is a maximum. Thus, the merchant should order z_0 units, where z_0 is the solution to the equation

$$F(z) = \frac{p - c}{p}$$

In the special case that F is the exponential distribution with parameter β, we find $z_0 = (\log p - \log c)/\beta$.   ////

REFERENCES

If X is any random variable, then the *characteristic function* of X is defined by $\phi(t) = E(e^{itX})$ for $-\infty < t < \infty$, where i denotes $\sqrt{-1}$. Characteristic functions have properties similar to those of moment-generating functions and have the advantage that every random variable has a characteristic function. On the other hand, they are more complicated since complex exponentials are involved. To readers interested in characteristic functions, we recommend Parzen (1960), chap. 9, for an elementary account and Feller (1966), chap. 15, for a more detailed treatment.

If X is a nonnegative random variable, then $M(-t) = E(e^{-tX})$, $t > 0$, is known as the *Laplace transform* of X. Laplace transforms have several interesting additional properties which are discussed in Feller (1966), chap. 13.

For a more complete account of stratified sampling and other interesting sampling schemes, see Cochran (1963).

Readers interested in inventory models (Example 8.8.1) should consult Arrow, Karlin, and Scarf (1958, 1962).

PROBLEMS

8.1 Let X have the Rayleigh distribution [with density $f(x) = 2\beta x \exp(-\beta x^2)$, $x > 0$, and $f(x) = 0$ for $x \leq 0$]. Find $E(X)$.

8.2 Let X have the bilateral exponential distribution [with density $f(x) = (\beta/2)e^{-\beta|x|}$, $x \in R$]. Find $E(X)$.

8.3 Let two balanced dice be rolled, and let X denote the total number of spots which appear. Find $E(X)$.

8.4 Two balanced dice are rolled. If X denotes the maximum number of spots which appear, find $E(X)$.

8.5 Would you be willing to play the following game repeatedly? Two balanced dice are rolled. You win 3 dollars if 2, 7, or 12 spots appear and lose 1 dollar otherwise.

8.6 Consider the following game. A fair coin is tossed until a head appears. If a head appears on the first toss, you lose 1 dollar. Otherwise, you win n dollars, where n is the number of tosses required to get a head. What is your expected gain?

8.7 Let $X_1, \ldots, X_n$ be independent random variables which are uniformly distributed over (0,1). Find the expected value of min $(X_1, \ldots, X_n)$ and max $(X_1, \ldots, X_n)$.

8.8 Let $X_1, \ldots, X_n$ be as in Problem 8.7. Find the expected value of Y_k, the kth smallest of $X_1, \ldots, X_n$.

8.9 Let X be any bounded random variable, that is, any random variable for which $\Pr(a < X \le b) = 1$ for some constants a and b. For $n = 1, 2, \ldots$, define a random variable X_n by

$$X_n = k2^{-n} \qquad \text{if } k - 1 < X2^n \le k$$

Show that lim $E(X_n) = E(X)$ as $n \to \infty$. *Hint:* This follows easily from the definition of the Riemann-Stieltjes integral.

8.10 Let X be any random variable for which $E(X)$ is defined, and for every integer $n = 1, 2, \ldots$ let $X_n = X$ if $|X| \le n$ and $X_n = 0$ if $|X| > n$. Show that lim $E(X_n) = E(X)$. *Hint:* Again, this follows easily from the definition of the Riemann-Stieltjes integral.

8.11 Let X have distribution function F, where $F(x) = 0$ for $x < 0$ and $F(x) = 1 - (1 + x)^{-2}$ for $x \ge 0$. Find $E(X)$.

8.12 Let X have the uniform distribution on (0,1). Find the expectations of cos $2\pi X$ and cos $(2\pi X)^2$.

8.13 Let X have the gamma distribution with parameter $\alpha = 2$ and $\beta = 1$. Find the expectation of $1/X$. Compare your answer with $1/E(X)$.

8.14 Let X be a random variable, and let D be an interval for which $\Pr(X \in D) = 1$. Let X have density f which is continuous on D, and let w be a continuously differentiable function for which $w'(x) \ne 0$ for any $x \in D$. Let $Y = w(X)$. Show directly that

$$E(Y) = \int_D w(x)f(x)\,dx$$

8.15 Prove Theorem 8.2.2 in the special case that X and Y have a joint density.

8.16 If k balls are placed in n cells according to Maxwell-Boltzmann statistics (every ball is equally likely to go into every cell), find:

(a) The expected number of balls in the first cell.

(b) The expected number of empty cells.

8.17 Repeat Problem 8.16 for Bose-Einstein statistics (see Section 1.6).

8.18 Derive Schwarz's inequality. If X and Y are any random variables for which $E(X^2)$ and $E(Y^2)$ are both finite, then $E(XY)^2 \le E(X^2)E(Y^2)$. *Hint:* $E[(X - tY)^2]$ is nonnegative for all $t \in R$. Find its minimum by differentiation.

8.19 Derive *Minkowski's inequality*. If X and Y are random variables for which $E(X^2) < \infty$ and $E(Y^2) < \infty$, then

$$E[(X + Y)^2]^{\frac{1}{2}} \le E(X^2)^{\frac{1}{2}} + E(Y^2)^{\frac{1}{2}}$$

Hint: Use Schwarz's inequality.

8.20 If (X, Y) has the uniform distribution over the unit square in R^2, find the expectation of XY. *Hint:* X and Y are independent.

8.21 N radioactive particles decay independently. If the time until decay of the ith particle has the exponential distribution with parameter α, find the expectation of the time at which the first decay is observed. *Hint:* See Section 7.6.

8.22 In Problem 8.21 find the expectation of the time at which the kth decay is observed.

8.23 In the notation of Problem 8.21, find the expected number of particles which decay in the time interval $(0,t)$.

8.24 Find the mean and variance of X when X has the bilateral exponential distribution [with density $f(x) = (\beta/2)e^{-\beta|x|}$, $-\infty < x < \infty$].

8.25 Find the mean and variance of X when X has Rayleigh distribution [with density $f(x) = \beta x \exp(-\beta x^2/2)$, $x > 0$].

8.26 Find the mean and variance of X when X has the Pareto distribution [with density $Cx^{\alpha-1}/(1 + x)^{\alpha+\beta}$, $x > 0$, where $C = \Gamma(\alpha + \beta)/\Gamma(\alpha)\Gamma(\beta)$].

8.27 Let $X_1,\ldots,X_n$ be independent random variables which are uniformly distributed over $(0,1)$. Let $Y = \max(X_1,\ldots,X_n)$. Find the mean and variance of Y.

8.28 Let $X_1,\ldots,X_n$ be as in Problem 8.27. Let $Y_1 = \min(X_1,\ldots,X_n)$, and let $Y_2 = \max(X_1,\ldots,X_n)$. Find the mean and variance of $Y_2 - Y_1$.

8.29 Let $X_1,\ldots,X_n$ be independent random variables with common mean μ and common variance σ^2. Find the mean and variance of $\bar{X} = (X_1 + \cdots + X_n)/n$.

8.30 Let $X_1,\ldots,X_n$ be as in Problem 8.29, and let

$$(n - 1)S^2 = \sum_{i=1}^{n} (X_i - \bar{X})^2$$

Find $E(S^2)$ in terms of μ and σ^2.

8.31◄ Let X and Y be independent random variables with finite positive variances σ^2 and τ^2, respectively. Find the number α which minimizes the variance of $Z = \alpha X + (1 - \alpha)Y$.

8.32 Let $X_1,\ldots,X_n$ be independent random variables with finite positive variances $\sigma_1^2,\ldots,$ σ_n^2, respectively. Find $\alpha_1,\ldots,\alpha_n$ which minimize the variance of $\alpha_1 X_1 + \cdots + \alpha_n X_n$, subject to the constraint that $\alpha_1 + \cdots + \alpha_n = 1$.

8.33 Let X be a positive random variable with finite mean μ and variance σ^2. Then, the ratio μ/σ is known as the *signal-to-noise ratio*. Find the signal-to-noise ratio for the (*a*) gamma; (*b*) Rayleigh; (*c*) Pareto; and (*d*) Poisson distributions. Your answer will involve the parameters of these distributions.

8.34◄ If X has the exponential distribution with variance $\sigma^2 = 1$, find $\Pr(X \le 1)$. *Hint:* Find β.

8.35 Consider three scales. If an object is weighed on any of the three scales, the result is a normally distributed random variable whose mean is the true weight of the object. The variances of the three machines are different. In fact, they are $\sigma_1^2 = 1$, $\sigma_2^2 = 2$, and $\sigma_3^2 = 3$. Would you obtain a more accurate estimate of the true weight of an object by weighing it on scale 1 or by weighing it on both scales 2 and 3 and using the average? (Assume the errors committed by the three scales to be independent.)

8.36 Consider two brands of light bulb. Brand A burns a normally distributed length of time with mean $\mu_A = 100$ hours and standard deviation $\sigma_A = 1$ hour. Brand B burns for a normally distributed length of time with mean $\mu_B = 102$ hours and standard deviation $\sigma_B = 10$ hours. Which brand has the larger probability of burning for more than 90 hours?

8.37 Let X be a random variable with finite expectation. Show that $E(|X - a|)$ is minimum when a is a median of X.

8.38 Suppose that mass is distributed over the interval $I = (0,1]$ according to a mass distribution F. That is, suppose that the amount of mass in the interval $(a,b]$ is $F(b) - F(a)$ for $0 \le a \le b \le 1$, where F is a nondecreasing, right-continuous function. The center of gravity is defined to be that number a which minimizes $\int_0^1 (x - a)^2 \, dF(x)$. Derive an expression for the *center of gravity*. Comment on any analogies with probability theory.

8.39 In Problem 8.38 let m denote the center of gravity. Then $\sigma^2 = \int_0^1 (x - m)^2 \, dF(x)$ is called the *moment of inertia*. Show that

$$\sigma^2 = \int_0^1 x^2 \, dF(x) - \left[\int_0^1 x \, dF(x) \right]^2 [F(1) - F(0)]^{-1}$$

8.40 If X is a random variable with moments $\mu_1, \mu_2, \ldots$, we define the *central moments* of X by $v_k = E[(X - \mu)^k]$, $k = 1, 2, \ldots$, where $\mu = \mu_1$ is the mean of X. Express the central moments as linear combinations of the ordinary moments $\mu_1, \mu_2, \ldots$.

8.41 The *skewness* and *kurtosis* of a random variable X are defined to be

$$s = v_3 \sigma^{-3} \quad \text{and} \quad k = v_4 \sigma^{-4} - 3$$

where v_k denote the central moments of X and $\sigma^2 = v_2$ denotes the variance of X. Derive the following two properties of skewness and kurtosis:
(a) If $Y = aX + b$, where a and b are constants, then Y has the same skewness and kurtosis as X.
(b) If X has a normal distribution then $s = 0 = k$.
Skewness and kurtosis may be thought of as measures of how much the distribution of X deviates from normality.

8.42 Compute the skewness and kurtosis of X when X has each of the following distributions:
(a) Binomial with parameters n and p.
(b) Poisson with parameter $\beta > 0$.
(c) Uniform on the interval (a,b) with $a < b$.
(d) Beta with parameters α and β.
(e) Exponential with parameter β.

8.43 Express the moments of the standard normal distribution in terms of the gamma function.

8.44 Let X have the bilateral exponential distribution [with density $f(x) = (\beta/2)e^{-\beta|x|}$, $-\infty < x < \infty$].
(a) Find the moment-generating function of X.
(b) Find the first four moments.

8.45 If X has moment-generating function $M(t) = (\sinh t)/t$ for $t \ne 0$ and $M(0) = 1$, what is the distribution of X?

8.46 If X has moment-generating function $M(t) = \cosh t$ for $-\infty < t < \infty$, what is the distribution of X?

8.47 If X has moment-generating function $M(t) = \exp(t + t^2)$, what is the distribution of X?

8.48 Prove assertions (*iii*) and (*iv*) of Theorem 8.4.4.

8.49 Let $X_1, \ldots, X_n$ be independent, geometrically distributed random variables with distinct parameters $p_1, \ldots, p_n$. Find the mass function of $S = X_1 + \cdots + X_n$.

8.50 Let X_1 and X_2 be independent, exponentially distributed random variables with distinct parameters β_1 and β_2. Find a density for $S = X_1 + X_2$.

8.51 Let $X_1, \ldots, X_n$ be independent, exponentially distributed random variables with distinct parameters $\beta_1, \ldots, \beta_n$. Find a density for $S = X_1 + \cdots + X_n$.

8.52 Let X have generating function A. Show that

$$E[(X)_k] = A^{(k)}(1-)$$

finite or infinite for $k = 1, 2, \ldots$. The numbers $f_k = E[(X)_k]$ are called the *factorial moments* of X.

8.53 Let X be an integer-valued random variable, and suppose that the factorial moments f_1, f_2, and f_3 are finite. Show that the usual moments μ_1, μ_2, and μ_3 are given by $\mu_1 = f_1$, $\mu_2 = f_2 + f_1$, and $\mu_3 = f_3 + 3f_2 + f_1$.

8.54 Find the factorial moments for (*a*) the binomial, (*b*) geometric, and (*c*) Poisson distribution.

8.55 Use the results of Problems 8.53 and 8.54 to find the mean and variance of (*a*) the binomial, (*b*) geometric, and (*c*) Poisson distribution.

8.56 Evaluate $\sum_{k=j}^{n} (k)_j$ for arbitrary n and j.

8.57 Let U have the uniform distribution on $(0,1)$, and let $X = \sin 2\pi U$ and $Y = \cos 2\pi U$. Show that var $(X + Y) = $ var $X + $ var Y. Are X and Y independent?

8.58 Let X and Y have joint density $f(x,y) = 120xy(1 - x - y)$ for $x > 0$, $y > 0$, and $x + y \leq 1$. Find the correlation between X and Y.

8.59 More generally, let X and Y have joint density $f(x,y) = Cx^{\alpha-1}y^{\beta-1}(1 - x - y)^{\gamma-1}$ for $x > 0$, $y > 0$, and $x + y \leq 1$, where $C = \Gamma(\alpha + \beta + \gamma)/\Gamma(\alpha)\Gamma(\beta)\Gamma(\gamma)$. Show that the correlation between X and Y is $-\sqrt{\alpha\beta/(\alpha + \gamma)(\beta + \gamma)}$.

8.60 Prove Equation (5.7).

8.61 Prove Lemma 8.5.1.

8.62 Let X and Y be independent random variables with mean μ and ν and variances σ^2 and τ^2, respectively. Let $W = X + Y$ and $Z = X - Y$. Find the correlation between X and Z.

8.63 Let $X_1, \ldots, X_n$ have the multivariate hypergeometric distribution (Example 6.3.1*a*). Find the covariance and correlation between X_i and X_j for $i \neq j$.

8.64 Let $X_1, \ldots, X_k$ have the multinomial distribution with parameter n and $p = (p_1, \ldots, p_k)$, as in Example 6.3.1*b*. Show that the covariance between X_i and X_j is $-p_i p_j$ for $i \neq j$.

8.65 If $A_1, \ldots, A_n$ are events with union $A = A_1 \cup \cdots \cup A_n$, then $1 - I_A = \prod_{k=1}^{n}(1 - I_{A_k})$.

8.66 Use the result of Problem 8.65 to give an independent proof of Theorem 2.4.1. *Hint:* $P(A) = E(I_A)$.

9

LIMIT THEOREMS

9.1 SOME USEFUL INEQUALITIES

There are several important inequalities which relate expectations and probabilities. Many are variations on the following basic inequality, known as *Markov's inequality*.

Theorem 9.1.1 *Let X be any random variable, and let ε and r be any positive real numbers. Then*

$$\Pr\left(|X| \geq \varepsilon\right) \leq \varepsilon^{-r}E(|X|^r) \qquad (1.1)$$

    PROOF   Let A denote the event that $|X| \geq \varepsilon$, and let I_A denote the indicator function of A. That is, let $I_A = 1$ if A occurs, and let $I_A = 0$ otherwise. Then $\varepsilon^r I_A \leq |X|^r$, since $|X| \geq \varepsilon$ if $I_A = 1$ and $I_A = 0$ otherwise. Thus, $E(|X|^r) \geq E(\varepsilon^r I_A) = \varepsilon^r E(I_A) = \varepsilon^r P(A) = \varepsilon^r \Pr\left(|X| \geq \varepsilon\right)$, as asserted.   ////

As a special case of Markov's inequality, we obtain *Chebyshev's inequality*.

Corollary 9.1.1 *Let X be a random variable with mean μ and variance σ^2. Then for every $\varepsilon > 0$*

$$\Pr\left(|X - \mu| \geq \varepsilon\right) \leq \frac{\sigma^2}{\varepsilon^2} \qquad (1.2)$$

PROOF   We apply Markov's inequality to $Y = X - \mu$ with $r = 2$ and find that $\Pr(|X - \mu| \geq \varepsilon) \leq \varepsilon^{-2} E\{|X - \mu|^2\} = \varepsilon^{-2}\sigma^2$.   ////

EXAMPLE 9.1.1

a Let X denote the number of heads which result from n independent tosses of a coin which has probability p of turning up heads on each toss, so that X has the binomial distribution with parameters n and p. Let $Y = X/n$ be the relative frequency of heads. Then

$$\Pr(|Y - p| \geq 0.1) = \Pr(|X - np| \geq 0.1n) \leq \frac{npq}{0.01n^2} = \frac{100pq}{n} \qquad (1.3)$$

since the mean and variance of X are np and npq, respectively. Moreover, since $pq = p(1 - p) \leq \frac{1}{4}$, $0 \leq p \leq 1$, (1.3) may be further bounded by $25/n$. Thus, if $n \geq 1000$, the probability that Y differs from p by more than 0.1 is at most 0.025 for every possible p.

b If X has the standard normal distribution, then $\Pr(|X| \geq 2) = 1 - \Pr(-2 < X < 2) = 0.046 \leq \frac{1}{20}$ by Example 5.6.1*a*. Chebyshev's inequality yields only that $\Pr(|X| \geq 2) \leq \frac{1}{4}$, however.   ////

As this example indicates, Chebyshev's inequality may badly overestimate $\Pr(|X - \mu| \geq \varepsilon)$. In fact, it does badly overestimate $\Pr(|X - \mu| \geq \varepsilon)$ for most distributions, although there are situations in which it is sharp (Problem 9.4). *The virtue of Chebyshev's inequality is its generality, not its sharpness.* Chebyshev's inequality is valid for any random variable with a finite variance, whereas the sharper estimate $\Pr(|X| \geq 2) \leq \frac{1}{20}$ of Example 9.1.1*b* depends on X having the standard normal distribution.

Chebyshev's inequality supports the interpretation of the variance as a measure of the tendency of a random variable to deviate from its mean (Section 8.3). Indeed, Chebyshev's inequality provides a bound on the probability that X deviates from its mean by more than ε in terms of ε and σ^2, the variance of X.

In the extreme case that $\sigma^2 = 0$ we have the following corollary.

Corollary 9.1.2   *Let X be a random variable with variance $\sigma^2 = 0$. Then, $\Pr(X = \mu) = 1$, where $\mu = E(X)$.*

PROOF   If $\sigma^2 = 0$, then $\Pr(|X - \mu| \geq \varepsilon) = 0$ for every $\varepsilon > 0$ by Chebyshev's inequality. Letting $\varepsilon \to 0$, we have then $\Pr(|X - \mu| > 0) = 0$, as asserted.   ////

Another useful variation on Markov's inequality, *Bernstein's inequality*, may be stated as follows.

Theorem 9.1.2 *Let S be any random variable with a moment-generating function M. Then for any s and any $t > 0$, we have*

$$\Pr (S > s) \leq e^{-st} M(t) \qquad (1.4)$$

PROOF   We apply Markov's inequality to the random variable $Y = e^S$ with $r = t$ and $\varepsilon = e^s$. We have

$$\Pr (S \geq s) = \Pr (e^S \geq e^s) \leq e^{-st} E(e^{tS}) = e^{-st} M(t)$$

as asserted.   ////

For a given s, the inequality (1.4) is valid for all $t > 0$. It seems natural to use the t which minimizes the right side of (1.4), but unfortunately this t can seldom be found explicitly. However, if $E(S) = 0$, it can be found approximately in the following manner. Let $m(t) = \log M(t)$, and recall that $m'(0) = E(S) = 0$ and $m''(0) = \sigma^2 = D(S)$. Thus, by Taylor's theorem, we can approximate $m(t)$ by $m'(0)t + \frac{1}{2} m''(0)t^2 = \frac{1}{2}\sigma^2 t^2$ for small t. It follows that we can approximate $M(t) = \exp m(t)$ by $\exp \frac{1}{2}\sigma^2 t^2$ and $e^{-st}M(t)$ by $\exp(-st + \frac{1}{2}\sigma^2 t^2)$. The minimum value of $\exp(\frac{1}{2}\sigma^2 t^2 - st)$ is easily seen to occur when $t = t_0$, where

$$t_0 = \frac{s}{\sigma^2} \qquad (1.5)$$

Thus, t_0 seems a reasonable choice of t in Equation (1.4) and yields the inequality

$$P(S \geq s) \leq e^{-s^2/\sigma^2} M\left(\frac{s}{\sigma^2}\right)$$

In the special case that S is the sum of independent random variables, Bernstein's inequality yields surprisingly low bounds.

Corollary 9.1.3 *Let $X_1, \ldots, X_n$ be independent random variables with common moment-generating function M_0, and let $S = X_1 + \cdots + X_n$. Then*

$$\Pr (S \geq ns) \leq e^{-nst} M_0(t)^n$$

for all $s > 0$ and all $t > 0$.

PROOF   The moment-generating function of S is $M(t) = M_0(t)^n$ by Theorem 8.4.3, so that

$$\Pr (S \geq ns) \leq e^{-nst} M_0(t)^n$$

for $s > 0$ and $t > 0$, as asserted.   ////

EXAMPLE 9.1.2   Let $X_1, \ldots, X_n$ be independent random variables with the common distribution

$$\Pr (X_i = \pm 1) = \frac{1}{2}$$

Table 12

			n	
ε	10	25	50	100
0.1	0.9512	0.8823	0.7785	0.6060
0.2	0.8177	0.6045	0.3655	0.1336
0.5	0.2728	0.0389	0.0015	

and let $S = X_1 + \cdots + X_n$. The moment-generating function of X_1 is

$$E(e^{tX_1}) = \tfrac{1}{2}e^t + \tfrac{1}{2}e^{-t} = \cosh t$$

for $-\infty < t < \infty$, so that the moment-generating function of S is $M(t) = (\cosh t)^n$ (Theorem 8.4.3). Also, the mean and variance of X_1 are 0 and 1, respectively, so that the mean and variance of S are 0 and n. Letting $s = n\varepsilon$, we now find that $t_0 = n\varepsilon/n = \varepsilon$, and therefore

$$\Pr(S \geq n\varepsilon) \leq (e^{-\varepsilon^2} \cosh \varepsilon)^n \qquad (1.6)$$

for $\varepsilon > 0$. The right side of (1.6) is given in Table 12 for several values of n and ε.

/////

9.2 THE WEAK LAW OF LARGE NUMBERS

In Section 8.1 we indicated that the mean μ of a random variable X may be regarded as the long-run empirical average of X over many repetitions of the experiment to which X refers. In this section we shall prove two theorems which support this interpretation of the mean. We shall consider independent random variables $X_1, \ldots, X_n$ with a common mean μ, and we shall think of $X_1, \ldots, X_n$ as the outcomes of n repetitions of an experiment. We shall show that the empirical average

$$\overline{X}_n = \frac{1}{n}(X_1 + \cdots + X_n)$$

converges to μ as $n \to \infty$, in a sense to be defined below.

We begin by remarking that if $X_1, \ldots, X_n$ are independent random variables with both a common mean μ and a common variance σ^2, then the mean and variance of the sum $S_n = X_1 + \cdots + X_n$ are

$$E(S_n) = n\mu \qquad \text{and} \qquad D(S_n) = n\sigma^2$$

by Theorem 8.3.1. Hence the mean and variance of $\overline{X}_n = S_n/n$ are

$$E(\overline{X}_n) = \frac{1}{n} E(S_n) = \mu \quad \text{and} \quad D(\overline{X}_n) = \left(\frac{1}{n}\right)^2 D(S_n) = \frac{\sigma^2}{n} \quad (2.1)$$

by Lemma 8.3.3. The point to be observed is that the variance of $\overline{X}_n$ is substantially less than that of each of the individual X_i if n is large. If we recall that the variance is a measure of the tendency of a random variable to deviate from its mean, then we see that the average $\overline{X}_n$ tends to deviate much less from μ than the X_i do. This simple remark is the basis for the results of this section.

We say that a sequence of random variables $Y_1, Y_2, \ldots$ *converges in probability* to another random variable Y as $n \to \infty$ if and only if

$$\lim_{n \to \infty} \Pr \left(|Y_n - Y| \geq \varepsilon \right) = 0 \quad (2.2)$$

for every $\varepsilon > 0$. Of course, (2.2) is equivalent to

$$\lim_{n \to \infty} \Pr \left(|Y_n - Y| < \varepsilon \right) = 1 \quad (2.2a)$$

for every $\varepsilon > 0$. That is, Y_n converges to Y in probability as $n \to \infty$ if and only if Y_n is arbitrarily close to Y with arbitrarily high probability for sufficiently large n.

Most use of this terminology will be in cases where Y is a constant.

EXAMPLE 9.2.1 Let $X_1, \ldots, X_n$ be independently random variables which are uniformly distributed over $(0,1)$, and let $Y_n = \max (X_1, \ldots, X_n)$ for $n = 1, 2, \ldots$. Then, $Y_n \to 1$ in probability as $n \to \infty$. Indeed, $\Pr (Y_n > 1) = 0$, and for $0 < \varepsilon < 1$

$$\Pr (Y_n \leq 1 - \varepsilon) = (1 - \varepsilon)^n$$

which tends to zero as $n \to \infty$ (see Example 7.2.1a). ////

We shall now state two theorems which assert the convergence of $\overline{X}_n$ to μ in probability.

Theorem 9.2.1 *Let $X_1, \ldots, X_n$ be independent random variables which have a common mean μ and a common (finite) variance σ^2, and let*

$$\overline{X}_n = \frac{X_1 + \cdots + X_n}{n} \quad n \geq 1$$

Then $\overline{X}_n \to \mu$ in probability as $n \to \infty$.

PROOF   Theorem 9.2.1 follows easily from Equation (2.1) and Chebyshev's inequality. Indeed, for $\varepsilon > 0$

$$\Pr\left(|\overline{X}_n - \mu| \geq \varepsilon\right) \leq \frac{1}{\varepsilon^2} D(\overline{X}_n) = \frac{\sigma^2}{n\varepsilon^2}$$

which tends to zero as $n \to \infty$.   ////

While Theorem 9.2.1 is adequate for many applications, it does have a defect in that the variance σ^2 is assumed to be finite. By assuming that the random variables $X_1, \ldots, X_n$ have a common distribution function, we can eliminate the condition that $\sigma^2 < \infty$. The hypothesis that $X_1, \ldots, X_n$ have the same distribution function is certainly reasonable if we think of $X_1, \ldots, X_n$ as the outcomes of independent trials of the same experiment. We shall say that $X_1, \ldots, X_n$ are *identically distributed* if they have the same distribution function.

Theorem 9.2.2   *Let $X_1, \ldots, X_n$ be independent random variables which have a common distribution function F. If the mean*

$$\mu = \int_{-\infty}^{\infty} x \, dF(x) \qquad (2.3)$$

is finite, then $\overline{X}_n$ converges to μ in probability as $n \to \infty$.

Theorem 9.2.2 is known as the _weak law of large numbers_. We shall prove this theorem below, but first we discuss some of its implications.

First, it does make precise the interpretation given to the mean in Section 8.1 as the limiting value of $\overline{X}_n$. In particular, it supports the frequentistic interpretation of probability by predicting the type of behavior on which the frequentistic interpretation is founded. Indeed, if A is an event which may occur on each of n independent trials, then the relative frequency with which A does occur is simply $f_n(A) = \overline{X}_n$, where $X_i = 1$ if A occurs on the ith trial and $X_i = 0$ otherwise, $i = 1, \ldots, n$. Since $X_1, \ldots, X_n$ are independent with a common distribution, Theorem 9.2.2 asserts the convergence of $\overline{X}_n$ to the mean $\mu = E(X_i)$, which is simply $P(A)$. Since $E(X_1^2)$ is finite in this example, the convergence of $\overline{X}_n$ to $\mu = P(A)$ also follows from Theorem 9.2.1.

Since it is almost an axiom of probability theory that any really interesting theorem should have applications to gambling, let us see what Theorem 9.2.2 has to say about gambling. Consider a gambler who plays n repetitions of a fixed game, and let X_i denote his reward (possibly negative) on the ith game, $i = 1, \ldots, n$. Then his total reward is simply $S_n = X_1 + \cdots + X_n$. If we now suppose that $X_1, \ldots, X_n$ are independent with common distribution function, then Theorem 9.2.2 has the following implications.

If the expected reward $\mu = E(X_i)$ on each play is positive, then

$$\text{Pr } (S_n \geq \tfrac{1}{2}n\mu) \geq \text{Pr } (|\overline{X}_n - \mu| \leq \tfrac{1}{2}\mu)$$

which tends to 1 as $n \to \infty$ by Theorem 9.2.2. That is, by playing the game often enough (n large), the gambler will win an arbitrarily large amount (at least $n\mu/2$) with arbitrarily high probability. In this case, $\mu > 0$, we say that the game is *favorable*. Similarly, if $\mu < 0$, the gambler will lose an arbitrarily large amount with arbitrarily high probability by playing the game often enough and we say that the game is *unfavorable*. If $\mu = 0$, it is tempting to call the game *fair*, and we shall do so, although this terminology is somewhat questionable in cases where the variance is infinite.[1]

We now turn to the proof of Theorem 9.2.2. Let $X_1, \ldots, X_n$ be as described in its hypotheses, independent random variables with a common distribution function F and a finite mean μ. We must show that, given arbitrary $\varepsilon > 0$ and $\delta > 0$, the inequality

$$\text{Pr } (|\overline{X}_n - \mu| \geq \varepsilon) \leq \delta$$

holds for all sufficiently large n.

Let $\varepsilon > 0$ and $\delta > 0$ be given; let $\gamma = E(|X_1|)$, which is finite by assumption; let $\alpha = \delta\varepsilon^2/8\gamma$; and define $Y_1, \ldots, Y_n$ by

$$Y_k = \begin{cases} X_k & \text{if } -\alpha n \leq X_k \leq \alpha n \\ 0 & \text{otherwise} \end{cases}$$

Then $Y_1, \ldots, Y_n$ are independent with common mean

$$\mu_n = \int_{-\alpha n}^{\alpha n} y \, dF(y)$$

and variance

$$\sigma_n^2 = \int_{-\alpha n}^{\alpha n} y^2 \, dF(y) - \mu_n^2 \tag{2.4}$$

Now, as $n \to \infty$, $\alpha n \to \infty$, so that $\mu_n \to \mu$ by definition of the improper Riemann–Stieltjes integral in (2.3). Hence there is an n_0 for which $|\mu_n - \mu| \leq \tfrac{1}{2}\varepsilon$ for every $n \geq n_0$. Therefore, for $n \geq n_0$ we have

$$\text{Pr } (|\overline{X}_n - \mu| \geq \varepsilon) \leq \text{Pr } (|\overline{X}_n - \mu_n| \geq \tfrac{1}{2}\varepsilon)$$

$$\leq \text{Pr } (|\overline{Y}_n - \mu_n| \geq \tfrac{1}{2}\varepsilon) + \text{Pr } (\overline{X}_n \neq \overline{Y}_n) \tag{2.5}$$

Now, by Chebyshev's inequality, $\text{Pr } (|\overline{Y}_n - \mu_n| \geq \tfrac{1}{2}\varepsilon) \leq 4\sigma_n^2/n\varepsilon$. Moreover, by (2.4),

$$\sigma_n^2 \leq \int_{-\alpha n}^{\alpha n} y^2 \, dF(y) \leq \alpha n \int_{-\alpha n}^{\alpha n} |y| \, dF(y) \leq \alpha n \gamma$$

---

[1] See Feller (1968), p. 249.

where (we recall) $\gamma = E(|X_1|)$. Therefore, by definition of α, we have

$$\Pr\left(|\overline{Y}_n - \mu_n| \geq \tfrac{1}{2}\varepsilon\right) \leq 4\alpha\gamma\varepsilon^{-2} = \tfrac{1}{2}\delta \qquad (2.6)$$

for every $n = 1, 2, \ldots$. Moreover,

$$\Pr\left(\overline{X}_n \neq \overline{Y}_n\right) \leq \sum_{k=1}^{n} \Pr\left(X_k \neq Y_k\right)$$

$$= n \Pr\left(|X_1| \geq \alpha n\right)$$

$$\leq n \int_{|x| \geq \alpha n} dF(x) \leq \alpha^{-1} \int_{|x| \geq \alpha n} |x| \, dF(x)$$

which tends to zero as $n \to \infty$, since it is the tail of a convergent integral. Thus, there is an n_1 for which $\Pr\left(\overline{X}_n \neq \overline{Y}_n\right) \leq \tfrac{1}{2}\delta$ for every $n \geq n_1$. Combining this information with (2.5) and (2.6), we now see that $\Pr\left(|\overline{X}_n - \mu| \geq \varepsilon\right) \leq \delta$ whenever $n \geq \max(n_0, n_1)$.

$/////$

9.3   VARIATIONS OF THE WEAK LAW OF LARGE NUMBERS

There are several variations on the laws of large numbers given in the previous section, and we present three of them in this section. We begin by remarking that Theorem 9.2.1 did not really use the independence of the X_i very strongly. It only used the fact that they were uncorrelated, which follows from independence but is a much weaker condition (see Example 8.5.3). In fact, all that is really necessary is that the X_i be asymptotically uncorrelated in the sense that

$$\lim C(X_i, X_j) = 0 \qquad (3.1)$$

where the limit is taken as $|i - j| \to \infty$.

Theorem 9.3.1   *Let $X_1, \ldots, X_n$ be jointly distributed random variables with common mean μ and bounded variances $\sigma_1^2, \ldots, \sigma_n^2$, say $\sigma_i^2 \leq b$, $i = 1, \ldots, n$, where b is independent of n. If (3.1) holds, then $\overline{X}_n \to \mu$ in probability as $n \to \infty$.*

PROOF   By Chebyshev's inequality, we have

$$\Pr\left(|\overline{X}_n - \mu| \geq \varepsilon\right) \leq \frac{1}{\varepsilon^2} D(\overline{X}_n)$$

for all $\varepsilon > 0$, so that it will suffice to show that

$$\lim_{n \to \infty} D(\overline{X}_n) = 0 \qquad (3.2)$$

Now

$$D(\overline{X}_n) = \left(\frac{1}{n}\right)^2 \sum_{i=1}^{n} \sum_{j=1}^{n} C(X_i, X_j)$$

by Theorem 8.5.2. Given $\delta > 0$, there is by (3.1) an integer m for which $|C(X_i, X_j)| \leq \delta/2$ provided only that $|i - j| \geq m$. Moreover, by Schwarz's inequality (Section 8.5), we also have

$$|C(X_i, X_j)| \leq \sigma_i \sigma_j \leq b$$

For all i and j. Therefore, for $n \geq 2mb/\delta$,

$$D(\overline{X}_n) \leq \left(\frac{1}{n}\right)^2 \left[\sum_{|i-j| \leq m} b + \sum_{|i-j| > m} |C(X_i, X_j)| \right]$$

$$\leq \frac{nmb}{n^2} + \frac{n(n-m)\delta}{2n^2} \leq \delta$$

and (3.2) follows. ////

If we require that the X_i in Theorem 9.3.1 actually be uncorrelated, the assumption that their variances are bounded may be relaxed.

Theorem 9.3.2 *Let $X_1, \ldots, X_n$ be uncorrelated random variables with common mean μ and variances $\sigma_1^2, \ldots, \sigma_n^2$, respectively. If there are constants $\alpha > 0$ and $\beta < 1$ such that*

$$\sigma_k^2 \leq \alpha k^\beta \qquad k = 1, \ldots, n \qquad n \geq 1$$

then $\overline{X}_n$ converges to μ in probability as $n \to \infty$.

We leave the proof of Theorem 9.3.2 as an exercise (Problem 9.13). Let us now consider some examples.

EXAMPLE 9.3.1 Let Y be uniformly distributed over the interval $(-1, 1)$, and let $X_k = \sin \pi k Y$, $k = 1, 2, \ldots$. Then

$$E(X_k) = \frac{1}{2} \int_{-1}^{1} \sin \pi k y \, dy = 0$$

$$E(X_k^2) = \frac{1}{2} \int_{-1}^{1} \sin (\pi k y)^2 \, dy = \tfrac{1}{2}$$

and $\quad E(X_j X_k) = \frac{1}{4} \int_{-1}^{1} \{\cos [(j - k)\pi y] - \cos [(j + k)\pi y]\} \, dy = 0$

for $j \neq k$. Therefore, the X_i are uncorrelated with common mean $\mu = 0$ and common variance $\sigma^2 = \frac{1}{2}$. It now follows from either Theorem 9.3.1 or 9.3.2 that $\overline{X}_n$ converges to $\mu = 0$ in probability, even though the X_k are highly dependent.   ////

EXAMPLE 9.3.2 Let $X_0, \ldots, X_{n-1}$ be uncorrelated with common mean $E(X_i) = 0$ and variance $E(X_i^2) = \sigma^2$, and define

$$Y_k = \sum_{j=1}^{k} 2^{-j} X_{k-j}$$

for $k = 1, \ldots, n$. The sequence $Y_1, \ldots, Y_n$ is called a *moving average* of the sequence $X_0, \ldots, X_{n-1}$. Then

$$E(Y_k) = \sum_{j=1}^{k} 2^{-j} E(X_{k-j}) = 0$$

$$E(Y_k^2) = \sum_{j=1}^{k} 4^{-j} E(X_{k-j}^2) = \sum_{j=1}^{k} 4^{-j} \sigma^2 = \tfrac{1}{3}(1 - 4^{-k})\sigma^2$$

for $k = 1, \ldots, n$. Moreover, letting $m = \min(j,k)$ gives

$$C(Y_j, Y_k) = E(Y_j Y_k) = 2^{-|k-j|} \sum_{i=1}^{m} 4^{-i}$$

which tends to zero as $|k - j| \to \infty$. Therefore, Theorem 9.3.1 applies and asserts that $\overline{Y}_n$ converges to 0 in probability as $n \to \infty$.   ////

Another application of the weak law of large numbers consists in using probabilistic methods to prove a famous theorem from analysis, the *Weierstrass approximation theorem*, which asserts that any continuous function on the closed interval [0,1] can be approximated uniformly to any desired degree of accuracy by a polynomial.

Theorem 9.3.3 *Let g be any continuous function defined on the closed interval* [0,1]. *Then there is a sequence of polynomials* g_n, $n = 1, 2, \ldots$, *for which* $\lim g_n(p) = g(p)$ *uniformly in* $0 \le p \le 1$ *as* $n \to \infty$.

PROOF Let S_n have the binomial distribution with parameters n and p, let $\overline{X}_n = S_n/n$, and define g_n by

$$g_n(p) = E[g(\overline{X}_n)]$$

for $0 \le p \le 1$. Each g_n is a polynomial, since

$$E[g(\overline{X}_n)] = \sum_{k=0}^{n} g\left(\frac{k}{n}\right) \binom{n}{k} p^k (1 - p)^{n-k}$$

$$= \sum_{k=0}^{n} \sum_{j=0}^{n-k} g\left(\frac{k}{n}\right) \binom{n}{k} \binom{n-k}{j} (-1)^j p^{k+j}$$

for $0 \leq p \leq 1$ and $n = 1, 2, \ldots$. Thus we need only show that g_n converges to g uniformly as n tends to infinity, that is, for every $\varepsilon > 0$, there is an integer n_0 for which $|g_n(p) - g(p)| \leq \varepsilon$ for all $n \geq n_0$ and all p, $0 \leq p \leq 1$. Since g is continuous, g is bounded and uniformly continuous. Hence, there is a constant b for which $|g(p)| \leq b$, $0 \leq p \leq 1$, and given $\varepsilon > 0$, there is a constant δ for which $|g(p_1) - g(p_2)| \leq \varepsilon/2$ whenever $|p_1 - p_2| \leq \delta$. Moreover, we have

$$\Pr\left(|\overline{X}_n - p| \geq \delta\right) \leq \frac{pq}{n\delta^2} \leq \frac{1}{4n\delta^2}$$

for all $n = 1, 2, \ldots$ by Chebyshev's inequality. Let A_n be the event that $|\overline{X}_n - p| < \delta$, and let I_{A_n} be the indicator function of A_n, so that $I_{A_n} = 1$ or 0 according as $|\overline{X}_n - p| < \delta$ or $|\overline{X}_n - p| \geq \delta$. Then $I_{A_n}|g(\overline{X}_n) - g(p)| \leq \varepsilon/2$ by choice of δ, so that

$$
\begin{aligned}
|g_n(p) - g(p)| &= |E[g(\overline{X}_n) - g(p)]| \\
&\leq E[I_{A_n}|g(\overline{X}_n) - g(p)|] \\
&\quad + E[(1 - I_{A_n})|g(\overline{X}_n) - g(p)|] \\
&\leq \frac{\varepsilon}{2}\Pr\left(|\overline{X}_n - p| < \delta\right) + 2b\,\Pr\left(|\overline{X}_n - p| \geq \delta\right) \\
&\leq \frac{\varepsilon}{2} + \frac{2b}{4n\delta^2}
\end{aligned}
$$

which is less than ε provided that $n \geq b/\varepsilon\delta^2$.   ////

The polynomials g_n are known as *Bernstein polynomials*, after S. Bernstein, to whom this proof is due.

9.4 THE CENTRAL-LIMIT THEOREM

In this section we present a theorem, known as the *central-limit theorem*, which simultaneously provides a simple, effective approximation to probabilities determined by sums of independent random variables and explains the great importance of the normal distribution in probability theory. Its precise statement is the following: let $X_1, \ldots, X_n$ be independent, identically distributed random variables with mean μ and positive finite variance σ^2. Further, let $S_n = X_1 + \cdots + X_n$ and

$$S_n^* = \frac{S_n - n\mu}{\sigma\sqrt{n}}$$

S_n^* is the number of standard deviations by which S_n differs from its mean. The mean and standard deviation of S_n^* are 0 and 1, respectively, by Lemma 8.3.3.

Theorem 9.4.1 *Let $X_1, \ldots, X_n$ be independent random variables with common distribution function F, mean μ, and positive finite variance σ^2. Then*

$$\lim_{n \to \infty} \Pr\,(S_n^* \leq a) = \Phi(a) \qquad (4.1)$$

for all a, $-\infty < a < \infty$, where Φ denotes the standard normal distribution function

$$\Phi(a) = \int_{-\infty}^{a} \frac{e^{-\frac{1}{2}x^2}}{\sqrt{2\pi}}\,dx$$

That is, if we subtract the mean $n\mu = E(S_n)$ from S_n and divide the difference by the standard deviation $\sigma\sqrt{n} = \sqrt{D(S_n)}$, we obtain a random variable S_n^* whose distribution function $\Pr\,(S_n \leq a)$ is approximately the standard normal distribution function $\Phi(a)$. It follows that

$$\Pr\,(S_n \leq a) = \Pr\left(S_n^* \leq \frac{a - n\mu}{\sigma\sqrt{n}}\right)$$

is approximately $\Phi[(a - n\mu)/(\sigma\sqrt{n})]$. Accordingly, we shall say that the distribution function of S_n is approximately normal with mean $n\mu$ and variance $n\sigma^2$. Observe that *the approximation to the distribution function of S_n depends on the common distribution of $X_1, \ldots, X_n$ only through the common mean μ and the common variance σ^2.*

It is even possible to place a bound on the rate of convergence in (4.1). The result is known as the *Berry-Esseen theorem.*

Theorem 9.4.2 *If, in addition to the hypotheses of Theorem 9.4.1, $\gamma = E(|X_i^3|)$ is finite, then*

$$|\Pr\,(S_n^* \leq a) - \Phi(a)| \leq \frac{5\gamma}{\sqrt{n}\,\sigma^3}$$

for all a, $-\infty < a < \infty$, and $n = 1, 2, \ldots$.

We shall give a plausibility argument for the central-limit theorem at the end of this section. We shall not prove the Berry-Esseen theorem.

Before turning to examples, let us remark that (4.1) implies that

$$\lim_{n \to \infty} \Pr\,(a < S_n^* \leq b) = \Phi(b) - \Phi(a) \qquad (4.2)$$

for all $a < b$, since $\Pr\,(a < S_n^* \leq b) = \Pr\,(S_n^* \leq b) - \Pr\,(S_n^* \leq a)$.

EXAMPLE 9.4.1 The central-limit theorem contains the DeMoivre-Laplace integral limit theorem of Sections 4.5 and 4.6 as a special case. Indeed, if S_n has the binomial distribution with parameters n and p, then S_n has the same distribution as $X_1 + \cdots + X_n$, where X_i are independent with common distribution $\Pr(X_i = 1) = p$ and $\Pr(X_i = 0) = q = 1 - p$. Since these random variables are identically distributed with common mean $\mu = p$ and common variance $\sigma^2 = pq$, (4.2) asserts that

$$\lim_{n \to \infty} \Pr\left(\alpha < \frac{S_n - np}{\sqrt{npq}} \le \beta\right) = \Phi(\beta) - \Phi(\alpha) \qquad (4.3)$$

for all $\alpha < \beta$. Of course, (4.3) is simply the DeMoivre-Laplace integral limit theorem, of which practical applications were given in Section 4.5.   ////

EXAMPLE 9.4.2   Suppose that n numbers are rounded to the nearest integer and then added. How large a difference may we expect to find between the sum of the rounded numbers and the sum of the original numbers; that is, how large will the total error due to rounding be?

Let X_i denote the error introduced by rounding the ith number, $i = 1, \ldots, n$, and suppose that $X_1, \ldots, X_n$ are independent with a common uniform distribution on the interval $(-\frac{1}{2}, \frac{1}{2}]$. Then $X_1, \ldots, X_n$ are identically distributed with

$$\mu = \int_{-\frac{1}{2}}^{\frac{1}{2}} x \, dx = 0 \qquad \text{and} \qquad \sigma^2 = \int_{-\frac{1}{2}}^{\frac{1}{2}} x^2 \, dx = \tfrac{1}{12}$$

Since the total error is $S_n = X_1 + \cdots + X_n$, the central-limit theorem provides an answer to our question. For example, if $n = 12$, it asserts that $\Pr(-1 < S_n \le 1) = \Pr(-1 < S_n^* \le 1)$ is approximately $\Phi(1) - \Phi(-1) = 0.683$. More generally, it shows that

$$\Pr\left(-\alpha \sqrt{\frac{n}{12}} < S_n \le \alpha \sqrt{\frac{n}{12}}\right) = \Pr(-\alpha < S_n^* \le \alpha)$$

is approximately $\Phi(\alpha) - \Phi(-\alpha)$, which is nearly 1 for $\alpha \ge 3$. Therefore, the rounding error grows like $\sqrt{n}$ rather than n.   ////

The central-limit theorem may be viewed as a supplement to the weak law of large numbers. Indeed, in the notation of the previous section, we have

$$\frac{\sqrt{n}}{\sigma}(\overline{X}_n - \mu) = \frac{S_n - n\mu}{\sigma\sqrt{n}} = S_n^*$$

Thus, if $X_1, \ldots, X_n$ are independent and identically distributed with mean μ and variance σ^2, then $\overline{X}_n$ converges to μ at the rate $1/\sqrt{n}$ in the sense that $(\sqrt{n}/\sigma)(\overline{X}_n - \mu)$ has a distribution which approaches normality.[1]

---

[1] For a related result, see Section 11.8.

The central-limit theorem also has implications concerning the classification of games as fair in Section 9.2. Indeed, if $X_1, \ldots, X_n$ denote the rewards in n plays of a game, if the expected gain $\mu = E(X_i)$ on each play is $\mu = 0$, and if the variance $\sigma^2 = E(X_i)$ is finite, then for every $\alpha > 0$ we have, for large n,

$$\Pr\left(S_n > a\sqrt{n}\right) = \Pr\left(S_n^* > \frac{a}{\sigma}\right)$$

$$\approx 1 - \Phi\left(\frac{a}{\sigma}\right)$$

$$= \Phi\left(-\frac{a}{\sigma}\right) \approx \Pr\left(S_n \leq -a\sqrt{n}\right)$$

so that the probability of winning at least $a\sqrt{n}$ is approximately the same as that of losing at least $a\sqrt{n}$.

In addition to the applications cited above, the central-limit theorem has some important implications for model building. Indeed, it states that any random variable which is determined as the sum of a large number of independent, identically distributed random variables with finite variance will have approximately a normal distribution. Moreover, the requirement that the summands be identically distributed may be relaxed to the requirement that each contribute negligibly to the sum (Section 9.4.1). Many naturally occurring phenomena may be thought of in this manner, that is, as the sum of many independent deviations, each of which contributes negligibly. For example, such attributes as the heights and intelligence quotients of individuals are determined by many independent or nearly independent genetic and environmental factors each of which contributes only a small amount. Likewise, many measurement and production errors are the sum of many independent smaller errors.

Suppose now that we wish to build a model for some phenomenon of the above type. We shall probably wish to make some assumptions about the distribution of the phenomenon, and in view of the central-limit theorem, normality seems to be the natural assumption. For example, in the examples mentioned above, heights, intelligence quotients, and measurement and production errors are commonly assumed to follow normal distributions by people who work with them.

Let us now indicate the proof of the central-limit theorem. Consider n independent, identically distributed random variables $X_1, \ldots, X_n$ with common mean μ and variance σ^2, $0 < \sigma^2 < \infty$, and suppose also that $X_1, \ldots, X_n$ have a common moment-generating function M which is defined (finite) on some open interval $(-h,h)$ containing zero. We shall show that the moment-generating function of $S_n^* = (S_n - n\mu)/\sigma\sqrt{n}$ converges to the moment-generating function of the standard normal distribution, namely,

$$M_0(t) = e^{\frac{1}{2}t^2} \qquad -\infty < t < \infty$$

(Example 8.4.6). This is admittedly somewhat different from the conclusion of Theorem 9.4.1, which asserts that the distribution function of S_n^* converges to the standard normal distribution function, but it should at least render the conclusion of Theorem 9.4.1 highly plausible.

The moment-generating function of S_n is $M(t)^n$ by Theorem 8.4.3, and hence the moment-generating function of S_n^* is

$$M_n(t) = \exp\left(-\mu t \frac{\sqrt{n}}{\sigma}\right) M\left(\frac{t}{\sigma\sqrt{n}}\right)^n$$

for $|t| < h\sigma\sqrt{n}$ by Lemma 8.4.1. In terms of logarithms, we have

$$\log M_n(t) = n\psi\left(\frac{t}{\sigma\sqrt{n}}\right) - \mu t \frac{\sqrt{n}}{\sigma} \tag{4.4}$$

where $\psi = \log M$. Let us expand ψ in a Taylor series about $t = 0$, recalling that $\psi'(0) = \mu$ and $\psi''(0) = \sigma^2$ (Corollary 8.4.1). We have

$$\psi\left(\frac{t}{\sigma\sqrt{n}}\right) = \mu\left(\frac{t}{\sigma\sqrt{n}}\right) + \frac{\sigma^2}{2}\left(\frac{t}{\sigma\sqrt{n}}\right)^2 + \tfrac{1}{6}\psi'''(t_1)\left(\frac{t}{\sigma\sqrt{n}}\right)^3$$

where $|t_1| \le |t|/\sigma\sqrt{n}$. Substituting the Taylor series expansion into (4.4), we find that

$$\log M_n(t) = \frac{t^2}{2} + \frac{\tfrac{1}{6}\psi'''(t_1)t^3}{\sigma^3\sqrt{n}} \tag{4.5}$$

Now, the last term in (4.5) contains the factor $1/\sqrt{n}$ and therefore tends to zero as n tends to infinity, so that $\lim\left[\log M_n(t)\right] = t^2/2$ as $n \to \infty$ or, equivalently,

$$\lim_{n\to\infty} M_n(t) = \lim_{n\to\infty} \exp\left[\log M_n(t)\right] = e^{\frac{1}{2}t^2}$$

as asserted.

9.4.1 The Lindeberg-Feller Theorem[1]

There is a more general version of the central-limit theorem which allows the random variables $X_1, \ldots, X_n$ to have different distributions. It is known as the *Lindeberg-Feller theorem* and may be stated as follows.

[1] This section treats a special topic and may be omitted without loss of continuity.

Theorem 9.4.3 *Let $X_1, \ldots, X_n$ be independent random variables with distribution functions $F_1, \ldots, F_n$, means $\mu_1, \ldots, \mu_n$, and finite variances $\sigma_1{}^2, \ldots, \sigma_n{}^2$. Let*

$$v_n = \mu_1 + \cdots + \mu_n \qquad \tau_n{}^2 = \sigma_1{}^2 + \cdots + \sigma_n{}^2$$

and $S_n^* = (S_n - v_n)/\tau_n$ for $n \geq 1$. If

$$\lim \tau_n{}^{-2} \sum_{k=1}^{n} \int_{|x| \geq \varepsilon \tau_n} (x - v_k)^2 \, dF_k(x) = 0 \qquad (4.6)$$

as $n \to \infty$ for every $\varepsilon > 0$, then for any a, $-\infty < a < \infty$,

$$\lim_{n \to \infty} \Pr (S_n^* \leq a) = \Phi(a) \qquad (4.7)$$

Condition (4.6) is known as the *Lindeberg-Feller condition*. In particular, it requires that $\tau_n{}^2 \to \infty$ as $n \to \infty$.

EXAMPLE 9.4.3

a If there is a constant c for which $\Pr (|X_k - \mu_k| \geq c) = 0$ for $k = 1, \ldots, n$, $n \geq 1$, and if $\tau_n{}^2 \to \infty$ as $n \to \infty$, then the Lindeberg-Feller condition is satisfied. Indeed, if $\varepsilon > 0$ is given, then there is an n_0 for which $\varepsilon \tau_n > c$ for $n \geq n_0$. Thus, for $n \geq n_0$, we have

$$\int_{|x| \geq \varepsilon \tau_n} (x - \mu_k)^2 \, dF_k(x) = 0$$

for $k = 1, \ldots, n$, so that the left side of (4.6) is actually equal to zero for $n \geq n_0$.
b Let $Y_1, \ldots, Y_n$ be independent random variables with common distribution function F, common mean μ, and common finite, positive variance σ^2, and let $X_k = kY_k$, $k = 1, \ldots, n$. Then the Lindeberg-Feller condition is satisfied.

For simplicity, we consider only the case that $\mu = 0$ and $\sigma^2 = 1$. In this case $\sigma_k{}^2 = k^2$, so that

$$\tau_n{}^2 = \sigma_1{}^2 + \cdots + \sigma_n{}^2 \sim \int_0^n x^2 \, dx = \tfrac{1}{3} n^3$$

as $n \to \infty$. Moreover, we have

$$\int_{|x| \geq \varepsilon \tau_n} x^2 \, dF_k = k^2 \int_{|x| \geq (\varepsilon/k)\tau_n} x^2 \, dF(x)$$

$$\leq k^2 \int_{|x| \geq (\varepsilon/n)\tau_n} x^2 \, dF(x)$$

for $k = 1, \ldots, n$, so that

$$\tau_n^{-2} \sum_{k=1}^{n} \int_{|x| \geq \varepsilon \tau_n} x^2 \, dF_k(x)$$

$$\leq \tau_n^{-2} \left(\sum_{k=1}^{n} k^2 \right) \int_{|x| \geq (\varepsilon/n)\tau_n} x^2 \, dF(x) = \int_{|x| \geq (\varepsilon/n)\tau_n} x^2 \, dF(x)$$

which tends to 0 as $n \to \infty$, since $(1/n)\tau_n \to \infty$ as $n \to \infty$. ////

The Lindeberg-Feller condition requires that *each summand contribute negligibly to the sum* $S_n = X_1 + \cdots + X_n$ in the following sense.

Lemma 9.4.1 *Let* $X_1, \ldots, X_n$ *be as in Theorem 9.4.3. If condition (4.6) is satisfied, then*

$$\lim_{k \leq n} (\max \tau_n^{-2} \sigma_k^2) = 0 \qquad (4.8)$$

as $n \to \infty$.

PROOF We may suppose that $\mu_k = 0$, $k = 1, \ldots, n$. If ε is given, $0 < \varepsilon < \frac{1}{2}$, then we have

$$\sigma_k^2 = \int_{-\infty}^{\infty} x^2 \, dF_k(x) = \int_{-\varepsilon\tau_n}^{\varepsilon\tau_n} x^2 \, dF_k(x) + \int_{|x| \geq \varepsilon\tau_n} x^2 \, dF_k(x)$$

$$\leq \varepsilon^2 \tau_n^2 + \sum_{j=1}^{n} \int_{|x| \geq \varepsilon\tau_n} x^2 \, dF_j(x) \qquad (4.9)$$

for $k = 1, \ldots, n$. Now, by (4.6), there is an n_0 for which the second term in the last line of (4.9) is at most $\varepsilon^2 \tau_n^2$ for $n \geq n_0$. Therefore, for $n \geq n_0$, we have

$$\tau_n^{-2} \max_{k \leq n} \sigma_k^2 \leq 2\varepsilon^2 \leq \varepsilon$$

The lemma follows. ////

EXAMPLE 9.4.4 Let $X_1, \ldots, X_n$ be independent, and let $\Pr(X_k = \pm 2^{k-1}) = \frac{1}{2}$, $k = 1, \ldots, n$. Then $\sigma_k^2 = 4^{k-1}$, so that

$$\tau_n^2 = \sum_{k=1}^{n} 4^{k-1} = \frac{1}{3}(4^n - 1)$$

for $n \geq 1$. Hence, $\sigma_n^2/\tau_n^2 \to \frac{3}{4} \neq 0$ as $n \to \infty$, and the Lindeberg-Feller condition is violated. ////

We observe that in this example X_n and S_n are both of the order of magnitude 4^n, so that X_n does not contribute negligibly to S_n.

The Lindeberg-Feller condition implies that S_n^* has an approximate normal distribution for large n and that each of the summands contributes negligibly to the sum. In fact, the Lindeberg-Feller condition is equivalent to these two statements. A complete statement of the Lindeberg-Feller theorem is the following.

Theorem 9.4.4 *Let $X_1, \ldots, X_n$ be as in the statement of Theorem 9.4.3. Then condition (4.6) is equivalent to (4.7) and (4.8).*

We omit the proof.

9.5 EXTREME-VALUE DISTRIBUTIONS[1]

In the previous section we presented a simple approximation to the distribution function of normalized sums of independent random variables. In this section we shall present a similar approximation to the distribution function of normalized maxima of independent random variables. The treatment of maxima is similar to that of sums but simpler.

Let $X_1, \ldots, X_n$ be independent random variables with common distribution function F, and let

$$M_n = \max (X_1, \ldots, X_n)$$

The distribution function of M_n is then F^n (Example 7.2.1). Thus, if $a_n > 0$ and b_n are constants, then the distribution function of

$$M_n^* = \frac{M_n - b_n}{a_n}$$

is

$$G_n(x) = F(a_n x + b_n)^n$$

for $-\infty < x < \infty$. We shall now show how to choose the constants a_n and b_n in such a manner that G_n approaches a limit as $n \to \infty$. For simplicity, we shall consider only the case that $F(x) < 1$ for all $-\infty < x < \infty$. The case that $F(b) = 1$ of some (finite) b is treated in the problems at the end of this chapter.

Let us consider some examples.

EXAMPLE 9.5.1 Suppose that $F(x) = 1 - x^{-\alpha}$ for $x \geq 1$. In this case we may let $a_n^\alpha = n$ and $b_n = 0$ to obtain

$$G_n(x) = \left(1 - \frac{x^{-\alpha}}{n}\right)^n$$

[1] This section treats a special topic and may be omitted.

for $a_n x \geq 1$, and it follows easily that

$$\lim_{n \to \infty} G_n(x) = \exp(-x^{-\alpha})$$

for all $x > 0$ (see Lemma 4.3.1).   ////

EXAMPLE 9.5.2   Let F be the exponential distribution function $F(x) = 1 - e^{-x}$ for $x \geq 0$. In this case we may let $a_n = 1$ and $b_n = \log n$ to obtain

$$G_n(x) = \left(1 - \frac{e^{-x}}{n}\right)^n$$

for $x \geq -\log n$, so that

$$\lim_{n \to \infty} G_n(x) = \exp(-e^{-x})$$

for $-\infty < x < \infty$.   ////

These examples are more general than might at first appear. In fact, we have the following theorem.

Theorem 9.5.1   *Suppose there exist positive constants c and α for which*

$$1 - F(x) \sim cx^{-\alpha}$$

as $x \to \infty$. Define a_n and b_n by $a_n^{\alpha} = cn$ and $b_n = 0$. Then

$$\lim_{n \to \infty} G_n(x) = \exp(-x^{-\alpha}) \qquad (5.1)$$

for all $x > 0$.

Theorem 9.5.2   *Suppose that there are constants α, β, c, and d for which c, d, and β are positive and*

$$1 - F(x) \sim cx^{\alpha} \exp(-dx^{\beta}) \qquad (5.2)$$

as $x \to \infty$. Define a_n and b_n by

$$db_n^{\beta} = \log n + \alpha\beta^{-1} \log(\log n) - \alpha\beta^{-1} \log d + \log c$$

$$d\beta b_n^{\beta-1} \overset{(a)}{a_n} = 1$$

Then

$$\lim_{n \to \infty} G_n(x) = \exp(-e^{-\alpha})$$

for all x, $-\infty < x < \infty$.

PROOF   The proof of Theorem 9.5.1 is similar to that of Example 9.5.1 and will be omitted. To prove Theorem 9.5.2, observe first that

$$b_n \sim \left(\frac{1}{d} \log n\right)^{1/\beta} \to \infty \qquad \text{and} \qquad a_n b_n^{-1} = \frac{1}{d\beta b_n^{\beta}} \to 0$$

as $n \to \infty$. In particular, for any x, $-\infty < x < \infty$,

$$a_n x + b_n \sim b_n \to \infty$$

as $n \to \infty$, so that

$$1 - F(a_n x + b_n) \sim c(a_n x + b_n)^\alpha \exp\left[-d(a_n x + b_n)^\beta\right] \tag{5.3}$$

as $n \to \infty$. If we now expand $(a_n x + b_n)^\beta$ in a Taylor series about b_n, we find that

$$(a_n x + b_n)^\beta = b_n^\beta + \beta c_n^{\beta-1} a_n x$$

where c_n is an intermediate value and $|b_n - c_n| \le |a_n x|$. In particular, since $a_n b_n^{-1} \to 0$, we must also have $c_n \sim b_n$ and $d\beta c_n^{\beta-1} a_n \to 1$ as $n \to \infty$ by definition of a_n. Exponentiating and using the definitions of a_n and b_n, we now find that

$$1 - F(a_n x + b_n) \sim c(a_n x + b_n)^\alpha \exp\left(-db_n^\beta - d\beta c_n^{\beta-1} a_n x\right)$$

where

$$\exp\left(-d\beta c_n^{\beta-1} a_n x\right) \to e^{-x}$$

and

$$c(a_n x + b_n)^\alpha \exp\left(-db_n^\beta\right)$$

$$= \frac{1}{n} c(a_n x + b_n)^\alpha \exp\left[\frac{-\alpha}{\beta} \log(\log n) + \frac{\alpha}{\beta} \log d - \log c\right] \sim \frac{1}{n}$$

as $n \to \infty$. Therefore, as $n \to \infty$,

$$1 - F(a_n x + b_n) \sim \frac{1}{n} e^{-x}$$

and

$$G_n(x) = \{1 - [1 - F(a_n x + b_n)]\}^n \sim \left(1 - \frac{1}{n} e^{-x}\right)^n \to \exp\left(-e^{-x}\right)$$

as asserted.   ////

EXAMPLE 9.5.3   If F is the standard normal distribution function, then

$$1 - F(x) \sim \frac{1}{x\sqrt{2\pi}} e^{-\frac{1}{2}x^2}$$

as $x \to \infty$ (see Lemma 4.4.2). This is of the form (5.2) with $c = 1/\sqrt{2\pi}$, $d = \frac{1}{2}$, $\alpha = -1$, and $\beta = 2$. Thus, the appropriate choices of a_n and b_n are

$$b_n = \sqrt{2 \log n} - \log(\log n) - \log 4\pi$$

$$a_n = b_n^{-1} \qquad\qquad ////$$

The limiting distribution functions of (5.1) and (5.3) are known as the *Weibull* and *double-exponential* distribution functions, respectively. They are both referred to as *extreme-value distributions*, since they arise in the context of sample extrema (maxima and minima).

REFERENCES

The usual proof of the central-limit theorem uses characteristic functions. Parzen (1960), chaps. 9 and 10, gives such a proof of Theorem 9.4.1 along with the preliminary results on characteristic functions. Feller (1966), chap. 15, gives proofs of Theorems 9.4.1, 9.4.3, and 9.4.4 via characteristic functions.

PROBLEMS

9.1 Let $X_1, \ldots, X_n$ be independent random variables which are uniformly distributed over $(-1, 1)$. Use Chebyshev's inequality to estimate $\Pr(|\bar{X}_n| \geq 0.05)$. How large must n be for the bound to be less than or equal to 0.05?

9.2 Let X have the binomial distribution with parameters n and p, $0 < p < 1$.
(a) Use Chebyshev's inequality to estimate $\Pr(|\bar{X} - p| \geq 0.1)$, where $\bar{X} = X/n$.
(b) How large must n be for the bound to be less than or equal to 0.05 for all p, $0 < p < 1$?

9.3 Let X and Y be jointly distributed random variables with finite variances and correlation coefficient r. If $|r| = 1$, then there are constants a and b for which $\Pr(Y = aX + b) = 1$. Show this and find the constants a and b. *Hint:* Find a and b for which $X - aY - b$ has zero variance.

9.4 Show that if X is a random variable which equals $\varepsilon > 0$ with probability p and equals 0 with probability $q = 1 - p$, then Markov's inequality is, in fact, an equality. Show that if X equals $\pm \varepsilon$ each with probability $p < \frac{1}{2}$ and $X = 0$ with probability $1 - 2p$, then Chebyshev's inequality is an equality.

9.5 Let $X_1, \ldots, X_n$ be independent with mean μ, variance σ^2, and finite fourth central moment $\gamma = E[(X_1 - \mu)^4]$. Show that $E[(\bar{X} - \mu)^4] \leq 3\sigma^4/n^2 + \gamma/n^3$.

9.6 Let $X_1, \ldots, X_n$ be as in Problem 9.5. Show that $\sum_{n=1}^{\infty} \Pr(|\bar{X}_n - \mu| \geq \varepsilon) < \infty$ for every $\varepsilon > 0$.

9.7 Let $X_1, \ldots, X_n$ be independent random variables with common distribution function F. Show that if there is a number b for which $F(b) = 1$ and $F(a) < 1$ for every $a < b$, then $\max(X_1, \ldots, X_n) \to b$ in probability as $n \to \infty$.

9.8 Let X_n and Y_n be jointly distributed random variables. Show that if $X_n \to X$ and $Y_n \to Y$ in probability as $n \to \infty$, then $X_n + Y_n \to X + Y$ in probability as $n \to \infty$.

9.9 If X_n and Y_n are jointly distributed random variables for which $X_n \to X$ and $Y_n \to Y$ in probability as $n \to \infty$, show that $X_n Y_n \to XY$ in probability as $n \to \infty$.

9.10   Let $X_1, \ldots, X_n$ be independent random variables with finite mean μ and variance σ^2. Show that

$$S_n^2 = \frac{1}{n} \sum_{i=1}^{n} (X_i - \bar{X}_n) \to \sigma^2$$

in probability as $n \to \infty$. *Hint:* Use Problem 9.8.

9.11   Let $X_1, \ldots, X_n$ be independent, identically distributed, nonnegative random variables with infinite expectation. Show that $\bar{X}_n \to \infty$ in probability in the sense that $\Pr(\bar{X}_n \geq a) \to 1$ as $n \to \infty$ for every finite constant a.

9.12→ Let g be a continuous function on R. Show that if $X_n \to X$ in probability, then $g(X_n) \to g(X)$ in probability. *Hint:* Given ε, $\varepsilon' > 0$, choose r such that $\Pr(|X| \geq r - 1) \leq \varepsilon'/2$; then since g is uniformly continuous on $[-r, r]$, there is a δ, $0 < \delta < 1$, such that $|x| \leq r$, $|y| \leq r$, and $|x - y| \leq \delta$ implies $|g(x) - g(y)| \leq \varepsilon$. It now follows that $\Pr(|g(X_n) - g(X)| \geq \varepsilon) \leq \Pr(|X_n - X| \geq \delta) + \Pr(|X| \geq r - 1)$, which is at most $\varepsilon'/2 + \varepsilon'/2 = \varepsilon'$ for n sufficiently large.

9.13   Prove Theorem 9.3.2.

9.14   Let $g(x) = \sin 2\pi x$, $0 < x < 1$, and let g_n be the Bernstein polynomial approximating g. Find an n for which $|g(x) - g_n(x)| \leq 0.05$ for all x, $0 < x < 1$.

9.15   Let $X_1, \ldots, X_n$ be independent random variables, and let X_k have the Poisson distribution with parameter $\sqrt{k}$, $k = 1, \ldots, n$. Describe the behavior of $\bar{X}_n$ as $n \to \infty$.

9.16   Let a balanced die be tossed 100 times, and let X denote the total sum of spots. Use the central-limit theorem to estimate the probability that $300 \leq X \leq 400$.

9.17   Repeat Problem 9.1 using the central-limit theorem in place of Chebyshev's inequality.

9.18   Repeat Problem 9.2 using the central-limit theorem in place of Chebyshev's inequality.

9.19   Let X_n have the Poisson distribution with parameter $\beta = n$, and let $Y_n = (X_n - n)/\sqrt{n}$. Show that the moment-generating function of Y_n converges to the moment-generating function of the standard normal distribution. What does this suggest?

9.20   If X has the Poisson distribution with parameter 100, estimate the probability that $85 \leq X \leq 110$.

9.21   If X has the gamma distribution with parameters $\alpha = 400$ and $\beta = 1$, estimate the probability that $390 \leq X \leq 450$.

9.22   Let $X_1, \ldots, X_n$ be independent random variables which are uniformly distributed over $(0,1)$, and let $Y_n = \max(X_1, \ldots, X_n)$. Show that the distribution function of $Z_n = n(1 - Y_n)$ approaches a limit, and evaluate that limit.

9.23   Let F be a distribution function, and let b be a real number for which $F(b) = 1$ and $F(x) < 1$ for $x < b$. Suppose also that there are positive numbers c and α for which $1 - F(x) \sim c(b - x)^\alpha$ as $x \to b$. Let $X_1, \ldots, X_n$ be independent with common distribution function F, and let G_n be the distribution function of $a_n(b - M_n)$, where $M_n = \max(X_1, \ldots, X_n)$ and $a_n^\alpha = n$. Find the limit of G_n as $n \to \infty$.

9.24   Let $X_1, \ldots, X_n$ be independent random variables which have the gamma distribution with parameters $\alpha = k$, a positive integer, and $\beta = 1$, and let $M_n = \max(X_1, \ldots, X_n)$. How should a_n and b_n be chosen for the distribution function of $M_n^* = (M_n - b_n)/a_n$ to approach a limit as $n \to \infty$?

9.25   Let $X_1, X_2, \ldots, X_n$ be independent random variables for which $X_k = \pm k^\alpha$, each with probability $\frac{1}{2}$, where $\alpha > 0$, $k = 1, \ldots, n$. Show that the Lindeberg-Feller condition is satisfied.

9.26   Let X_k be uniformly distributed on the interval $(-a_k, a_k)$, where $a_k > 0$, $k = 1, \ldots, n$. Suppose also that $a_k \leq 1$, $k = 1, \ldots, n$, $n \geq 1$. Show that the Lindeberg-Feller condition is satisfied if and only if $a_1^2 + \cdots + a_n^2 \to \infty$ as $n \to \infty$.

10

CONDITIONAL DISTRIBUTIONS AND EXPECTATION

10.1 CONDITIONAL DENSITIES AND MASS FUNCTIONS

Let X and Y be jointly distributed, discrete random variables with joint mass function f, and let g and h denote the (marginal) mass functions of X and Y, respectively. Further, let D be the set of $x \in R$ for which $g(x) = \text{Pr}\,(X = x) > 0$. Then for $x \in D$ we define the *conditional mass function* of Y given that $X = x$ by

$$h(y \mid x) = \frac{f(x,y)}{g(x)} \qquad -\infty < y < \infty \qquad (1.1a)$$

That is, we define $h(y \mid x)$ to be the conditional probability of the event that $Y = y$ given that $X = x$. Similarly, if $h(y) > 0$, we define the conditional mass function of X given that $Y = y$ by the formula

$$g(x \mid y) = \frac{f(x,y)}{h(y)} \qquad -\infty < x < \infty \qquad (1.1b)$$

We remark that for every $x \in D$, the function $h(\cdot \mid x)$ does define a mass function. Indeed, letting E be the finite or countably infinite set for which $h(y) > 0$, we find that $f(x,y) \leq h(y) = 0$ for $y \notin E$, so that $h(y \mid x) = 0$ if $y \notin E$. Moreover,

$$\sum_{y \in E} h(y \mid x) = \frac{1}{g(x)} \sum_{y \in E} f(x,y) = \frac{g(x)}{g(x)} = 1$$

for $x \in D$ by Equation (2.3) of Chapter 6. Moreover, if X and Y are *independent* random variables, then $f(x,y) = g(x)h(y)$ for all x and y, so that

$$h(y \mid x) = h(y) \qquad (1.2)$$

for $-\infty < y < \infty$ and $x \in D$. In any case (even if X and Y are not independent), we have the factorization

$$f(x,y) = h(y \mid x)g(x) \qquad (1.3)$$

for $-\infty < y < \infty$ and $x \in D$. Of course, the roles of X and Y may be reversed in the above discussion.

EXAMPLE 10.1.1

a Let a balanced die be rolled n times, let X denote the number of aces which appear, and let Y denote the number of sixes. Then

$$f(x,y) = \binom{n}{x,\, y,\, n - x - y} \left(\frac{1}{6}\right)^x \left(\frac{1}{6}\right)^y \left(\frac{4}{6}\right)^{n-x-y}$$

for all nonnegative integers x and y for which $x + y \le n$ (see Theorem 4.1.2). Also,

$$g(x) = \binom{n}{x} \left(\frac{1}{6}\right)^x \left(\frac{5}{6}\right)^{n-x}$$

for $x = 0, \ldots, n$, by Theorem 4.1.1. It follows that

$$h(y \mid x) = \binom{n - x}{y} \left(\frac{1}{5}\right)^y \left(\frac{4}{5}\right)^{n-x-y}$$

$y = 0, \ldots, n - x$ and $x = 0, \ldots, n$. Thus, the conditional distribution of Y given that $X = x$ is binomial with parameters $n - x$ and $\frac{1}{5}$. We may interpret this result as follows: given that there were x aces, Y has the distribution of the number of sixes in $n - x$ tosses of a five-sided die (no aces).

b Let X and Y be independent random variables which have binomial distributions, say

$$\Pr(X = x) = \binom{m}{x} p^x q^{m-x} \qquad x = 0, \ldots, m$$

$$\Pr(Y = y) = \binom{n}{y} p^y q^{n-y} \qquad y = 0, \ldots, n$$

and let $Z = X + Y$. Then, the conditional distribution of X, given that $Z = z$ where $z = 0, \ldots, m + n$, is hypergeometric. Indeed, we have

$$\Pr(Z = z) = \binom{m + n}{z} p^z q^{m+n-z}$$

for $z = 0, \ldots, m + n$, so that

$$
\begin{aligned}
g(x \mid z) &= \frac{\Pr(X = x, Z = z)}{\Pr(Z = z)} \\
&= \frac{\Pr(X = x) \Pr(Y = z - x)}{\Pr(Z = z)} \\
&= \frac{\binom{m}{x} p^x q^{m-x} \binom{n}{z-x} p^{z-x} q^{n-z+x}}{\binom{m+n}{z} p^z q^{m+n-z}} = \frac{\binom{m}{x}\binom{n}{z-x}}{\binom{m+n}{z}}
\end{aligned}
$$

for $x = 0, \ldots, m$ and $z = 0, \ldots, m + n$. [Recall that $\binom{n}{k} = 0$ if $k < 0$ or $k > n$.]   ////

Let us now consider the absolutely continuous case. Thus, let X and Y have joint density f, and let g and h denote the marginal densities of X and Y, respectively. Let D be the set of $x \in R$ for which $g(x) > 0$. Then for $x \in D$ we define the *conditional density* of Y given that $X = x$ by the formula

$$
h(y \mid x) = \frac{f(x, y)}{g(x)} \qquad -\infty < y < \infty \qquad (1.4a)
$$

and similarly, if $h(y) > 0$ we define the conditional density of X given that $Y = y$ by the formula

$$
g(x \mid y) = \frac{f(x, y)}{h(y)} \qquad -\infty < x < \infty \qquad (1.4b)
$$

As in the discrete case, $h(\cdot \mid x)$ does define a density for each $x \in D$, and

$$
f(x, y) = h(y \mid x) g(x) \qquad (1.5)
$$

for $-\infty < y < \infty$ and $x \in D$. Moreover, if X and Y are *independent*, then $f(x, y) = g(x)h(y)$ defines a density for X and Y, in which case

$$
h(y \mid x) = h(y) \qquad (1.6)
$$

for $-\infty < y < \infty$ and $x \in D$. Of course, the roles of X and Y may be reversed in the above discussion.

There are two novel features in the absolutely continuous case. First, $h(y \mid x)$ no longer gives the conditional probability of the event $Y = y$ given the event $X = x$. Both events have probability zero. Moreover, there is some ambiguity in the definition of $h(y \mid x)$ and $g(x \mid y)$, since there is some ambiguity in the choice of f (recall that a density may be changed on a finite set without affecting its integral). We should really refer to $h(\cdot \mid x)$ and $g(\cdot \mid y)$ as conditional densities with respect to f, but the qualifying phrase will be omitted.

EXAMPLE 10.1.2   Let (X,Y) have the uniform distribution over the unit disk. That is, let X and Y have joint density

$$f(x,y) = \pi^{-1} \qquad x^2 + y^2 \le 1$$

and $f(x,y) = 0$ for other values of x and y. Then, the marginal density of X is

$$g(x) = 2\pi^{-1}\sqrt{1 - x^2} \qquad -1 < x < 1$$

and $g(x) = 0$ for other values of x (see Example 6.2.3). It follows that for $-1 < x < 1$,

$$h(y \mid x) = \frac{1}{2\sqrt{1 - x^2}} \qquad -\sqrt{1 - x^2} \le y \le \sqrt{1 - x^2}$$

Thus, the conditional distribution of Y given $X = x$ is uniform on the interval $(-a,a)$, where $a = \sqrt{1 - x^2}$. ////

EXAMPLE 10.1.3   Let X and Y be independent, exponentially distributed random variables with the same parameter $\beta > 0$. Then the conditional distribution of X given that $Z = X + Y = z$ is uniform on the interval $(0,z)$ for $z > 0$.

   To see this we must first find a joint density for X and Z and the marginal density of Z. A joint density for X and Y is

$$d(x,y) = \beta^2 e^{-(x+y)}$$

for $x > 0$ and $y > 0$, and $d(x,y) = 0$ for other values of x and y by independence. Thus, by Theorem 7.4.1, a joint density for X and Z is $f(x,z) = d(x, z - x)$, which simplifies to

$$f(x,z) = \beta^2 e^{-\beta z}$$

for $0 < x < z$ and $f(x,z) = 0$ for other values of x and z. The marginal density of Z can now be computed by a straightforward integration (it was also found in Sections 7.3 and 8.4) as

$$h(z) = \beta^2 z e^{-\beta z}$$

for $z > 0$. Thus,

$$g(x \mid z) = \frac{1}{z}$$

for $0 < x < z$, as asserted. ////

EXAMPLE 10.1.4   Let X and Y have the standard bivariate normal distribution with parameter (correlation coefficient) r, $-1 < r < 1$. Then the conditional distribution of Y given that $X = x$ is normal with mean rx and variance $1 - r^2$. To see this recall from Example 6.2.3

$$f(x,y) = C_r e^{-\frac{1}{2}Q(x,y)} \qquad -\infty < x < \infty$$
$$-\infty < y < \infty$$

$$g(x) = \frac{1}{\sqrt{2\pi}} e^{-\frac{1}{2}x^2} \qquad -\infty < x < \infty$$

where $C_r^{-1} = 2\pi\sqrt{1 - r^2}$ and

$$Q(x,y) = \frac{x^2 - 2rxy + y^2}{1 - r^2}$$

We may write $Q(x,y) = x^2 + z^2$, where

$$z = \frac{y - rx}{\sqrt{1 - r^2}}$$

so that

$$h(y \mid x) = \sqrt{2\pi}\, C_r e^{-\frac{1}{2}z^2} = \frac{1}{\sqrt{2\pi(1 - r^2)}} \exp\left[-\frac{1}{2} \frac{(y - rx)^2}{1 - r^2} \right]$$

for $-\infty < x < \infty$ and $-\infty < y < \infty$, as asserted. Of course, the roles of X and Y may be reversed in this example. ////

Let us return briefly to Equations (1.3) and (1.5). For example, (1.5) states that if X and Y have joint density f, then

$$f(x,y) = h(y \mid x)g(x) \qquad (1.5)$$

for $-\infty < y < \infty$ and $x \in D$, where g and h denote the marginal density of X and the conditional density of Y given X, respectively, and D denotes the set of $x \in R$ for which $g(x) > 0$. That is, *we can determine a joint density for X and Y by specifying a marginal density for X and a conditional density for Y given X.* In many problems this is the most natural way to introduce a joint density. The marginal density of Y and the conditional density of X given $Y = y$ can then be computed from

$$h(y) = \int_D h(y \mid x)g(x)\, dx \qquad -\infty < y < \infty \qquad (1.7)$$

$$g(x \mid y) = \frac{h(y \mid x)g(x)}{h(y)} \qquad (1.8)$$

for $x \in D$ and $h(y) > 0$. Equations (1.7) and (1.8) may be regarded as extensions of Bayes' theorem to the absolutely continuous case, and the general discussion of

Bayes' theorem (Section 3.2) applies to (1.7) and (1.8). The corresponding formulas in the discrete case,

$$f(x,y) = h(y \mid x)g(x) \qquad x \in D \qquad (1.3)$$

$$h(y) = \sum_{x \in D} h(y \mid x)g(x) \qquad (1.7a)$$

$$g(x \mid y) = \frac{h(y \mid x)g(x)}{h(y)} \qquad (1.8)$$

for $h(y) > 0$ are, in fact, simply restatements of Bayes' theorem.

EXAMPLE 10.1.5 Let X have the uniform distribution on the interval $(0,1)$, and conditionally, given $X = x$, let Y have the uniform distribution on the interval $(0,x)$. That is, let

$$g(x) = 1 \qquad 0 < x < 1$$

and

$$h(y \mid x) = \frac{1}{x} \qquad 0 < y < x$$

Then

$$h(y) = \int_y^1 \frac{1}{x} \, dx = -\log y$$

for $0 < y < 1$, and

$$g(x \mid y) = -\frac{1}{x} \log y$$

for $y < x < 1$. It is interesting to remark that the conditional distribution of X given $Y = y$ is not uniform, even though the conditional distribution of Y given $X = x$ is uniform.   ////

10.1.1   Mixed Distributions[1]

In this section we shall extend the notions of the previous section to the case of *mixed distributions*, by which we mean joint distributions where one variable is discrete and the other is absolutely continuous. Thus, let X and Y be jointly distributed random variables, and suppose that X is discrete with mass function g and that Y is absolutely continuous with density h. Further, let D denote the finite or countably infinite set of $x \in R$ for which $g(x) > 0$, and let E be an interval for which $h(y) > 0$ when $y \in E$ and $h(y) = 0$ when $y \notin E$. Then, for any $x \in D$ and $y \in R$,

---

[1] This section treats a special topic and may be omitted.

we can compute

$$H(y \mid x) = \Pr(Y \le y \mid X = x) = \frac{\Pr(Y \le y, X = x)}{\Pr(X = x)}$$

and we shall call $H(\cdot \mid x)$ the *conditional distribution function of* Y given $X = x$. Accordingly, we define the *conditional density* of Y given $X = x$ to be the derivative

$$h(y \mid x) = \frac{d}{dy} H(y \mid x)$$

for $-\infty < y < \infty$ and $x \in D$, provided that the derivative exists.[1] Further, we define the *conditional mass function* of X given $Y = y$ by a variation on Bayes' theorem. That is, we define

$$g(x \mid y) = \frac{h(y \mid x)g(x)}{h(y)} \qquad (1.9)$$

for $x \in D$ and $y \in E$. It then follows easily that

$$h(y \mid x) = \frac{g(x \mid y)h(y)}{g(x)} \qquad (1.10)$$

for $x \in D$ and $y \in E$.

As in the discrete and absolutely continuous cases, the relations

$$g(x) = \int_E g(x \mid y)h(y) \, dy \qquad x \in D \qquad (1.11)$$

$$h(y) = \sum_{x \in D} h(y \mid x)g(x) \qquad y \in E \qquad (1.12)$$

can be obtained (see Problem 10.21). It follows easily that $h(\cdot \mid x)$ is a density for every $x \in D$ and that $g(\cdot \mid y)$ is a mass function for every $y \in E$. Moreover, if X and Y are *independent*, then $\Pr(Y \le y \mid X = x) = \Pr(Y \le y)$ for $-\infty < y < \infty$ and $x \in D$, so that

$$h(y \mid x) = h(y) \qquad \begin{matrix} y \in E \\ x \in D \end{matrix} \qquad (1.13)$$

by differentiation, and

$$g(x \mid y) = g(x) \qquad \begin{matrix} x \in D \\ y \in E \end{matrix} \qquad (1.14)$$

by (1.9).

EXAMPLE 10.1.6 Let Z be an absolutely continuous random variable, and suppose that Z has a continuous density f which is everywhere positive. We shall compute $\Pr(Z > 0 \mid |Z| = y)$ for $y > 0$. Let X be the indicator of the event $Z > 0$.

---

[1] It is sufficient that the derivative exist at all but a finite number of points.

That is, let $X = 1$ if $Z > 0$, and let $X = 0$ if $Z \leq 0$. Also, let $Y = |Z|$. Then, X has mass function

$$g(0) = F(0) \quad \text{and} \quad g(1) = 1 - F(0)$$

where F denotes the distribution function of Z; and Y has density

$$h(y) = f(y) + f(-y) \quad y > 0$$

by Example 7.1.3b. Let us find $h(y \mid 0)$. Now,

$$\begin{aligned} H(y \mid 0) &= \Pr\,(Y \leq y \mid X = 0) \\ &= \Pr\,(|Z| \leq y \mid Z \leq 0) \\ &= \frac{\Pr\,(-y \leq Z \leq 0)}{\Pr\,(Z \leq 0)} = \frac{F(0) - F(-y)}{F(0)} \end{aligned}$$

for $y > 0$. Therefore,

$$h(y \mid 0) = \frac{f(-y)}{F(0)} \quad y > 0$$

by differentiation and similarly,

$$h(y \mid 1) = \frac{f(y)}{1 - F(0)} \quad y > 0$$

$\Pr\,(Z > 0 \mid |Z| = y) = \Pr\,(X = 1 \mid Y = y) = g(1 \mid y)$ can now be found from Equation (1.9). Indeed, we have

$$g(1 \mid y) = \frac{h(y \mid 1)r(1)}{h(y)} = \frac{f(y)}{f(y) + f(-y)}$$

for $y > 0$. In particular, $g(1 \mid y) = \frac{1}{2}$ for all $y > 0$ if f is symmetric, that is, $f(x) = f(-x)$ for all x. ////

It is sometimes natural to describe a joint mixed distribution by specifying the unconditional distribution of one variable and the conditional distribution of the other. In such cases many interesting probabilities and conditional probabilities can be computed directly from (1.9) to (1.12). We shall illustrate this procedure with some examples.

EXAMPLE 10.1.7

a Let a point Y be chosen from the unit interval according to the uniform distribution, and then let a coin with probability Y of turning up heads be tossed until a head appears. Let X denote the number of tosses required to get a head. Then it seems most natural to describe a joint distribution for X and Y by specifying first the marginal density of Y and then the conditional mass function of X given $Y = y$. In fact, we are given that Y has the uniform distribution, so that

$$h(y) = 1 \qquad 0 < y < 1$$

Moreover, given $Y = y$, X is simply the number of tosses required to get a head, so that X should have the geometric distribution with parameter y. That is,

$$g(x \mid y) = y(1 - y)^{x-1}$$

for $x = 1, 2, \ldots$ and $0 < y < 1$. The unconditional mass function of X can now be computed from (1.11) to be

$$g(x) = \int_0^1 y(1 - y)^{x-1} \, dy = \int_0^1 (1 - u)u^{x-1} \, du = \frac{1}{x(x + 1)}$$

for $x = 1, 2, \ldots$. (Here we made the change of variable $u = 1 - y$.) The conditional density of Y given that $X = x$ can now be computed from (1.10) to be

$$h(y \mid x) = \frac{g(x \mid y)h(y)}{g(x)} = x(x + 1)y(1 - y)^{x-1}$$

for $0 < y < 1$ and $x = 1, 2, \ldots$.

b Now let Y have the beta distribution with parameters $\alpha > 0$ and $\beta > 0$, and let X be the number of heads in n independent tosses of a coin which has probability Y of coming up heads on each toss. In this case we are given that Y has density

$$h(y) = \frac{\Gamma(\alpha + \beta)}{\Gamma(\alpha)\Gamma(\beta)} y^{\alpha-1}(1 - y)^{\beta-1}$$

for $0 < y < 1$, and that X has conditional mass function

$$g(x \mid y) = \binom{n}{x} y^x(1 - y)^{n-x}$$

for $x = 0, \ldots, n$ and $0 < y < 1$. Letting $\alpha' = \alpha + x$ and $\beta' = \beta + x$, it follows that

$$g(x) = \int_0^1 \binom{n}{x} \frac{\Gamma(\alpha + \beta)}{\Gamma(\alpha)\Gamma(\beta)} y^{\alpha'-1}(1 - y)^{\beta'-1} \, dy$$

$$= \binom{n}{x} \frac{\Gamma(\alpha + \beta)}{\Gamma(\alpha)\Gamma(\beta)} \frac{\Gamma(\alpha')\Gamma(\beta')}{\Gamma(\alpha' + \beta')}$$

for $x = 0, \ldots, n$ and that

$$h(y \mid x) = \frac{\Gamma(\alpha' + \beta')}{\Gamma(\alpha')\Gamma(\beta')} y^{\alpha'-1}(1 - y)^{\beta'-1}$$

for $0 < y < 1$ and $x = 0, \ldots, n$. Thus, the conditional distribution of Y given $X = x$ is again beta, but with new parameters, $\alpha' = \alpha + x$ and $\beta' = \beta + (n - x)$. ////

10.2 CONDITIONAL PROBABILITY

If X and Y are discrete random variables, if x is a real number for which $g(x) = \Pr(X = x) > 0$, and if B is a finite or countably infinite subset of R, then by Theorem 3.1.1, we have

$$\Pr(Y \in B \mid X = x) = \sum_{y \in B} \Pr(Y = y \mid X = x) = \sum_{y \in B} h(y \mid x) \qquad (2.1)$$

where $h(\cdot \mid x)$ denotes the conditional mass function of Y given $X = x$. However, if X and Y are (jointly) absolutely continuous, then the conditional probability that $Y \in B$ given that $X = x$ is undefined because the latter event has probability zero. We shall now define the notation $\Pr(Y \in B \mid X = x)$ in the absolutely continuous case by a formula analogous to (2.1). Let X and Y be jointly absolutely continuous, let g denote the marginal density of X, and let $h(\cdot \mid x)$ denote the conditional density of Y given that $X = x$. If B is a subset of R and $g(x) > 0$, then we define

$$\Pr(Y \in B \mid X = x) = \int_B h(y \mid x) \, dy \qquad (2.2)$$

provided that the integral on the right side of (2.2) exists.

We may also define the notation $\Pr(Y \in B \mid X = x)$ in the case that X and Y have a mixed distribution. In fact, if X is absolutely continuous and Y is discrete, we define $\Pr(Y \in B \mid X = x)$ by (2.1); and if X is discrete and Y is absolutely continuous, we define $\Pr(Y \in B \mid X = x)$ by Equation (2.2). The only difference is that the conditional mass function or density is computed as in Section 10.1.1. In any of the four cases, we define the conditional distribution function of Y given $X = x$ by the formula[1]

$$H(y \mid x) = \Pr(Y \le y \mid X = x) \qquad (2.3)$$

for $-\infty < y < \infty$ provided that $g(x) > 0$.

Of course, the roles of X and Y may be reversed in the above discussion to yield the definition of $P(X \in B \mid Y = y)$.

Let us now consider some examples.

---

[1] That this definition of H is consistent with the one given in Section 10.1.1 follows from (1.9) and (1.10).

EXAMPLE 10.2.1

a Let X and Y have the uniform distribution on the unit disk in R^2, as in Example 10.1.2. Then, for $-1 < x < 1$, the conditional distribution of Y given $X = x$ is uniform on the interval $(-a,a)$, where $a = \sqrt{1 - x^2}$, so that

$$\Pr (Y > 0 \mid X = x) = \frac{1}{2a} \int_0^a dy = \tfrac{1}{2}$$

for $-1 < x < 1$.

b Let X and Y have the standard, bivariate normal distribution with correlation coefficient r, $-1 < r < 1$. Then, by Example 10.1.4, the conditional distribution of Y given that $X = x$ is normal with mean rx and variance $1 - r^2$. Hence,

$$\Pr (Y \le y \mid X = x) = \Phi \left(\frac{y - rx}{\sqrt{1 - r^2}} \right)$$

where Φ denotes the standard normal distribution function.   ////

We shall now develop some general properties of conditional probability. For simplicity, the results will be stated and proved for jointly absolutely continuous variables only. The analogous results for discrete and mixed variables can be obtained by exchanging the words "density" and "mass function" and the symbols $\int$ and $\sum$ in appropriate places in both the statements and the proofs of Theorems 10.2.1 to 10.2.3. Theorem 10.2.4 is interesting only in the case of absolutely continuous and mixed variables.

The first item of business is to show that conditional probability obeys the axioms of probability.

Theorem 10.2.1   *Let X and Y be jointly absolutely continuous, and let g denote the marginal density of X. If $g(x) > 0$, then $0 \le \Pr (Y \in B \mid X = x) \le 1$ for all intervals $B \subset R$, and*

$$\Pr (Y \in A \cup B \mid X = x) = \Pr (Y \in A \mid X = x) + P(Y \in B \mid X = x)$$

whenever A and B are disjoint intervals.

The theorem is an obvious consequence of the definition (2.2). The third axiom of probability is also true in the discrete case (see Theorem 3.1.1).

Theorem 10.2.2   *Let X and Y be independent, absolutely continuous random variables, and let g denote the marginal density of X. If $g(x) > 0$, then*

$$\Pr(Y \in B \mid X = x) = \Pr(Y \in B)$$

for all intervals $B \subset R$.

PROOF   In fact, if X and Y are independent, and $g(x) > 0$, then $h(y \mid x) = h(y)$ for all y by Equation (1.6), where $h(\cdot \mid x)$ and h denote the conditional density of Y given $X = x$ and the marginal density of Y, respectively. Therefore,

$$\Pr(Y \in B \mid X = x) = \int_B h(y \mid x)\, dy = \int_B h(y)\, dy = \Pr(Y \in B) \qquad (2.4)$$

for all B, as asserted.   ////

As in Section 3.2, we can use conditional probabilities as tools in the computation of unconditional probabilities.

Theorem 10.2.3   *Let X and Y be jointly absolutely continuous; let g denote the marginal density of X; and let D denote the set of $x \in R$ for which $g(x) > 0$. If B is a subregion of R^2, then*

$$\Pr((X,Y) \in B) = \int_D \Pr(Y \in B_x \mid X = x) g(x)\, dx$$

where for each $x \in D$, B_x denotes the set of $y \in R$ for which $(x,y) \in B$.

PROOF   Clearly, $\Pr(X \in D) = 1$, so that

$$\Pr((X,Y) \in B) = \Pr((X,Y) \in B,\ X \in D)$$

Moreover (see Section 6.4),

$$\Pr((X,Y) \in B,\ X \in D) = \int_D \left[\int_{B_x} f(x,y)\, dy \right] dx$$

$$= \int_D \left[\int_{B_x} h(y \mid x)\, dy \right] g(x)\, dx$$

$$= \int_D \Pr(Y \in B_x \mid X = x) g(x)\, dx$$

as asserted.   ////

Theorem 10.2.3 has some interesting corollaries.

Corollary 10.2.1 *If A and B are intervals, then*

$$\Pr\left(X \in A, Y \in B\right) = \int_{AD} \Pr\left(Y \in B \mid X = x\right) g(x)\, dx \qquad (2.5)$$

In particular,

$$\Pr\left(Y \in B\right) = \int_{D} \Pr\left(Y \in B \mid X = x\right) g(x)\, dx \qquad (2.6)$$

PROOF   Since (2.6) follows from (2.5) on taking $A = D$, it will suffice to prove (2.5). Now $\Pr\left(X \in A, Y \in B\right) = \Pr\left((X,Y) \in A \times B\right)$, where $A \times B$ denotes the cartesian product of A and B. Moreover, $(A \times B)_x = B$ if $x \in A$ and $(A \times B)_x = \varnothing$ if $x \notin A$. Thus (2.5) follows directly from Theorem 10.2.3. ////

Corollary 10.2.2 *If $\Pr\left(Y \in B \mid X = x\right) = \Pr\left(Y \in B\right)$ for all $x \in D$ and every interval B, then X and Y are independent.*

PROOF   By Corollary 10.2.1

$$\Pr\left(X \in A, Y \in B\right) = \int_{AD} \Pr\left(Y \in B \mid X = x\right) g(x)\, dx$$

$$= \Pr\left(Y \in B\right) \int_{AD} g(x)\, dx$$

$$= \Pr\left(X \in AD\right) \Pr\left(Y \in B\right)$$

$$= \Pr\left(X \in A\right) \Pr\left(Y \in B\right)$$

for all intervals A and B. Hence X and Y are independent. ////

Example 10.2.1a shows that it is possible to have $\Pr\left(Y \in B \mid X = x\right) = \Pr\left(Y \in B\right)$ for all $x \in D$ for a particular B, even if X and Y are dependent.

EXAMPLE 10.2.2   As an application of Theorems 10.2.2 and 10.2.3, we shall rederive the convolution formula of Section 7.3. Let X and Y be independent random variables with densities g and h, respectively. We shall first compute the probability that $Z = X + Y \le z$ for arbitrary z. Let B denote the set of (x,y) for which $x + y \le z$. Then B_x is simply the interval $(-\infty, z - x]$, and $\Pr\left(Y \in B_x \mid X = x\right) = \Pr\left(Y \in B_x\right) = H(z - x)$ by Theorem 10.2.2. Here H denotes the distribution function of Y. Therefore, by Theorem 10.2.3, we have

$$\Pr\left(Z \le z\right) = \Pr\left((X,Y) \in B\right) = \int_{D} H(z - x) g(x)\, dx$$

where D denotes the set of $x \in R$ for which $g(x) > 0$. Differentiation now shows that Z has density

$$f(z) = \int_D h(z - x)g(x)\, dx$$

for $-\infty < z < \infty$. ////

In the absolutely continuous case, where $\Pr(X = x) = 0$, it is natural to expect $\Pr(Y \in B \mid X = x)$ to be the limit as $\varepsilon \to 0$ of $\Pr(Y \in B \mid |X - x| \le \varepsilon) = \Pr(Y \in B, |X - x| \le \varepsilon)/\Pr(|X - x| \le \varepsilon)$. We shall now show that this is in fact the case under some modest regularity conditions.

Theorem 10.2.4 *Let X and Y be jointly absolutely continuous random variables, let g denote the marginal density of X, and let D denote the set of $x \in R$ for which $g(x) > 0$. Further, let $B \subset R$ and define the function w on D by*

$$w(x) = \Pr(Y \in B \mid X = x) \qquad x \in D$$

If $a \in D$, and if both w and g are continuous at a, then

$$w(a) = \lim \Pr(Y \in B \mid |X - a| \le \varepsilon)$$

as $\varepsilon \to 0$.

PROOF Since g is continuous at a and $g(a) > 0$, we have $(a - \varepsilon, a + \varepsilon) \subset D$ for sufficiently small $\varepsilon > 0$. Moreover, for such ε, we have

$$\Pr(Y \in B, |X - a| \le \varepsilon) = \int_{a-\varepsilon}^{a+\varepsilon} w(x)g(x)\, dx$$

by Theorem 10.2.3. Also,

$$\Pr(|X - a| \le \varepsilon) = \int_{a-\varepsilon}^{a+\varepsilon} g(x)\, dx$$

by definition of a density. Now, by the fundamental theorem of calculus,

$$\lim_{\varepsilon \to 0} \frac{1}{2\varepsilon} \int_{a-\varepsilon}^{a+\varepsilon} g(x)\, dx = g(a)$$

so that $\Pr(|X - a| \le \varepsilon)/2\varepsilon \to g(a)$ as $\varepsilon \to 0$. Similarly, $\Pr(Y \in B, |X - a| \le \varepsilon)/2\varepsilon \to w(a)g(a)$ as $\varepsilon \to 0$, again by the fundamental theorem of calculus. Therefore,

$$\Pr(Y \in B \mid |X - a| \le \varepsilon) \to \frac{w(a)g(a)}{g(a)} = w(a)$$

as $\varepsilon \to 0$. ////

Theorem 10.2.4 is also valid if one variable is absolutely continuous and the other is discrete. We illustrate with an example.

EXAMPLE 10.2.3 Let us reconsider Example 10.1.6. Thus, let Z be an absolutely continuous random variable with continuous, everywhere positive density f; let $Y = |Z|$; and let X be the indicator of the event $Z > 0$ (that is, $X = 1$ if $Z > 0$ and $X = 0$ if $Z \leq 0$). Let us compute $\Pr(X = 1 \mid Y = y)$ by Theorem 10.2.4. For $a > 0$ and $\varepsilon \leq a/2$, we have

$$\Pr(X = 1, |Y - a| \leq \varepsilon) = \Pr(|Z - a| \leq \varepsilon) = \int_{a-\varepsilon}^{a+\varepsilon} f(z)\, dz$$

so that $\Pr(X = 1, |Y - a| \leq \varepsilon)/2\varepsilon \to f(a)$ as $\varepsilon \to 0$. Also,

$$\Pr(|Y - a| \leq \varepsilon) = \Pr(|Z - a| \leq \varepsilon) + \Pr(|Z + a| \leq \varepsilon)$$

so that $\Pr(|Y - a| \leq \varepsilon)/2\varepsilon \to f(a) + f(-a)$. Therefore,

$$\Pr(X = 1 \mid |Y - a| \leq \varepsilon) = \frac{\Pr(X = 1, |Y - a| \leq \varepsilon)}{\Pr(|Y - a| \leq \varepsilon)} \to \frac{f(a)}{f(a) + f(-a)} \qquad ////$$

10.3 CONDITIONAL EXPECTATION

Let X and Y be jointly distributed random variables, and let Z be a random variable which is determined as a function of X and Y, say

$$Z = w(X, Y)$$

We shall define the *conditional expectation of Z given X = x*.

Suppose first that X and Y are discrete with joint mass function f, and let g and h denote the marginal mass functions of X and Y, respectively. Further, let D and E denote the set of $x \in R$ for which $g(x) > 0$ and the set of $y \in R$ for which $h(y) > 0$. For $x \in D$ we define the conditional expectation of Z given $X = x$ to be

$$E(Z \mid X = x) = \sum_{y \in E} w(x, y) h(y \mid x) \qquad (3.1)$$

provided that the summation converges absolutely. Here $h(\cdot \mid x)$ denotes the conditional mass function of Y given $X = x$.

Similarly, if X and Y are jointly absolutely continuous, if g denotes the marginal density of X, and if D denotes the set of $X \in R$ for which $g(x) > 0$, then we define the conditional expectation of Z given $X = x$ for $x \in D$ by

$$E(Z \mid X = x) = \int_{-\infty}^{\infty} w(x, y) h(y \mid x)\, dy \qquad (3.2)$$

provided that the integral converges absolutely. Here $h(\cdot \mid x)$ denotes the conditional density of Y given $X = x$.

If X and Y have a mixed distribution, then we can also define the conditional expectation of Z given $X = x$ by an appropriate version of (3.1) or (3.2). In fact, if X is absolutely continuous and Y is discrete, then we define $E(Z \mid X = x)$ by (3.1); and if X is discrete and Y is absolutely continuous, then we define $E(Z \mid X = x)$ by (3.2).

The four cases may be subsumed under one equation by writing

$$E(Z \mid X = x) = \int_{-\infty}^{\infty} w(x, y) \, dH(y \mid x) \tag{3.3}$$

where $H(\cdot \mid x)$ denotes the conditional distribution function of Y given $X = x$. The conditions under which $E(Z \mid X = x)$ is defined may also be stated succinctly as follows: $H(\cdot \mid x)$ must be defined, and the integral appearing in (3.3) must converge absolutely.

An important special case occurs when we take $Z = Y$. Thus,

$$E(Y \mid X = x) = \int_{-\infty}^{\infty} y \, dH(y \mid x) \tag{3.4}$$

subject to the conditions stated above.

EXAMPLE 10.3.1

a   Let X and Y have the standard, bivariate normal distribution with correlation coefficient r, $-1 < r < 1$. Then, by Example 10.1.4, the conditional distribution of Y given $X = x$ is normal with mean rx and variance $1 - r^2$. Hence, $E(Y \mid X = x) = rx$.

b   Similarly, if X and Y denote the number of aces and sixes in n tosses of a fair die, then the conditional distribution of Y given $X = x$ is binomial with parameters $\frac{1}{5}$ and $n - x$, $x = 0, \dots, n$ (see Example 10.1.1a). Hence, $E(Y \mid X = x) = (n - x)/5$, since the mean of a binomial distribution with parameters n and p is np.

c   Let Y have the uniform distribution on $(0,1)$; let a coin with probability Y of turning up heads be tossed until a head appears; and let X denote the number of tosses required, as in Example 10.1.7a. Then, by Example 10.1.7a,

$$h(y \mid x) = x(x + 1)y(1 - y)^{x-1}$$

for $0 < y < 1$ and $x = 1, 2, \dots$. Therefore,

$$E(Y \mid X = x) = x(x + 1) \int_0^1 y^2 (1 - y)^{x-1} \, dy$$

which reduces to $2/(x + 2)$ after some manipulations.

d Let *B* be a region of R^2 and let $Z = I_B(X,Y)$. Thus, $Z = 1$ if $(X,Y) \in B$ and $Z = 0$ if $(X,Y) \notin B$. Further, let B_x denote the set of $y \in R$ for which $(x,y) \in B$. Then $I_B(x,y) = 1$ if $y \in B_x$ and 0 if $y \notin B_x$, so that

$$E(Z \mid X = x) = \int_{-\infty}^{\infty} I_B(x,y)\, dH(y \mid x)$$

$$= \int_{B_x} 1\, dH(y \mid x) = \Pr(Y \in B_x \mid X = x)$$

for all *x* for which $H(\cdot \mid x)$ is defined. /////

Like ordinary (unconditional) expectation, conditional expectation enjoys several useful and interesting properties, which we shall now develop. The first two are linearity and monotonicity.

Theorem 10.3.1 *Let X and Y be jointly distributed; let* $Z_1 = w_1(X,Y)$ *and* $Z_2 = w_2(X,Y)$; *and let* α_1 *and* α_2 *be real numbers. If* $E(Z_1 \mid X = x)$ *and* $E(Z_2 \mid X = x)$ *are both defined, then*

$$E(\alpha_1 Z_1 + \alpha_2 Z_2 \mid X = x) = \alpha_1 E(Z_1 \mid X = x) + \alpha_2 E(Z_2 \mid X = x)$$

Moreover, if $w_1(x,y) \le w_2(x,y)$ *for all* $y \in R$, *then* $E(Z_1 \mid X = x) \le E(Z_2 \mid X = x)$. *In particular,* $|E(Z_1 \mid X = x)| \le E(|Z_1| \mid X = x)$.

PROOF The properties claimed for conditional expectation are well-known properties of the summation and integration operations which define conditional expectation (compare Theorem 8.2.2). /////

Our next theorem is also to be anticipated.

Theorem 10.3.2 *Let X and Y be independent, and let* $Z = v(Y)$, *where v is a function on R. If* $E(|Z|) < \infty$, *then*

$$E(Z \mid X = x) = E(Z)$$

for all x for which conditional expectation is defined.

The proof is left as an exercise. Our next theorem asserts that given $X = x$, functions of *X* act like scalars in conditional expectations.

Theorem 10.3.3 *Let X and Y be jointly distributed, and let* $Z = u(X)w(X,Y)$, *where u and w are functions on R and* R^2, *respectively. If* $E[w(X,Y) \mid X = x]$

is defined, then so is $E(Z \mid X = x)$ and

$$E(Z \mid X = x) = u(x)E[w(X,Y) \mid X = x]$$

PROOF   We shall give the proof in the absolutely continuous case. In this case

$$E(Z \mid X = x) = \int_{-\infty}^{\infty} u(x)w(x,y)h(y \mid x)\,dy$$

$$= u(x)\int_{-\infty}^{\infty} w(x,y)h(y \mid x)\,dy$$

$$= u(x)E[w(X,Y) \mid X = x]$$

as asserted.   ////

Our next theorem and its corollary are the most important results of this section.

Theorem 10.3.4   *Let X and Y be jointly distributed random variables, and let D be a subset of R for which $\Pr(X \in D) = 1$. Also, let $Z = w(X,Y)$ be a random variable for which $E(Z \mid X = x)$ is defined for every $x \in D$. If $E(Z)$ is finite, then*

$$E(Z) = \int_{D} E(Z \mid X = x)\,dG(x) \qquad (3.5)$$

where G denotes the distribution function of X, provided that the integral in (3.5) converges absolutely.

PROOF   We shall prove the theorem in the discrete case only, although it is true in the generality stated. Let f denote the joint mass function of X and Y and g denote the marginal mass function of X, and let D be the set of $x \in R$ for which $g(x) > 0$. Further let E be any finite or countably infinite set for which $\Pr(Y \in E) = 1$. Then

$$E(Z) = \sum_{(x,y) \in D \times E} w(x,y)f(x,y)$$

$$= \sum_{(x,y) \in D \times E} w(x,y)h(y \mid x)g(x)$$

$$= \sum_{x \in D} [\sum_{y \in E} w(x,y)h(y \mid x)]g(x)$$

$$= \sum_{x \in D} E(Z \mid X = x)g(x) = \int_{D} E(Z \mid X = x)\,dG(x)$$

as asserted.   ////

Combining Theorems 10.3.3 and 10.3.4 produces the following corollary.

Corollary 10.3.1 *Let Z be as in Theorem* 10.3.4, *and let* $U = u(X)$, *where u is a function on R. Then*

$$E(UZ) = \int_D u(x)E(Z \mid X = x)\, dG(x)$$

provided that the integral converges absolutely.

PROOF Simply observe that $E(UZ \mid X = x) = u(x)E(Z \mid X = x)$, $x \in D$, and apply Theorem 10.3.4 to the random variable UZ.   /////

Theorem 10.3.4 is most naturally stated in a slightly different notation. Let X, Y, and Z be as in the statement of Theorem 10.3.4, and for each $x \in D$, let $\mu(x) = E(Z \mid X = x)$. Then, μ is a well-defined function on D, so that $\mu(X)$ is a random variable. We shall denote this random variable by $E(Z \mid X)$. Thus,

$$E(Z \mid X) = \mu(X) \qquad (3.6)$$

where $\mu(x) = E(Z \mid X = x)$ for each $x \in D$. The result of Theorem 10.3.4 can now be stated

$$E(Z) = E[E(Z \mid X)] \qquad (3.5')$$

for the right side of (3.5) is simply the expectation of the random variable $\mu(X) = E(Z \mid X)$.

EXAMPLE 10.3.2

a Let X have the uniform distribution on $(0,1)$, and conditionally given $X = x$, let Y have the uniform distribution on $(0,x)$, $0 < x < 1$, as in Example 10.1.5. Then, $h(y \mid x) = 1/x$, $0 < y < x < 1$, so that

$$E(Y \mid X = x) = \frac{1}{x}\int_0^x y\, dy = \tfrac{1}{2}x$$

$0 < x < 1$. That is, $E(Y \mid X) = \tfrac{1}{2}X$. Since $E(X) = \tfrac{1}{2}$, it now follows that

$$E(Y) = E[E(Y \mid X)] = E(\tfrac{1}{2}X) = \tfrac{1}{4}$$

b Let Y have the beta distribution with parameters $\alpha > 0$ and $\beta > 0$, and conditionally given $Y = y$, let X have the binomial distribution with parameters n and y, $0 < y < 1$. Then, $E(X \mid Y = y) = ny$, $0 < y < 1$, and $E(Y) = \alpha/(\alpha + \beta)$ by Example 8.4.1a. Therefore, $E(X) = E[E(X \mid Y)] = nE(Y) = n\alpha/(\alpha + \beta)$.   /////

We shall call the function μ of Equation (3.6) the *conditional mean* of Z given X and the function σ^2 defined by

$$\sigma^2(x) = E\{[Z - \mu(x)]^2 \mid X = x\}$$

for $x \in D$ the *conditional variance* of Z given $X = x$, provided, of course, that it is finite for all $x \in D$. It is then easily verified that

$$\sigma^2(x) = E(Z^2 \mid X = x) - \mu(x)^2 \qquad (3.7)$$

for $x \in D$. Our final corollary relates the unconditional variance of Z to its conditional mean and variance. In it, we have denoted the random variable $\sigma^2(X)$ by $D(Z \mid X)$.

Corollary 10.3.2 *Let Z have finite variance, and let the conditional mean and variance of Z be well defined. Then*

$$D(Z) = E[D(Z \mid X)] + D[E(Z \mid X)]$$

In words, the variance of Z is the expectation of its conditional variance plus the variance of its conditional expectation.

PROOF   We may suppose that $E(Z) = 0$, in which case $D(Z) = E(Z^2) = E[E(Z^2 \mid X)]$. Moreover, by (3.7) $E(Z^2 \mid X) = \sigma^2(X) + \mu(X)^2$, so that

$$E(Z^2) = E[\sigma^2(X)] + E[\mu(X)^2] \qquad (3.8)$$

Finally, $E[\mu(X)] = E(Z) = 0$, by Theorem 10.3.4, so that $E[\mu(X)^2] = D[\mu(X)]$. The theorem follows.   ////

10.4   HIGHER DIMENSIONS[1]

The notions of the preceding three sections extend easily from the case of two random variables to the case of several. In fact, *the definitions and theorems of Sections 10.1 to 10.3 remain valid when either X or Y, or both, are random vectors, provided that proper allowance is made for the dimension of the domain of the conditional and marginal densities and mass functions which appear in Sections 10.1 to 10.3*. We shall sketch these extensions only in the absolutely continuous case. The discrete case is similar and generally simpler, and the mixed is also similar.

Let $X_1, \ldots, X_m$ and $Y_1, \ldots, Y_n$ be jointly distributed random variables. Further, let f be a joint density for the random vector $(X_1, \ldots, X_m, Y_1, \ldots, Y_n)$, and let g and h denote the marginal densities of the random vectors $X = (X_1, \ldots, X_m)$ and $Y = (Y_1, \ldots, Y_n)$, respectively. Finally, let D denote the set of $x = (x_1, \ldots, x_m) \in R^m$

---

[1] This section treats a special topic and may be omitted without loss of continuity.

for which $g(x) > 0$. For $x \in D$, we define the conditional density of Y given that $X = x$ by

$$h(y \mid x) = \frac{f(x,y)}{g(x)} \qquad y \in R^n \qquad (4.1)$$

Moreover, if B is a region of R^n, and if $x \in D$, then we define the conditional probability that $Y \in B$ given that $X = x$ by

$$\Pr (Y \in B \mid X = x) = \int_B h(y \mid x)\, dy \qquad (4.2)$$

And finally, if $Z = w(X,Y)$ is a random variable which is determined as a function of X and Y, then we define the conditional expectation of Z given that $X = x$ for $x \in D$ by

$$E(Z \mid X = x) = \int_{R^n} w(x,y)h(y \mid x)\, dy \qquad (4.3)$$

provided that the (n-dimensional) integral appearing in (4.3) converges absolutely. Alternatively, the conditional expectation of Z given that $X = x$ can be computed from the formula

$$E(Z \mid X = x) = \int_{-\infty}^{\infty} z\, dK(z \mid x) \qquad (4.4)$$

where $K(\cdot \mid x)$ denotes the conditional distribution function of Z given $X = x$. That is, $K(z \mid x) = \Pr (Z \le z \mid X = x)$ for $z \in R$ and $x \in D$. The equivalence of (4.3) and (4.4) can be established by applying Theorem 8.2.1 to the conditional distribution of Y given $X = x$ and the function $Z = w(x,Y)$ for each $x \in D$.

It is now easily verified that Theorems 10.2.1 to 10.2.4 and 10.3.1 to 10.3.4 and their corollaries remain valid with the extended definitions of conditional probability and expectation, provided only that proper allowance is made for the dimensions of the domains of the conditional densities and mass functions which appear in them. Both the statements and the proofs of these results in the higher-dimensional case are so similar to those of the two-dimensional case that they need not be reproduced here.

Bayes' theorem is also valid for multivariate conditional densities. That is, in the notation of (4.1) to (4.4),

$$h(y) = \int_D h(y \mid x)g(x)\, dx \qquad (4.5)$$

and

$$g(x \mid y) = \frac{h(y \mid x)g(x)}{h(y)} \qquad (4.6)$$

if $h(y) > 0$.

EXAMPLE 10.4.1

a Let $X_1, \ldots, X_m$ and $Y_1, \ldots, Y_n$ have the multinomial distribution, say

$$f(x_1, \ldots, x_m, y_1, \ldots, y_n)$$
$$= \binom{N}{x_1, \ldots, x_m, y_1, \ldots, y_n} p_1^{x_1} \cdots p_m^{x_m} q_1^{y_1} \cdots q_n^{y_n}$$

for nonnegative integers $x_1, \ldots, y_n$ with $x_1 + \cdots + y_n = N$. Here $p_1, \ldots, p_m$, $q_1, \ldots, q_n$ are nonnegative, and $p_1 + \cdots + p_m + q_1 + \cdots + q_n = 1$. Suppose also that $q_1 + \cdots + q_n > 0$. By Problem 6.25 the marginal mass function of $X_1, \ldots, X_m$ is

$$g(x_1, \ldots, x_m) = \binom{N}{x_1, \ldots, x_m, k} p_1^{x_1} \cdots p_n^{x_m} r^k$$

for $x_1 + \cdots + x_m \leq N$, where $k = N - x_1 - \cdots - x_m$ and $r = q_1 + \cdots + q_n$. The conditional mass function of $Y_1, \ldots, Y_n$ given $X_1, \ldots, X_m$ is

$$h(y_1, \ldots, y_n \mid x_1, \ldots, x_m) = \frac{f(x_1, \ldots, x_m, y_1, \ldots, y_n)}{g(x_1, \ldots, x_m)}$$

which simplifies to

$$\binom{k}{y_1, \ldots, y_n} \left(\frac{q_1}{r}\right)^{y_1} \cdots \left(\frac{q_n}{r}\right)^{y_n}$$

Thus, the conditional distribution of $Y_1, \ldots, Y_n$ is multinomial with new parameters

$$k = N - x_1 - \cdots - x_m \quad \text{and} \quad q_i' = \frac{q_i}{r}$$

for $i = 1, \ldots, n$.

b Similarly, if $X_1, \ldots, X_m$ and $Y_1, \ldots, Y_n$ have the multinomial hypergeometric distribution, say

$$f(x_1, \ldots, x_m, y_1, \ldots, y_n) = \frac{\binom{s_1}{x_1} \cdots \binom{s_m}{x_m} \binom{r_1}{y_1} \cdots \binom{r_1}{y_1}}{\binom{s_1 + \cdots + s_m + r_1 + \cdots + r_n}{k}}$$

for nonnegative integers $x_1, \ldots, x_m, y_1, \ldots, y_n$ with $x_1 + \cdots + y_n = k$, then

$$h(y_1, \ldots, y_n \mid x_1, \ldots, x_m) = \frac{\binom{r_1}{y_1} \cdots \binom{r_n}{y_n}}{\binom{r_1 + \cdots + r_n}{k'}}$$

where $k' = k - x_1 - \cdots - x_m$.   ////

A new phenomenon in higher dimensions is the notion of *conditional independence*. Thus, let $X = (X_1, \ldots, X_m)$ and $Y = (Y_1, \ldots, Y_n)$ be jointly distributed, jointly absolutely continuous random vectors with densities g and h, respectively. Further, let D be the set of $x \in R^n$ for which $g(x) > 0$, and for $x \in D$ let $h(\cdot \mid x)$ and $h_1(\cdot \mid x), \ldots,$ $h_n(\cdot \mid x)$ denote the conditional densities of Y and $Y_1, \ldots, Y_n$ given $X = x$. If

$$h(y_1, \ldots, y_n \mid x) = \prod_{i=1}^{n} h_i(y_i \mid x)$$

for all $y = (y_1, \ldots, y_n) \in R^n$ and all $x \in D$, then we shall say that $Y_1, \ldots, Y_n$ *are conditionally independent given X*. In this case, the marginal density of Y will be

$$h(y_1, \ldots, y_n) = \int_D \prod_{i=1}^{n} h_i(y_i \mid x) g(x) \, dx$$

so that $Y_1, \ldots, Y_n$ need not be unconditionally independent. The notion of conditional independence leads us, in fact, to a new class of models.

EXAMPLE 10.4.2 Let X have the exponential distribution with parameter $\beta = 1$, and conditionally given $X = x > 0$, let $Y_1, \ldots, Y_n$ be independent, exponentially distributed random variables with parameter $\beta = x$ (in this case $m = 1$). That is, let

$$g(x) = e^{-x} \qquad x > 0$$

and
$$h(y_1, \ldots, y_n \mid x) = x^n e^{-x(y_1 + \cdots + y_n)}$$

for $y_i > 0$, $i = 1, \ldots, n$. The marginal density of $Y = (Y_1, \ldots, Y_n)$ is then

$$h(y_1, \ldots, y_n) = \int_0^{\infty} x^n e^{-x(y_1 + \cdots + y_n)} e^{-x} \, dx$$

$$= \int_0^{\infty} x^n e^{-x(1 + y_1 + \cdots + y_n)} \, dx = \frac{n!}{(1 + y_1 + \cdots + y_n)^{n+1}}$$

for $y_i > 0$, $i = 1, \ldots, n$. [The final equality follows from the change of variables $x' = x(1 + y_1 + \cdots + y_n)$ and the definition of the gamma function.] Letting $z = y_1 + \cdots + y_n$, it now follows from (4.6) that the conditional density of X given $Y = y = (y_1, \ldots, y_n)$ is

$$g(x \mid y) = \frac{(1 + z)^{n+1} x^n e^{-(1+z)x}}{n!}$$

for $x > 0$ and $y_i > 0$, $i = 1, \ldots, n$. That is, the conditional distribution of X given $Y = y$ is gamma with parameters $\alpha = n + 1$ and $\beta = 1 + z$. Hence,

$$E(X \mid Y = y) = \frac{n + 1}{1 + z} \qquad \text{////}$$

We shall now consider an extension of Theorem 10.3.4. Let X, Y, and Z be jointly absolutely continuous random vectors, and let f, h, and g denote the joint density of X, Y, and Z, the marginal density of X and Y, and the marginal density of X, respectively. Further, let

$$W = w(X, Y, Z)$$

be a random variable which is determined as a function of X, Y, and Z, and suppose that the conditional expectations

$$\mu(x) = E(W \mid X = x)$$

$$v(x, y) = E(W \mid X = x, Y = y)$$

exist whenever $g(x) > 0$ and $h(x, y) > 0$. As in the previous section, we shall denote the random variables $\mu(X)$ and $v(X, Y)$ by $E(W \mid X)$ and $E(W \mid X, Y)$, respectively, so that Theorem 10.3.4 (as extended to higher dimensions) asserts that

$$E(W) = E[E(W \mid X)] = E[E(W \mid X, Y)] \tag{4.7}$$

Theorem 10.4.1 *With the notations and assumptions of the previous paragraph, we have*

$$E[E(W \mid X, Y) \mid X = x] = E(W \mid X = x)$$

whenever $g(x) > 0$. That is, $E[E(W \mid X, Y) \mid X] = E(W \mid X)$.

PROOF Let $k(\cdot \mid x, y)$ and $l(\cdot, \cdot \mid x)$ denote the conditional density of Z given $X = x$ and $Y = y$ and the conditional density of Y and Z given $X = x$, respectively. Then

$$k(z \mid x, y) = \frac{f(x, y, z)}{h(x, y)} = \frac{f(x, y, z)/g(x)}{h(x, y)/g(x)} = \frac{l(y, z \mid x)}{h(y \mid x)}$$

whenever $g(x) > 0$ and $h(x, y) > 0$. Therefore,

$$v(x, y) = \int w(x, y, z) k(z \mid x, y)\, dz$$

$$= \frac{1}{h(y \mid x)} \int w(x, y, z) l(y, z \mid x)\, dz$$

if $h(y \mid x) > 0$. Moreover,

$$E[E(W \mid X, Y) \mid X = x]$$

$$= E[v(X, Y) \mid X = x] = \int v(x, y) h(y \mid x)\, dy$$

where the integral extends over y for which $h(y \mid x) > 0$. Combining these expressions, we find

$$E[E(W \mid X,Y) \mid X = x] = \int w(x,y,z) l(y,z \mid x) \, dy \, dz$$

$$= E(W \mid X = x)$$

as asserted.   ////

10.5   DECISION THEORY[1]

In this section we shall consider a mathematical model for the problem of making decisions in the face of uncertainty. The development of this model relies heavily on the notions of the preceding four sections and may be regarded as an application of them. Of course, we shall be able only to scratch the surface of this rich area, and we refer interested readers to the references at the end of this chapter.

Our model involves the following elements. First, we shall suppose that the state of nature is unknown to us but that there is a known set D of possible states of nature. We shall consider here only the case that D is an interval of real numbers, though the theory may be extended to the case that D is a region of higher-dimensional euclidean space. We shall denote the elements of D, that is, the possible states of nature, by x. Also, we suppose that we are required to take one of a specified set of actions A and that if we take action $a \in A$ when the state of nature is actually $x \in D$, then we incur a loss $L(a,x)$, where L is a continuous[2] function on the product space $A \times D$. Finally, we suppose that before taking any action we are allowed to perform an experiment in order to learn about the unknown state of nature. The outcome of this experiment is assumed to be a random variable or vector Y whose distribution depends on the unknown state of nature x. The question we wish to answer is the following: If we observe $Y = y$, what action should we take?

In order to answer this question, we shall specify a density g which represents our opinion about the state of nature prior to the experimentation. That is, we shall regard the unknown state of nature as a random variable X with density g, where g is so selected that

$$\Pr \, (X \in B) = \int_B g(x) \, dx$$

gives our subjective probability that X belongs to any subinterval $B \subset D$ prior to the experiment. We shall refer to g as the *prior density,* and we shall suppose that g

---

[1] This section treats a special topic and may be omitted without loss of continuity.
[2] If A is finite, this means that $L(a,x)$ should be continuous in x for each a.

is positive on D and vanishes off of D. We now agree to treat X and Y as jointly distributed variables in the following manner. The conditional density or mass function of Y given that $X = x$, say $h(\cdot \mid x)$, is supposed known for each $x \in D$, and the marginal density of X is g. The marginal density or mass function of Y will then be

$$h(y) = \int_D h(y \mid x)g(x)\, dx$$

After the experiment is performed and the value of Y observed, we can compute the conditional density of X given Y. The latter density then describes our new opinion about X, the unknown state of nature, after the experiment has been performed and is often referred to as the *posterior density* of X. By Bayes' theorem it is

$$g(x \mid y) = \frac{h(y \mid x)g(x)}{h(y)}$$

for $x \in D$ and $h(y) > 0$.

Let E be the set of y for which $h(y) > 0$, so that $\Pr(Y \in E) = 1$. We define a *decision policy* to be a function, δ say, from E into A, the action space. A decision policy is a rule which tells us to take action $\delta(y)$ when we observe the outcome $Y = y$. In this case our expected loss is

$$R(\delta) = E[L(\delta(Y),X)]$$

where the expectation is taken with respect to the joint distribution of X and Y. Of course, we must assume that the policy δ is sufficiently regular for the expectation defining $R(\delta)$ to exist. We shall call such policies *regular policies*, and we shall consider only regular policies.

A regular policy δ_0 will be called optimal if it minimizes the expected loss. That is, δ_0 is *optimal* if and only if

$$R(\delta_0) \le R(\delta)$$

for every other regular policy δ. The obvious question is then: How can we determine an optimal policy? The answer is provided by the following theorem.

Theorem 10.5.1 *If the regular policy δ_0 has the property that*

$$E[L(\delta_0(y),X) \mid Y = y] = \min_{a \in A} E[L(a,X) \mid Y = y] \qquad (5.1)$$

for every $y \in E$, then δ_0 is optimal. That is, the optimal policy can be determined by letting $\delta_0(y)$ be that action $a \in A$ which minimizes the conditional expected loss given $Y = y$ for each $y \in E$.

PROOF The proof of the theorem is easy. Indeed, if δ is any regular policy, then it follows from (5.1) that

$$E[L(\delta_0(y),X) \mid Y = y] \le E[L(\delta(y),X) \mid Y = y]$$

for every $y \in E$. Therefore, by Theorem 10.3.4,

$$R(\delta_0) = \int_E E[L(\delta_0(y),X) \mid Y = y]h(y)\, dy$$

$$\leq \int_E E[L(\delta(y),X) \mid Y = y]h(y)\, dy = R(\delta) \qquad (5.2)$$

as asserted. [If Y is discrete, the integral in (5.2) must be replaced by a summation, but the result is the same.]   ////

EXAMPLE 10.5.1 Suppose that we wish to determine the probability that a coin will turn up heads. Here we may take the state of nature to be the probability in question, in which case $D = (0,1)$, the open unit interval. Moreover, since we are required to guess the state of nature, we may take the action space to be $A = D = (0,1)$. For the loss function L, it seems natural to take

$$L(a,x) = c(x - a)^2 \qquad (5.3)$$

or possibly

$$L(a,x) = c|x - a|$$

where c is a positive constant. We shall consider only the loss function (5.3) in this example, leaving the other loss function for a problem. In order to learn about the unknown state of nature, we might toss the coin several times and count the number of heads. If we toss the coin n times and let Y denote the number of heads, then the conditional distribution of Y given $X = x$ will be binomial with parameters n and x. That is, we shall have

$$h(y \mid x) = \binom{n}{y} x^y (1 - x)^{n-y}$$

for $y = 0, \ldots, n$ and $0 < x < 1$.

Finally, we must specify the prior density g. For reasons of mathematical tractability, we shall suppose that our prior opinion is adequately represented by a beta density, say

$$g(x) = \frac{\Gamma(\alpha + \beta)}{\Gamma(\alpha)\Gamma(\beta)} x^{\alpha-1} (1 - x)^{\beta-1}$$

for $0 < x < 1$, where $\alpha > 0$ and $\beta > 0$. The parameters α and β may still be selected to represent our prior opinion. For example, the choice $\alpha = \beta = 6$ might be appropriate if we had a strong belief that X is close to $\frac{1}{2}$, while the choice $\alpha = \beta = 1$ (the uniform distribution) might be appropriate if we had very little prior opinion about X.

Having specified the problem completely, we shall now solve it. The first step is to find the conditional distribution of X given Y. By Example 10.1.7b, this is beta with parameters $\alpha' = \alpha + y$ and $\beta' = \beta + n - y$. That is,

$$g(x \mid y) = \frac{\Gamma(\alpha' + \beta')}{\Gamma(\alpha')\Gamma(\beta')} x^{\alpha'-1}(1 - x)^{\beta'-1}$$

for $0 < x < 1$ and $y = 0, \ldots, n$. Next, we must minimize the conditional expected loss

$$E[L(a,X) \mid Y = y] = c \int_0^1 (x - a)^2 g(x \mid y) \, dx$$

with respect to a. By Lemma 8.3.1 we know that this is done by taking

$$a = E(X \mid Y = y) = \int_0^1 x g(x \mid y) \, dx$$

Finally, by Example 8.4.1a, we know that the expectation of a beta density is $\alpha/(\alpha + \beta)$. Therefore, the optimal policy is

$$\delta_0(y) = E(X \mid Y = y) = \frac{\alpha + y}{\alpha + \beta + n}$$

We observe that the estimate $\delta_0(y)$ is different from the actual frequency of heads y/n.

The fact that the optimal decision policy was to let $\delta_0(y) = E(X \mid Y = y)$, $y \in E$, in Example 10.5.1 depended only on the loss function (5.3) and not on the other specifications of the problem.   ////

EXAMPLE 10.5.2 Let us suppose that we wish to decide whether an unknown quantity x is positive or negative. More precisely, we suppose that all real values of x are possible, in which case $D = R$, and that we are required to take one of the two actions a_0 and a_1, where a_0 represents the decision that $x \le 0$ and a_1 the decision that $x > 0$. We shall also suppose that there is no loss for a correct decision and that the loss for an incorrect decision is proportional to $|x|$. That is, we take our loss function to be

$$L(a,x) = \begin{cases} 0 & \text{if } a = a_0 \text{ and } x \le 0 \text{ or } a = a_1 \text{ and } x > 0 \\ c|x| & \text{otherwise} \end{cases}$$

where c is a positive constant. Finally, let us suppose that we are allowed to make n measurements on x, say $Y_1, \ldots, Y_n$, which are subject to measurement error. In fact, let us suppose that given $X = x$, $Y_1, \ldots, Y_n$ are independent and have the normal distribution with mean x and variance τ^2. Finally, let us suppose that our prior opinion about X is adequately described by a normal distribution with mean μ and variance σ^2.

Let $L_0(a,x) = L(a,x) - L(a_0,x)$, $a \in A$, $x \in R$. Then we minimize $E[L(\delta(Y),X)]$ with respect to δ is and only if we minimize

$$E[L_0(\delta(Y),X)] = E[L(\delta(Y),X)] - E[L(a_0,X)]$$

with respect to δ, for the difference is independent of δ. Now $L(a_0,x) = 0$ for all $x \in R$, and $L_0(a_1,x) = -cx$ for all $x \in R$. Therefore,

$$E[L_0(a_0,X) \mid Y = y] = 0 \tag{5.4a}$$

$$E[L_0(a_1,X) \mid Y = y] = -cE(X \mid Y = y) \tag{5.4b}$$

for all $y \in R^n$. By Theorem 10.5.1, an optimal policy is to let $\delta(y) = a_1$ if and only if (5.4b) is less than (5.4a). That is,

$$\delta_0(y) = \begin{cases} a_1 & \text{if } E(X \mid Y = y) > 0 \\ a_0 & \text{if } E(X \mid Y = y) \le 0 \end{cases} \tag{5.5}$$

[Actually, either decision is admissible if $E(X \mid Y = y) = 0$.]

All that is left is to find $E(X \mid Y = y)$, and this will be left as an exercise (Problem 10.41). The answer is

$$E(X \mid Y = y) = \frac{\mu\sigma^{-2} + nz\tau^{-2}}{\sigma^{-2} + n\tau^{-2}}$$

where $z = \bar{y} = (y_1 + \cdots + y_n)/n$.

As in the previous problem, the result (5.5), the general form of the optimal policy, depends only on the loss function and not on the other specifications of the problem. ////

10.6 BRANCHING PROCESSES[1]

In this section we shall consider a model for population growth. We suppose that at the beginning of the first generation a population has X_0 members. During the first generation each of these X_0 members has a random number of progeny, and at the end of the first generation all the original X_0 members die or leave the population. Let Z_{1i} be the number of progeny of the ith of the original X_0 members. Then the number of progeny at the end of the first generation is

$$X_1 = \sum_{i=1}^{X_0} Z_{1i} \tag{6.1}$$

In later generations this process repeats itself. If there are X_{n-1} members in the population at the end of the $(n - 1)$st generation, where $n \ge 2$, and if the ith of

---

[1] This section treats a special topic and may be omitted without loss of continuity.

these has Z_{ni} progeny during the nth generation, then the population size at the end of the nth generation is

$$X_n = \sum_{i=1}^{X_{n-1}} Z_{ni} \qquad (6.2)$$

We interpret X_n as zero if $X_{n-1} = 0$.

If $X_n = 0$ for some n, then $X_m = 0$ for all $m \geq n$ by (6.2). In this case we shall say that the population becomes *extinct*. We wish to compute the probability that the population becomes extinct. In order to do so, we shall have to make some assumptions about the evolution of the population. We shall suppose that *for each n, $Z_{n1}, \ldots, Z_{nk}$ are conditionally independent given $X_0 = k_0$, $X_1 = k_1, \ldots, X_{n-1} = k_{n-1}$. We shall also suppose that the conditional mass function of Z_{ni} does not depend on n, i, or $k_0, \ldots, k_{n-1}$.* Thus,

$$f(j) = \Pr (Z_{ni} = j \mid X_0 = k_0, \ldots, X_{n-1} = k_{n-1})$$

is the probability that a member of the population has exactly j progeny for $j = 0$, $1, 2, \ldots$, and this probability is assumed to be independent of the generation n, the population member i, and the sizes of previous generations $k_0, \ldots, k_{n-1}$. We shall suppose that $f(0) > 0$, since otherwise the probability of extinction is trivially zero.

It is easy to compute $E(X_n)$. Let μ denote the expected number of progeny from a single individual. That is, let

$$\mu = \sum_{j=0}^{\infty} jf(j)$$

Lemma 10.6.1 $E(X_n) = X_0 \mu^n$ for $n = 1, 2, \ldots$.

PROOF By (6.1), we have $E(X_1) = X_0 \mu$. Moreover, by (6.2) and the conditional independence of the Z_{ni}, we have $E(X_n \mid X_0, \ldots, X_{n-1}) = X_{n-1}\mu$. Therefore, by Theorem 10.3.4, we have $E(X_n) = \mu E(X_{n-1}) = \mu^2 E(X_{n-2}) = \cdots = \mu^{n-1} E(X_1) = X_0 \mu^n$. ////

A similar technique can be used to compute the generating function of X_n. Let G_n denote the generating function of X_n,

$$G_n(t) = E(t^{X_n}) = \sum_{j=0}^{\infty} \Pr (X_n = j)t^n$$

for $-1 \leq t \leq 1$. Also, let F denote the generating function of the Z_{ni},

$$F(t) = E(t^{Z_{ni}}) = \sum_{j=0}^{\infty} f(j)t^j$$

for $-1 \leq t \leq 1$. Then $G_1(t) = F(t)^{X_0}$ by (6.1), Theorem 8.4.7, and the independence of $Z_{11}, \ldots, Z_{1X_0}$. Moreover,

$$E(t^{X_n} \mid X_0, \ldots, X_{n-1}) = F(t)^{X_{n-1}}$$

by (6.2) and the conditional independence of $Z_{n1}, \ldots, Z_{nX_{n-1}}$ given X_{n-1}.

Lemma 10.6.2 *For $n = 1, 2, \ldots,$ we have $G_n(t) = G_{n-1} \circ F(t) = G_{n-1}(F(t))$ for $-1 \leq t \leq 1$.*

PROOF   By Theorem 10.3.4, we have

$$G_n(t) = E(t^{X_n}) = E[E(t^{X_n} \mid X_0, \ldots, X_{n-1})]$$
$$= E[F(t)^{X_{n-1}}] = G_{n-1}(F(t))$$

for $-1 \leq t \leq 1$.   ////

Let us define F_n recursively by $F_1(t) = F(t)$, $F_2(t) = F \circ F(t)$, and $F_n(t) = F_{n-1} \circ F(t)$ for $-1 \leq t \leq 1$. Since composition is associative, we can write $F_n(t) = F \circ F \cdots \circ F$, the composition of F with itself n times, and it follows that $F_n(t) = F \circ F_{n-1}(t)$ for $-1 \leq t \leq 1$ and $n \geq 1$.

Corollary 10.6.1 *For $n = 1, 2, \ldots$ and $-1 \leq t \leq 1$, $G_n(t) = F_n(t)^{X_0}$.*

PROOF   When $n = 1$, this relation has already been observed; and if it is true when $n = m - 1$, then $G_m(t) = G_{m-1} \circ F(t) = [F_{m-1} \circ F(t)]^{X_0} = F_m(t)^{X_0}$. The corollary follows by induction.   ////

Let $\alpha_n = \Pr(X_n = 0)$. Then, since $X_n = 0$ implies $X_{n+1} = 0$, we must have $\alpha_n \leq \alpha_{n+1}$ for every $n \geq 1$. It follows that

$$\alpha = \lim_{n \to \infty} \alpha_n$$

exists. We shall call α the *probability of extinction*.

Theorem 10.6.1 *If $\mu \leq 1$, then $\alpha = 1$; and if $\mu > 1$, then $\alpha = \beta^{X_0}$, where β is the smallest positive solution of the equation*

$$\beta = F(\beta) \qquad (6.3)$$

PROOF   Let $\beta_n = F_n(0)$, so that $\alpha_n = G_n(0) = F_n(0)^{X_0} = \beta_n^{X_0}$. Also, let $\beta = \lim \beta_n$ as $n \to \infty$, so that $\alpha = \beta^{X_0}$. Then $\beta_n = F_n(0) = F(F_{n-1}(0)) = F(\beta_{n-1})$, and Equation (6.3) can be obtained by letting $n \to \infty$. To see that β is the smallest positive solution to (6.3), let γ be any other positive solution. Then, since F is a nondecreasing function, we must have $\beta_1 = F(0) \leq F(\gamma) = \gamma$. By induction, we then have $\beta_n = F(\beta_{n-1}) \leq F(\gamma) = \gamma$, and therefore $\beta = \lim \beta_n \leq \gamma$ as $n \to \infty$.

It remains to show that $\beta = 1$ if $\mu \leq 1$. We claim that if $\mu \leq 1$, then $F'(t) < 1$ for $0 < t < 1$. To see this write

$$F'(t) = \sum_{j=1}^{\infty} jf(j)t^{j-1}$$

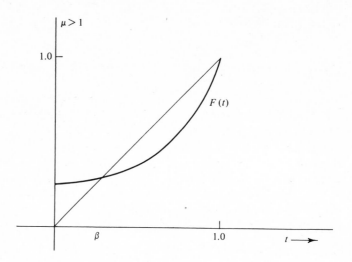

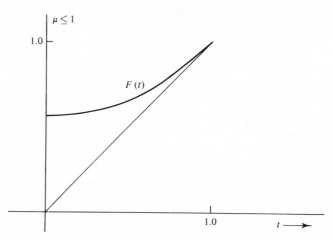

FIGURE 14
The equation $\beta = F(\beta)$.

for $0 < t < 1$. If $f(j) > 0$ for some $j \geq 2$, then $f(j)t^{j-1} < f(j)$ and consequently $F'(t) < F'(1-) = \mu \leq 1$ for $0 < t < 1$; and if $f(j) = 0$ for all $j \geq 2$, then $F'(t) = f(1)$, which is less than 1 because $f(0) > 0$. Now, if $\beta < 1$, then $1 - \beta = 1 - F(\beta) = F'(\delta)(1 - \beta)$ with $\beta < \delta < 1$ by the mean-value theorem. It follows that $F'(\delta) = 1$, contradicting the assumption that $\mu \leq 1$ (see Figure 14).   ////

REFERENCES

A completely general treatment of conditional probability and expectation requires the abstract theory of measure. Readers who are interested in this approach may consult Neveu (1965), chap. 4.

For a more detailed account of decision theory (Section 10.5), see DeGroot (1970) or Blackwell and Girshick (1954). For a more detailed treatment of branching processes (Section 10.6), see Karlin (1966), chap. 11.

The type of dependence exhibited by branching processes is a special case of *markovian dependence*. We shall not give a systematic exposition of this subject. Instead, we refer interested readers to Karlin (1966), chaps. 2 to 5, and Feller (1968), chaps. 15 and 16.

PROBLEMS

10.1   Let a random sample of size k be drawn without replacement from an urn which contains r red balls, b black balls, and w white balls ($k \le n = r + b + w$). Also, let X and Y denote the number of red and white balls in the sample, respectively. Find the conditional mass function of Y given $X = x$ for all possible values of x. Interpret your results.

10.2   Let X and Y be independent random variables which have the Poisson distribution with parameters $\alpha > 0$ and $\beta > 0$, respectively. Also, let $Z = X + Y$. Show that the conditional mass function of X given $Z = z$ is binomial with parameters $n = z$ and $p = \alpha/(\alpha + \beta)$ for $z = 0, 1, 2, \ldots$.

10.3   Let X and Y be independent random variables which are geometrically distributed with the same parameter p, $0 < p < 1$, and let $Z = X + Y$. For $z = 2, 3, \ldots$, find the conditional mass function of X given $Z = z$.

10.4   Let two balanced dice be tossed, and let X and Y be the sum and maximum number of spots appearing on the two dice, respectively. For $y = 1, \ldots, 6$, find the conditional mass function of X given $Y = y$.

10.5   Generalize Problem 10.4 to three dice.

10.6   Let X and Y have the bivariate Cauchy density $f(x,y) = 1/[2\pi\sqrt{(1 + x^2 + y^2)^3}]$, $-\infty < x, y < \infty$. Find the conditional density of X given $Y = y$ for $-\infty < y < \infty$.

10.7   In Problem 10.6 show that Y and $Z = X/\sqrt{1 + Y^2}$ are independent.

10.8   Let X and Y have the bivariate Dirichlet density $f(x,y) = cx^{\alpha-1}y^{\beta-1}(1 - x - y)^{\gamma-1}$ for $0 < x, y < 1$ and $x + y < 1$, where α, β, and γ are positive and $c = \Gamma(\alpha + \beta + \gamma)/\Gamma(\alpha)\Gamma(\beta)\Gamma(\gamma)$. Find the conditional density of X given $Y = y$ for $0 < y < 1$.

10.9   In Problem 10.8 show that Y and $Z = X/(1 - Y)$ are independent.

10.10   Let X and Y be independent, standard normal random variables, and let $Z = X + Y$. Find the conditional density of:
(a) X given $Z = z$ for $z \in R$.
(b) Z given $X = x$ for $x \in R$.

10.11 Let $X_1, \ldots, X_n$ be independent random variables which have a common density f, and let $Y = \min(X_1, \ldots, X_n)$ and $Z = \max(X_1, \ldots, X_n)$. Find the conditional density of Z given $Y = y$ for all possible values of y. Interpret your results.

10.12 Conditionally, given that $X = x$, $0 < x < 1$, let Y have the geometric distribution with parameter x, and let X have the beta distribution with parameters $\alpha > 0$ and $\beta > 0$. Find the (unconditional) mass function of Y in terms of factorials and gamma functions.

10.13 Conditionally, given $X = x$, let Y have the normal distribution with mean x and variance τ^2. Also, let X have the normal distribution with mean μ and variance σ^2. Show that the marginal distribution of Y is normal with mean μ and variance $\sigma^2 + \tau^2$.

10.14 Conditionally, given that $X = x > 0$, let Y have the Poisson distribution with parameter x, and let X have the gamma distribution with parameters $\alpha > 0$ and $\beta > 0$. Find the mass function of Y, and simplify your result in the special case that $\alpha = 1$ and $\beta = 1$.

10.15 Conditionally, given $X = x > 0$, let Y have the gamma distribution with parameters $\alpha > 0$ and x, and let X have standard exponential distribution. Find the density of Y.

10.16 Conditionally, given that $X = x$, let Y have the uniform distribution on $(0,x)$, and let X have density $g(x) = 1/x^2$ for $x \geq 1$. Find the density of Y.

10.17 Find the conditional distribution of X given $Y = y$ in Problems 10.14 and 10.15.

10.18 Find the conditional distribution of X given $Y = y$ in Problem 10.16.

10.19 Let X have the normal distribution with mean μ and variance σ^2. Find the conditional distribution of X, given $X^2 = z > 0$. Simplify your result in the special case that $\mu = 0$.

10.20 Let X have the uniform distribution on $(0,1)$. Find the conditional distribution of X given that $\sin 4\pi X = \frac{1}{2}$.

10.21 Derive Equations (1.11) and (1.12).

10.22 Let X and Y have the bivariate Cauchy distribution (Problem 10.6). Find the conditional expectation of $|Y|$ given $X = x$ for $x \in R$.

10.23 Let X and Y be independent, absolutely continuous random variables. Let f denote a density for X, let G denote the distribution function of Y, and suppose that $f(x) > 0$ if and only if $x > 0$. Let $Z = Y/X$. Show that the conditional distribution function of Z given $X = x > 0$ is $\Pr(Z \leq z \mid X = x) = G(xz)$ for $z \in R$. Use this result and Theorem 10.2.3 to derive the distribution function and density of Z.

10.24 How would your answers to Problem 10.23 change if $f(x)$ were assumed to be positive for all x, $-\infty < x < \infty$?

10.25 Let X and Y have the standard bivariate normal distribution with parameter r, $0 < r < 1$. How large must x be in order that $\Pr(Y \geq 0 \mid X = x) \geq 0.95$?

10.26 Let X have the uniform distribution on $(0,1)$, and conditionally given $X = x$, $0 < x < 1$, let Y have the geometric distribution with parameter x. Find $\Pr(X > \frac{1}{2} \mid Y = y)$ for $y = 1, 2, \ldots$.

10.27 Let X and Y have the uniform distribution on the unit disk in R^2. Find $E(Y \mid X = x)$ and $E(Y^2 \mid X = x)$ for $-1 < x < 1$.

10.28 Let X and Y have the bivariate Dirichlet distribution with parameters α, β, and γ (see Problem 10.8). Find the conditional mean and variance of Y given $X = x$ for $0 < x < 1$.

10.29 Let X and Y have the bivariate hypergeometric distribution (Example 6.1.4). Find $E(Y \mid X = x)$ for all possible values of x.

10.30 Let X and Y be independent, binomially distributed random variables with parameters m, n, and (the same) p. Further, let $Z = X + Y$. Find $E(X \mid Z = z)$ for $z = 0, \ldots, m + n$.

10.31 Conditionally, given $X = x$, let Y have the Poisson distribution with parameter x, and let X have the gamma distribution with parameters $\alpha > 0$ and $\beta > 0$ (as in Problem 10.14). Find the unconditional mean and variance of Y.

10.32 If the conditional distribution of Y given $X = x$ is exponential with parameter x, and if the unconditional distribution of X is gamma with parameters $\alpha > 2$ and $\beta > 0$, find the unconditional mean and variance of Y.

10.33 Let X have the beta distribution with parameters α and β, and conditionally given $X = x$, let Y have the binomial distribution with parameters n and x, as in Example 10.1.7. Find the unconditional mean and variance of Y.

10.34 Let X and Y be independent with means both zero and common variance σ^2. Let $Z = X + Y$. Show that $E(Z^2 \mid X = x) = x^2 + \sigma^2$ for all x for which the conditional expectation is defined.

10.35 Let X and Y be discrete random variables, and let $Z = w(X, Y)$. If $E(Z \mid X = x)$ is defined, then $E(Z \mid X = x) = \sum z \, \Pr(Z = z \mid X = x)$, where the summation extends over all z for which $\Pr(Z = z \mid X = x) > 0$.

10.36 Let X and Y be jointly distributed random variables, and suppose that $E(Y \mid X = x)$ is defined for all $x \in D$, where D is an interval for which $P(X \in D) = 1$. Suppose also that $E(Y \mid X = x) = ax + b$, $x \in D$, where a and b are constants. Express a and b in terms of the means and variances of X and Y and the correlation between X and Y.

10.37 If X and Y are jointly distributed random variables for which $E(Y) = 0$ and $E(Y^2) = E[E(Y \mid X)^2]$, what can be said about the joint distribution of X and Y?

10.38 In Example 10.4.1a, find the conditional mean and variance of Y_1, given $X_i = x_i$, $i = 1, \ldots, m$.

10.39 In Example 10.4.1a find the conditional mean and variance of $Y_1 + Y_2$ given $X_i = x_i$, $i = 1, \ldots, m$.

10.40 In Example 10.4.1b, find the conditional mean and variance of $Y_1 + Y_2$, given $X_i = x_i$, $i = 1, \ldots, m$.

10.41 Let X have the normal distribution with mean μ and variance σ^2, and conditionally given $X = x$, let $Y_1, \ldots, Y_n$ be independent, normally distributed random variables with mean x and variance τ^2. Show that the conditional distribution of X, given $Y_i = y_i$, $i = 1, \ldots, n$ is normal and mean $\mu' = (\mu\sigma^{-2} + z\tau^{-2})/(\sigma^{-2} + n\tau^{-2})$ and variance $1/(\sigma^{-2} + n\tau^{-2})$, where $z = y_1 + \cdots + y_n$.

10.42 Let X_1, X_2, and X_3 be independent random variables which are uniformly distributed over $(0,1)$, and let Y_1, Y_2, and Y_3 denote the ordered values of X_1, X_2, and X_3. Find the conditional density of Y_1 and Y_3 given $Y_2 = y$ for $0 < y < 1$.

10.43 Let $X_1, \ldots, X_n$ be independent with common density f, and let $Y_1, \ldots, Y_n$ denote the ordered values of $X_1, \ldots, X_n$. Find the conditional density of $Y_2, \ldots, Y_{n-1}$ given $Y_1 = y_1$ and $Y_n = y_n$ for all possible values of y_1 and y_n.

10.44 In Problem 10.43 let $1 < k < n$, and find the conditional density of $Y_1, \ldots, Y_{k-1}$, $Y_{k+1}, \ldots, Y_n$ given $Y_k = y$ for all possible values of y. Comment on your result.

10.45 Prove the following result: if $X_1, \ldots, X_n$ are independent with a common distribution function F, and if $S = X_1 + \cdots + X_n$, then $E(X_i \mid S) = (1/n)S$ for $i = 1, \ldots, n$.

10.46 In order to estimate the intensity $x > 0$ with which a radioactive substance decays, the substance is observed for $t > 0$ units of time and the number of emissions Y is recorded. Suppose that the conditional mass function of Y given $X = x$ is $h(y \mid x) = (1/y!)(tx)^y e^{-tx}$ for $y = 0, 1, 2, \ldots$ (see Section 7.6) and that the prior distribution of X is gamma with parameters $\alpha > 0$ and $\beta > 0$. If the loss for estimating x by a is $(x - a)^2$, find the optimal policy and the total expected loss incurred by using the optimal policy.

10.47 In Problem 10.46 let $\alpha = 1$ and $\beta = 1$, and suppose you must decide whether $x \leq 1$ or $x > 1$. If there is unit loss for a wrong decision and no loss for a correct decision, and if it is observed that $Y = 0$, which decision would you make?

10.48 In order to estimate the probability X with which a coin falls heads, the coin is tossed until a head appears and the number of tosses Y is recorded. If the prior distribution of X is uniform on the interval $(0,1)$, and if the loss for estimating X by a is $(X - a)^2$, how would you estimate X?

10.49 In Problem 10.48 suppose that we wish to decide whether $X \leq \frac{1}{2}$ or $X > \frac{1}{2}$ and that the loss for a wrong decision is $|X - \frac{1}{2}|$ with no loss for a correct decision. Describe the optimal policy.

10.50 Conditionally, given $X = x$ let $Y_1, \ldots, Y_n$ be normally distributed with mean x and variance 1, and let X be normally distributed with mean μ and variance σ^2. If the loss incurred by estimating X by a is $|X - a|$, find the optimal estimate of X.

10.51 In Problem 10.50 find the expected loss which is incurred when the optimal policy is used.

10.52 Show that if Y and Z are conditionally independent given X, then $E[w(Z) \mid X = x, Y = y] = E[w(Z) \mid X = x]$ for all choices of x and y for which the conditional expectations are defined.

11

RANDOM WALKS[1]

11.1 INFINITE SEQUENCE OF RANDOM VARIABLES

In the remainder of this book we shall be concerned with infinite sequences of random variables. That is, we shall consider random variables $X_1, X_2, \ldots$, all of which are defined on the same probability space $(S, \mathscr{S}, P)$. In this case $X_1, \ldots, X_n$ will have a joint distribution for every $n = 1, 2, \ldots$. We shall say that random variables $X_1, X_2, \ldots$ are *independent* if and only if $X_1, \ldots, X_n$ are (mutually) independent for every n. That is, $X_1, X_2, \ldots$ are independent if and only if

$$\Pr(X_1 \in I_1, \ldots, X_n \in I_n) = \prod_{i=1}^{n} \Pr(X_i \in I_i)$$

for every choice of the intervals $I_1, \ldots, I_n$ for every $n = 1, 2, \ldots$. Also, we shall say that random variables $X_1, X_2, \ldots$ are *indentically distributed* if they all have the same distribution function.

If $X_1, X_2, \ldots$ are both independent and identically distributed, we shall call the sequence of partial sums $S_0, S_1, S_2, \ldots$, defined by $S_0 = 0$ and

$$S_n = X_1 + \cdots + X_n$$

---

[1] This chapter treats a special topic and may be omitted.

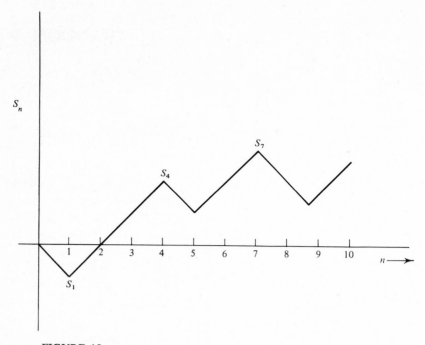

FIGURE 15
Linear interpolation of a simple random walk.

for $n = 1, 2, \ldots$, a *random walk*. We may regard the sequence $S_0, S_1, S_2, \ldots$ as the successive heights of a particle which moves a vertical distance X_k at every integral time k, and it is this interpretation which inspires the name random walk (see Figure 15). We may also regard $S_0, S_1, S_2, \ldots$ as the cumulative winnings of a gambler who plays a sequence of independent games and wins X_k on the kth game for every $k = 1, 2, \ldots$.

In the special case that the common distribution of $X_1, X_2, \ldots$ is given by

$$\Pr(X_k = 1) = p \quad \text{and} \quad \Pr(X_k = -1) = q$$

where $0 < p < 1$ and $q = 1 - p$, the random walk will be called *simple*. In this case the random walk may move only by unit jumps. We have graphed a possible realization of a simple random walk in Figure 15.

In this chapter we shall study random walks in some detail. We begin with two simple, useful observations.

Lemma 11.1.1 *Let $S_0, S_1, S_2, \ldots$ be a random walk; let n be any positive integer; and define $S'_0, S'_1, \ldots$ by*

$$S'_k = S_{n+k} - S_n$$

for $k = 0, 1, 2, \ldots$. Then $S'_0, S'_1, S'_2, \ldots$ is again a random walk, and $(S'_1, \ldots, S'_k)$ has the same distribution as $(S_1, \ldots, S_k)$ for every k. Moreover, $(S'_1, \ldots, S'_k)$ is independent of $(S_0, \ldots, S_n)$ for every k.

PROOF By assumption, $S_k = X_1 + \cdots + X_k$, where $X_1, X_2, \ldots$ are independent with a common distribution function, F say. Let $X'_k = X_{n+k}$, $k = 1, 2, \ldots$. Then $X'_1, X'_2, \ldots$ are again independent with common distribution function F. Moreover, $S'_k = S_{n+k} - S_n = X_{n+1} + \cdots + X_{n+k} = X'_1 + \cdots + X'_n$ for $k = 1, 2, \ldots$, so that $S'_0, S'_1, S'_2, \ldots$ is a random walk. Moreover, $(S'_1, \ldots, S'_k)$ has the same distribution as $(S_1, \ldots, S_k)$, since $(X'_1, \ldots, X'_k)$ has the same distribution as $(X_1, \ldots, X_k)$. Finally, $(S'_1, \ldots, S'_k)$ is determined by $(X'_1, \ldots, X'_k) = (X_{n+1}, \ldots, X_{n+k})$ and is therefore independent of $(S_1, \ldots, S_n)$, which is determined by $(X_1, \ldots, X_n)$. ////

Since, by definition, $S_{n+k} = S'_k + S_n$ for $k = 1, 2, \ldots$, the result of Lemma 11.1.1 can be paraphrased by saying that *at every integral time n, the random walk starts anew but starts from the position S_n.*

Lemma 11.1.2 *Let $S_0, S_1, S_2, \ldots$ be a random walk, and let n be a positive integer. Define*

$$S''_k = S_n - S_{n-k}$$

for $k = 1, \ldots, n$. Then $(S''_1, \ldots, S''_n)$ has the same distribution as $(S_1, \ldots, S_n)$, and $(S''_1, \ldots, S''_n)$ is independent of $(S'_1, \ldots, S'_k)$ for every $k = 1, 2, \ldots$.

PROOF Lemma 11.1.2 follows from the observation that $S''_k = X_n + \cdots + X_{n-k+1}$ by an argument similar to that given in the proof of Lemma 11.1.1. ////

Let us turn briefly to a technical point. We have not shown how to construct a sample space on which a sequence of independent random variables can be defined. We shall not give this construction because the details would lead us into the realm of abstract measure theory and away from the behavior of random walks. We ask the reader to accept without proof the following fact. Given any sequence $F_1, F_2, \ldots$ of univariate distribution functions, there is a sequence of independent random variables $X_1, X_2, \ldots$ with distribution functions $F_1, F_2, \ldots$, respectively. That is, sequences of independent random variables do exist. In fact, more is true and may be found in Problems 11.1 to 11.6.

11.2 THE GAMBLER'S-RUIN PROBLEM

Consider the simple random walk of the previous section. That is, let $X_1, X_2, \ldots$ be independent random variables with common distribution given by

$$\text{Pr}\,(X_1 = 1) = p \quad \text{and} \quad \text{Pr}\,(X_1 = -1) = q \tag{2.1}$$

where $q = 1 - p$, and let $S_0 = 0$ and $S_n = X_1 + \cdots + X_n$ for $n = 1, 2, \ldots$. In this section we shall regard $S_0, S_1, \ldots$ as the cumulative winnings of a gambler who wins a dollar with probability p and loses a dollar with probability q on each of a sequence of independent games. The gambler's opponent will be called the *house*. We suppose that the gambler starts with a dollars and the house starts with b dollars, where a and b are nonnegative integers. The total capital $c = a + b$ is a fixed positive integer which does not change from play to play. Finally, we suppose that the gambler and the house agree to continue playing until one of them has won all the money, and we are asked for the probability that the gambler eventually wins all the house's money.

We shall state the problem mathematically. For $n = 0, 1, 2, \ldots$, let B_n^a be the event

$$-a < S_k < b \quad \text{for } k = 0, \ldots, n - 1 \quad \text{and} \quad S_n = b$$

that the gambler wins all the house's money after exactly n plays of the game. We require the probability of the event

$$B_a = \bigcup_{n=0}^{\infty} B_n^a$$

that the gambler wins after an unspecified number of plays (after exactly n plays for some $n = 0, 1, 2, \ldots$). Let π_a denote the probability in question. Then

$$\pi_a = P(B_a) = \sum_{n=0}^{\infty} P(B_n^a) \tag{2.2}$$

since the events $B_1^a, B_2^a, \ldots$ are mutually exclusive. In particular, we have

$$\pi_0 = 0 \quad \text{and} \quad \pi_c = 1 \tag{2.3}$$

since B_0^c is certain and B_n^0 is impossible for every n. For $0 < a < c$, we shall compute π_a by the following novel method. We shall derive a difference equation which the π_a must satisfy, and then we shall solve the difference equation subject to the boundary conditions (2.3).

Lemma 11.2.1 *For $0 < a < c = a + b$, we have $\pi_a = p\pi_{a+1} + q\pi_{a-1}$.*

PROOF The idea is quite simple. B_a is the event that a gambler who starts with a dollars eventually wins. Moreover, if $X_1 = 1$, then the gambler effectively starts over with $a + 1$ dollars (see Lemma 11.1.1). Thus,

$$\Pr\,(B_a \mid X_1 = 1) = P(B_{a+1}) = \pi_{a+1}$$

and similarly, $\Pr\,(B_a \mid X_1 = -1) = P(B_{a-1}) = \pi_{a-1}$. Therefore,

$$
\begin{aligned}
\pi_a &= P(B_a) \\
&= \Pr\,(B_a \mid X_1 = 1)\,\Pr\,(X_1 = 1) \\
&\quad + \Pr\,(B_a \mid X_1 = -1)\,\Pr\,(X_1 = -1) \\
&= p\pi_{a+1} + q\pi_{a-1}
\end{aligned}
$$

Since infinitely many random variables enter into the definition of B_a, the statement that $\Pr\,(B_a \mid X_1 = 1) = \pi_{a+1}$ does require more justification than we have given. The details of this justification are sketched in Problems 11.13 and 11.14.

$////$

We shall now solve the difference equation of Lemma 11.2.1.

Theorem 11.2.1 Let $\beta = q/p$. If $\beta = 1$ $(p = q)$, then $\pi_a = a/(a + b)$; and if $\beta \neq 1$, then

$$\pi_a = \frac{1 - \beta^a}{1 - \beta^{a+b}} \qquad (2.4)$$

PROOF It follows from Lemma 11.2.1 and induction that $\pi_{a+1} - \pi_a = \beta(\pi_a - \pi_{a-1}) = \beta^a(\pi_1 - \pi_0)$ for $a = 0, \ldots, c = a + b$. Moreover, $\pi_0 = 0$ by (2.3), so that

$$\pi_a = \pi_a - \pi_0 = \sum_{k=0}^{a-1} (\pi_{k+1} - \pi_k) = \pi_1 \sum_{k=0}^{a-1} \beta^k \qquad (2.5)$$

for $a = 0, \ldots, c$. Therefore, if $\beta = 1$, then $\pi_a = a\pi_1$ for $a = 0, \ldots, c$, and since $\pi_c = 1$ by (2.3), it follows that $\pi_1 = 1/c$ and $\pi_a = a/c = a/(a + b)$ for $a = 0, \ldots, c$. If $\beta \neq 1$, then (2.5) yields

$$\pi_a = \pi_1 \frac{1 - \beta^a}{1 - \beta}$$

for $a = 0, \ldots, c$. Moreover, we again have $\pi_c = 1$, so that $\pi_1 = (1 - \beta)/(1 - \beta^c)$. Equation (2.4) now follows by substitution. $////$

We shall now consider the fortune of a gambler who plays against an infinitely rich opponent. Let $\sigma(a,b) = 1 - \pi_a$. We shall show below that $\sigma(a,b)$ is the probability that the gambler loses all his money to the house (goes broke) when the house starts with b dollars and the gambler starts with a dollars. That is, we shall show that the probability that the game terminates is 1 when both players start with finite capital. We expect the probability that the gambler goes broke when playing against an infinitely rich opponent to be the limit as $b \to \infty$ of $\sigma(a,b)$. Since $\beta^{a+b} \to 0$ or ∞ as $b \to \infty$ according as $\beta < 1$ or $\beta > 1$, it follows easily from Theorem 11.2.1 that

$$\sigma_a = \lim_{b \to \infty} \sigma(a,b) = \begin{cases} 1 & \text{if } p \leq \tfrac{1}{2} \\ \beta^a & \text{if } p > \tfrac{1}{2} \end{cases} \tag{2.6}$$

where (2.6) defines σ_a. That is, an unskillful gambler ($p \leq \tfrac{1}{2}$) is certain to lose all his money to an infinitely rich house, but a skillful gambler will lose with probability β^a, where $\beta = q/p$ and a is the gambler's initial fortune.

Table 13 gives the value of σ_a for selected values of p and a.

For example, a gambler who starts with $a = 12$ dollars and wins with probability $p = 0.6$ is virtually certain to prosper against an infinitely rich house.

We shall now show that (2.6) does, in fact, give the desired probability. The first item of business is to show that $\sigma(a,b)$ is the probability that the gambler loses when the house starts with b dollars.

Lemma 11.2.2 *Let a and b be positive integers, and let D be the event that $-a < S_n < b$ for every $n = 1, 2, \ldots$. Then $P(D) = 0$.*

    PROOF For $n = 1, 2, \ldots$, let D_n be the event that $-a < S_k < b$ for $k = 1, \ldots, n$. Then the occurrence of D implies the occurrence of D_n for every n, so that $P(D) \leq P(D_n)$ for every $n = 1, 2, \ldots$. Let $c = a + b$. Then

$$\Pr\left(|S_c| \geq c\right) = p^c + q^c > 0$$

Let $Z_k = S_{kc} - S_{kc-c}$ for $k = 1, 2, \ldots$. Then $Z_1, Z_2, \ldots$ are independent and identically distributed (Lemma 11.1.1), and $\Pr\left(|Z_k| \geq c\right) = p^c + q^c = d$, say, for every $k = 1, 2, \ldots$. Therefore,

$$\begin{aligned} P(D_{nc}) &\leq \Pr\left(-a < S_k < b, k = 1, \ldots, nc\right) \\ &\leq \Pr\left(-a < S_{kc} < b, k = 1, \ldots, n\right) \\ &\leq \Pr\left(-c < Z_k < c, k = 1, \ldots, n\right) \\ &= \prod_{k=1}^{n} \Pr\left(-c < Z_k < c\right) = (1 - d)^n \end{aligned}$$

for every $n = 1, 2, \ldots$. It follows that $P(D) \leq (1 - d)^n$ for every $n = 1, 2, \ldots$, and consequently, that $P(D) = 0$.   ////

For $a > 0$ and $b > 0$, let C_{ab} be the event for some $n = 1, 2, \ldots$

$$-a < S_k < b \quad \text{for } k = 1, \ldots, n \quad \text{and} \quad S_n = -a$$

Thus, C_{ab} is the event that the gambler loses when the house starts with b dollars.

Corollary 11.2.1 $P(C_{ab}) = \sigma(a,b) = 1 - \pi_a$.

PROOF   Let B_a and D be as in Theorem 11.2.1 and Lemma 11.2.2, respectively. Then B_a, C_{ab}, and D are mutually exclusive and exhaustive events, so that $P(B_a) + P(C_{ab}) + P(D) = 1$. Moreover, $P(D) = 0$ by Lemma 11.2.2, so that $P(C_{ab}) = 1 - P(B_a) = 1 - \pi_a$, as asserted.   ////

We shall now prove that σ_a gives the probability of losing to an infinitely rich house. Let C_a be the event that

$$S_n = -a \qquad \text{for some } n = 1, 2, \ldots$$

Thus, C_a is the event the gambler loses to an infinitely rich house.

Theorem 11.2.2 $P(C_a) = \sigma_a$, where σ_a is defined by (2.6).

PROOF   Define the events C_{ab} as in the previous corollary. Then for fixed a, C_{ab} implies $C_{a(b+1)}$ for every b, so that $C_{a1}, C_{a2}, \ldots$ is an increasing sequence of events. Moreover, the union of $C_{a1}, C_{a2}, \ldots$ is simply C_a. Therefore, by Theorem 2.5.1, we have

$$P(C_a) = \lim_{b \to \infty} P(C_{ab}) = \lim_{b \to \infty} \sigma(a,b) = \sigma_a \qquad ////$$

It is possible to view the result of Theorem 11.2.2 in another way. We shall say that the random walk $S_0, S_1, S_2, \ldots$ *passes through* or *visits* an integer a if

$$S_n = a \qquad \text{for some } n = 1, 2, \ldots$$

Since the gambler loses to an infinitely rich house if and only if $S_0, S_1, S_2, \ldots$ passes through $-a$, where a is the gambler's initial fortune, it follows from Theorem 11.2.2 that if a is a positive integer and $p \leq \frac{1}{2}$, then the probability that $S_0, S_1, S_2, \ldots$ passes through $-a$ is 1. By symmetry, if a is a positive integer and $p \geq \frac{1}{2}$, then the

Table 13

			a		
p	2	4	6	8	12
0.600	0.444	0.198	0.088	0.039	0.008
0.667	0.250	0.063	0.016	0.004	
0.750	0.111	0.012	0.001		

probability that $S_0, S_1, S_2, \ldots$ passes through a is 1. In particular, if $p = \frac{1}{2}$ and $a \neq 0$, then the probability that $S_0, S_1, S_2, \ldots$ passes through a is 1.

We shall say that the random walk *returns to the origin* if and only if $S_n = 0$ for some $n = 1, 2, \ldots$. Observe that if the random walk passes through both 1 and -1, then it must return to the origin. Thus, if $p = \frac{1}{2}$, then the probability that the random walk returns to the origin is 1.

We summarize the above discussion.

Theorem 11.2.3 *Let $S_0, S_1, S_2, \ldots$ be a simple random walk. If $p \geq \frac{1}{2}$ and a is a positive integer, then the probability that the random walk passes through a is 1. If $p = \frac{1}{2}$, then the probability that the random walk returns to the origin is 1.*

11.3 THE BOREL-CANTELLI LEMMAS

If $A_1, A_2, \ldots$ is any infinite sequence of events, we can form a new sequence $B_1, B_2, \ldots$ by letting

$$B_n = \bigcup_{k=n}^{\infty} A_k \qquad (3.1)$$

for $n = 1, 2, \ldots$. Thus, B_n is the event that A_k occurs for some $k \geq n$. Therefore, the event

$$B = \bigcap_{n=1}^{\infty} B_n \qquad (3.2)$$

is the event that infinitely many of the events $A_1, A_2, \ldots$ occur, for B occurs if and only if A_k occurs for some $k \geq n$ for *every* $n = 1, 2, \ldots$. We shall call B the event that A_n occurs *infinitely often*, and we shall write $B = \{A_n, \text{i.o.}\}$. The terminology $B = \limsup A_n$ is also used.

In this section we shall prove two theorems which relate the probability of B to the probabilities of the events $A_1, A_2, \ldots$. These theorems are known as the *Borel-Cantelli lemmas*.

Theorem 11.3.1 *Let $A_1, A_2, \ldots$ be any infinite sequence of events, and let $B = \{A_n, \text{i.o.}\}$. If*

$$\sum_{n=1}^{\infty} P(A_n) < \infty \qquad (3.3)$$

then $P(B) = 0$.

PROOF   For every n, we have $B_n = A_n \cup B_{n+1} \supset B_{n+1}$, so that B_1, $B_2, \ldots$ is a decreasing sequence. It follows from Theorem 2.5.1 that

$$P(B) = \lim_{n \to \infty} P(B_n)$$

Moreover,

$$P(B_n) \le \sum_{k=n}^{\infty} P(A_k)$$

for every n, so that the convergence of the series (3.3) implies that $\lim P(B_n) = 0$ as $n \to \infty$. The theorem follows.   ////

EXAMPLE 11.3.1   Let $S_0, S_1, S_2, \ldots$ be a simple random walk, and let $p = \Pr(S_1 = 1)$. If A_n denotes the event that $S_{2n} = 0$, then

$$P(A_n) = \binom{2n}{n} p^n q^n \sim \frac{1}{\sqrt{\pi n}} (4pq)^n$$

as $n \to \infty$ by Stirling's formula (Section 1.8). If $p \ne \frac{1}{2}$, then $4pq < 1$, so that $P(A_1) + P(A_2) + \cdots < \infty$. That is, if $p \ne \frac{1}{2}$, then the probability that the random walk $S_0, S_1, S_2, \ldots$ returns to 0 infinitely often is zero. If $p = \frac{1}{2}$, then $4pq = 1$ and the series $P(A_1) + P(A_2) + \cdots$ diverges. In fact, if $p = \frac{1}{2}$, then

$$\Pr(S_n = 0, \text{ i.o.}) = 1$$

as we shall show in the next section.   ////

There is a converse to Theorem 11.3.1. If $A_1, A_2, \ldots$ is any infinite sequence of events, then we shall say that $A_1, A_2, \ldots$ are *independent* if and only if $A_1, \ldots, A_n$ are (mutually) independent for every $n = 1, 2, \ldots$. It is easily verified that if X_1, $X_2, \ldots$ are independent random variables, and if A_n is determined by X_n, then A_1, $A_2, \ldots$ are independent events.

Theorem 11.3.2   *If $A_1, A_2, \ldots$ are independent events, and if*

$$\sum_{n=1}^{\infty} P(A_n) = \infty \qquad (3.4)$$

then $\Pr(A_n, \text{ i.o.}) = 1$.

PROOF   We shall prove Theorem 11.3.2 by showing that its hypotheses imply $P(B') = 0$, where B is defined by (3.2) and the prime denotes complement. As in the proof of Theorem 11.3.1, we have $P(B') = \lim P(B'_n)$ as $n \to \infty$, and so it will suffice to show that $P(B_n') = 0$ for every $n = 1, 2, \ldots$. Now

$$B_n' = \bigcap_{k=n}^{\infty} A_k'$$

so that

$$B_n' \subset \bigcap_{k=n}^{n+m} A_k'$$

for every n and m. Therefore,

$$P(B_n') \le \prod_{k=n}^{n+m} P(A_k') = \prod_{k=n}^{n+m} [1 - P(A_k)]$$

For every real number x, we have the inequality $1 - x < e^{-x}$, since the second term in the Taylor series expansion of e^x about $x = 0$ is positive. Therefore,

$$P(B_n') \le \prod_{k=n}^{n+m} e^{-P(A_k)} = \exp\left[-\sum_{k=n}^{n+m} P(A_k)\right] \qquad (3.5)$$

for every n and every m. Finally, if the series in (3.4) diverges, then the exponent in (3.5) must diverge to $-\infty$ as $m \to \infty$ for every n. Since the inequality (3.5) is valid for every n and every m, it follows that

$$P(B_n') \le \lim_{m \to \infty} \exp\left[-\sum_{k=n}^{n+m} P(A_k)\right] = 0$$

for every n, as required.   ////

EXAMPLE 11.3.2 Let $X_1, X_2, \ldots$ be independent random variables which have a common exponential density

$$f(x) = e^{-x}$$

for $x > 0$ and $f(x) = 0$ for $x \le 0$. Let A_n be the event that $X_n > a \log n$, where $a > 0$. Then A_n occurs infinitely often with probability 1 if $a \le 1$ and A_n occurs infinitely often with probability 0 if $a > 1$. In fact, $A_1, A_2, \ldots$ are independent (since A_n is determined by X_n), and

$$P(A_n) = \Pr(X_n > a \log n) = \exp(-a \log n) = n^{-a}$$

for $n = 1, 2, \ldots$. It is well known that the series $1^{-a} + 2^{-a} + 3^{-a} + \cdots$ is finite or infinite according as $a > 1$ or $a \le 1$.   ////

It is interesting that if $A_1, A_2, \ldots$ are independent events, then $\Pr(A_n, \text{ i.o.})$ is either 0 or 1, since the series $P(A_1) + P(A_2) + \cdots$ is either finite or infinite.

11.4 RECURRENCE

In this section we shall justify the statement of Example 11.3.1 that a simple, symmetric ($p = \frac{1}{2}$) random walk returns to 0 infinitely often. In fact, we shall show that a simple symmetric random walk visits every integer infinitely often.

Theorem 11.4.1 *Let $S_0, S_1, S_2, \ldots$ be a simple, symmetric random walk. Then*

$$\Pr (S_n = a, \text{i.o.}) = 1 \qquad (4.1)$$

for every integer a.

PROOF The idea is the following. We know from Section 11.2 that the random walk will pass through a at least once (with probability 1). Moreover, if it first passes through a at time n, then $S'_k = S_{n+k} - S_n$, $k = 0, 1, 2, \ldots$, will again be a simple symmetric random walk which must therefore pass through 0 (with probability 1). Since $S_n = a$, this means that the random walk must visit a at least twice. Continuing in this manner, we are forced to the conclusion that the random walk passes through a arbitrarily often.

We can make this idea precise as follows. For positive integers n and j, let B_{nj} be the event that the random walk passes through a for the jth time after exactly n moves. That is, let B_{nj} be the event that $S_n = a$ and exactly $j - 1$ of $S_1, \ldots, S_{n-1}$ are equal to a. Further, let

$$B_j = \bigcup_{n=1}^{\infty} B_{nj} \qquad \text{and} \qquad B = \bigcap_{j=1}^{\infty} B_j$$

Thus, B_j is the event that the random walk passes through a at least j times, and B is the event that $S_n = a$ for infinitely many values of n.

We shall show that $P(B) = 1$. We know from Section 11.2 that a simple symmetric random walk passes through any integer a with probability 1. Thus, $P(B_1) = 1$. Suppose inductively that $P(B_j) = 1$, and let us demonstrate that $P(B_{j+1}) = 1$. Since B_{j+1} implies B_j, we must have

$$P(B_{j+1}) = \sum_{n=1}^{\infty} P(B_{nj} \cap B_{j+1})$$

Moreover, since B_{nj} implies $S_n = a$, B_{nj} and B_{j+1} will occur simultaneously if and only if B_{nj} occurs and

$$S'_k = S_{n+k} - S_n = 0 \qquad \text{for some } k = 1, 2, \ldots \qquad (4.2)$$

Let C_n be the event defined by (4.2). Then $P(C_n) = 1$ by Lemma 11.1.1 and the results of Section 11.2. Moreover, $B_{nj} \cap B_{j+1} = B_{nj} \cap C_n$, so that

$$P(B_{nj} \cap B_{j+1}) = P(B_{nj} \cap C_n) = P(B_{nj}) - P(B_{nj} \cap C'_n) = P(B_{nj})$$

since $P(B_{nj} \cap C_n') \leq P(C_n') = 0$. Therefore,

$$P(B_{j+1}) = \sum_{n=1}^{\infty} P(B_{nj}) = P(B_j) = 1$$

where the final step follows from the induction hypothesis. Therefore, $P(B_j) = 1$ for all $j = 1, 2, \ldots$ by mathematical induction.

It now follows easily that $P(B) = 1$. In fact, $P(B') \leq P(B_1') + P(B_2') + \cdots = 0 + 0 + \cdots = 0$, so that $P(B) = 1$.   /////

11.5   CONVERGENCE WITH PROBABILITY 1

In this section we shall introduce and study a new mode of convergence. Let X, $X_1, X_2, \ldots$ be an infinite sequence of random variables, all of which are defined on the same probability space $(S, \mathscr{S}, P)$. We shall say that X_n *converges to X with probability* 1 as $n \to \infty$ if and only if

$$\Pr\left(\lim_{n \to \infty} X_n = X\right) = 1 \qquad (5.1)$$

That is, X_n converges to X with probability 1 if and only if $P(C) = 1$, where C denotes the set of $s \in S$ for which $\lim X_n(s) = X(s)$ as $n \to \infty$. Equivalently, X_n converges to X with probability 1 if and only if $P(D) = 0$, where $D = C'$ denotes the set[1] of $s \in S$ for which $X_n(s)$ fails to converge to $X(s)$ as $n \to \infty$.

Theorem 11.5.1   *Let $X, X_1, X_2, \ldots$ be random variables which are defined on the same probability space. Then X_n converges to X with probability 1 as $n \to \infty$ if and only if*

$$\Pr(|X_n - X| \geq \varepsilon, \text{ i.o.}) = 0 \qquad (5.2)$$

for every $\varepsilon > 0$.

PROOF   For any s, $X_n(s)$ will fail to converge to $X(s)$ if and only if there is an $\varepsilon = \varepsilon(s) > 0$ for which $|X_n(s) - X(s)| \geq \varepsilon$ for infinitely many values of n, and we may restrict our attention to ε of the form $1/j$, where j is a positive integer. Thus, the set of $s \in S$ for which $X_n(s)$ fails to converge to $X(s)$ as $n \to \infty$ is

$$D = \bigcup_{j=1}^{\infty} D_j$$

where D_j denotes the event that $|X_n - X| \geq 1/j$ infinitely often. Since $1/j > 1/(j+1)$ for every $j = 1, 2, \ldots$, we must have $D_j \subset D_{j+1}$ for every j. That

---

[1] That C and D are events is shown in the proof of Theorem 11.5.1.

is, $D_1, D_2, \ldots$ is an increasing sequence of events. It follows from Theorem 2.5.1 that

$$P(D_j) \leq P(D_{j+1}) \qquad \text{and} \qquad P(D) = \lim_{j \to \infty} P(D_j)$$

Thus, $P(D) = 0$ if and only if $P(D_j) = 0$ for every $j = 1, 2, \ldots$, and this is equivalent to (5.2).   ////

Theorem 11.5.1 has several interesting corollaries. Let $X, X_1, X_2, \ldots$ be random variables which are defined on the same probability space, and let $\varepsilon > 0$. Then the event that $|X_n - X| \geq \varepsilon$ for infinitely many values of n is

$$D_\varepsilon = \bigcap_{n=1}^{\infty} B_n$$

where B_n is the event that $|X_k - X| \geq \varepsilon$ for some $k \geq n$. Moreover, $B_1, B_2, \ldots$ is a decreasing sequence of events, so that $P(D_\varepsilon) = \lim P(B_n)$ as $n \to \infty$. Therefore, we have the following corollary.

Corollary 11.5.1 *Let $X, X_1, X_2, \ldots$ be random variables which are defined on the same probability space. Then X_n converges to X as $n \to \infty$ if and only if*

$$\lim_{n \to \infty} \Pr\left(|X_k - X| \geq \varepsilon \text{ for some } k \geq n\right) = 0 \qquad (5.3)$$

for every $\varepsilon > 0$.

Comparison of convergence with probability 1 and convergence in probability is now easy. If $X, X_1, X_2, \ldots$ are random variables which are defined on the same probability space, then X_n converges to X in probability if and only if

$$\lim_{n \to \infty} \Pr\left(|X_n - X| \geq \varepsilon\right) = 0 \qquad (5.4)$$

for every $\varepsilon > 0$ (Section 9.2). Since (5.3) implies (5.4), we have another corollary.

Corollary 11.5.2 *Let $X, X_1, X_2, \ldots$ be random variables which are defined on the same probability space. If X_n converges to X with probability 1 as $n \to \infty$, then X_n converges to X in probability as $n \to \infty$.*

An example of a sequence $X_1, X_2, \ldots$ for which X_n converges to 0 in probability and X_n does not converge to 0 with probability 1 as $n \to \infty$ will be given below.

Theorem 11.5.1 allows us to use the Borel-Cantelli lemmas to decide questions of convergence with probability 1.

EXAMPLE 11.5.1 Let $X_1, X_2, \ldots$ be independent and identically distributed random variables, and let F denote the common distribution function of $X_1, X_2, \ldots$. Then X_n/n converges to 0 with probability 1 as $n \to \infty$ if and only if

$$m = \int_{-\infty}^{\infty} |x| \, dF(x) < \infty$$

To see this observe that $X_n/n \to 0$ with probability 1 as $n \to \infty$ if and only if $\Pr(|X_n| \geq \varepsilon n, \text{i.o.}) = 0$ for every $\varepsilon > 0$ by Theorem 11.5.1. Moreover since X_1, $X_2, \ldots$ are independent, the latter condition is equivalent to

$$\sum_{n=1}^{\infty} \Pr(|X_n| \geq \varepsilon n) < \infty \qquad (5.5)$$

by Theorems 11.3.1 and 11.3.2. Thus, we need only show that (5.5) is equivalent to the finiteness of m.

Let Y be the greatest integer which is less than or equal to $|X_1 \varepsilon^{-1}|$. Then $0 \leq |X_1 \varepsilon^{-1}| - Y < 1$, so that $E(Y)$ is finite if and only if $m = E(|X_1|)$ is finite. Now

$$\Pr(|X_n| \geq \varepsilon n) = \Pr(|X_1| \geq \varepsilon n) = \Pr(Y \geq n)$$

so that

$$\sum_{n=1}^{\infty} \Pr(|X_n| \geq \varepsilon n) = \sum_{n=1}^{\infty} \Pr(Y \geq n) = E(Y)$$

by Corollary 8.1.2. The equivalence of (5.5) and the finiteness of m follows.   ////

EXAMPLE 11.5.2 Let $X_1, X_2, \ldots$ be independent and identically distributed random variables for which $E(|X_1|) = \infty$. Then X_n/n does not converge to zero with probability 1 as $n \to \infty$, by the previous example. However, $\Pr(|X_n| \geq \varepsilon n) = \Pr(|X_1| \geq \varepsilon n)$, which does tend to 0 as $n \to \infty$ for every $\varepsilon > 0$. Therefore, X_n/n converges to zero in probability as $n \to \infty$.   ////

11.6   SOME INEQUALITIES

In the next section we shall show that the convergence in the law of large numbers is, in fact, convergence with probability 1. In this section we shall develop some relevant inequalities. First, we shall show how Chebyshev's inequality can be improved in the presence of independence. The result is known as *Kolmogorov's inequality*.

Theorem 11.6.1 *Let $X_1, \ldots, X_n$ be independent random variables with means $E(X_i) = 0$, $i = 1, \ldots, n$, and finite variances $\sigma_i^2 = E(X_i^2)$, $i = 1, \ldots, n$.*

Then for every ε > 0,

$$\Pr\left(\max_{k \leq n} |S_k| \geq \varepsilon\right) \leq \frac{\sigma^2}{\varepsilon^2}$$

where $\sigma^2 = \sigma_1^2 + \cdots + \sigma_n^2$ is the variance of S_n.

Observe that Chebyshev's inequality gives the same bound for the probability of the smaller event, $|S_n| \geq \varepsilon$.

PROOF   For $k = 1, \ldots, n$, let A_k be the event that $|S_k| \geq \varepsilon$ and $|S_j| < \varepsilon$ for $j = 1, \ldots, k - 1$. Then $A_1, \ldots, A_n$ are mutually exclusive, and the union $A = A_1 \cup \cdots \cup A_n$ is the event that $|S_k| \geq \varepsilon$ for some $k = 1, \ldots, n$. Therefore,

$$\Pr\left(\max_{k \leq n} |S_k| \geq \varepsilon\right) = \sum_{k=1}^{n} P(A_k) \qquad (6.1)$$

Let I_{A_k} denote the indicator function of the event A_k, $k = 1, \ldots, n$. That is, let $I_{A_k} = 1$ if A_k occurs, and let $I_{A_k} = 0$ if A_k does not occur. Then $\varepsilon^2 I_{A_k} \leq S_k^2 I_{A_k}$ for all possible realizations of $X_1, \ldots, X_n$. Therefore,

$$P(A_k) = E(I_{A_k}) \leq \varepsilon^{-2} E(S_k^2 I_{A_k}) \qquad (6.2)$$

for $k = 1, \ldots, n$. We now claim that

$$E(S_k^2 I_{A_k}) \leq E(S_n^2 I_{A_k}) \qquad (6.3)$$

for $k = 1, \ldots, n$. To see this observe that A_k is determined by $X_1, \ldots, X_k$ and is therefore independent of $S_n - S_k = X_{k+1} + \cdots + X_n$. Hence,

$$E[I_{A_k} S_k (S_n - S_k)] = E(I_{A_k} S_k) E(S_n - S_k) = 0$$

Therefore,

$$E(I_{A_k} S_n^2) = E(I_{A_k} S_k^2) + E[I_{A_k}(S_n - S_k)^2] \geq E(I_{A_k} S_k^2)$$

for $k = 1, \ldots, n$, as asserted.

If we now combine (6.1) to (6.3) and use the fact that $I_{A_1} + \cdots + I_{A_n} = I_A \leq 1$, we find that

$$\Pr\left(\max_{k \leq n} |S_k| \geq \varepsilon\right) \leq \varepsilon^{-2} \sum_{k=1}^{n} E(S_n^2 I_{A_k}) = \varepsilon^{-2} E(S_n^2 I_A) \leq \varepsilon^{-2} \sigma^2$$

as asserted.   ////

There is another interesting inequality which relates the distribution of $\max(S_1, \ldots, S_n)$ to that of S_n in the special case that $X_1, \ldots, X_n$ have distributions which are symmetric about 0. The result, known as *Levy's inequality*, will now be presented.

If F is a distribution function, we say that F is *symmetric about* 0 if and only if

$$F(x) = 1 - F(x-) \qquad (6.4a)$$

for all x, $-\infty < x < \infty$. If X is a random variable with distribution function F, then (6.4a) is equivalent to

$$\Pr (X \le x) = \Pr (X \ge -x) \qquad (6.4b)$$

for all x, $-\infty < x < \infty$. Thus, X has a symmetric (about 0) distribution function if and only if X and $-X$ have the same distribution function.

EXAMPLE 11.6.1   If F has a density f for which $f(x) = f(-x)$ for all x, $-\infty < x < \infty$, then F is symmetric. In fact,

$$F(x) = \int_{-\infty}^{x} f(y) \, dy = \int_{-x}^{\infty} f(y) \, dy = 1 - F(-x)$$

for $-\infty < x < \infty$, and F is continuous. In particular, the standard normal and standard Cauchy distributions are symmetric about zero.   ////

EXAMPLE 11.6.2   If $X_1, \ldots, X_n$ are independent random variables, all of which have distributions which are symmetric about 0, then $(X_1, \ldots, X_n)$ and $(-X_1, \ldots, -X_n)$ have the same distributions. Consequently, $S = X_1 + \cdots + X_n$ and $-S = -X_1 - \cdots - X_n$ have the same distributions. That is, S has a distribution which is symmetric about 0.   ////

Theorem 11.6.2   *Let $X_1, \ldots, X_n$ be independent random variables whose distributions are all symmetric about 0. Further, let $S_k = X_1 + \cdots + X_k$ for for $k = 1, \ldots, n$. Then*

$$\Pr (\max_{k \le n} S_k \ge \varepsilon) \le 2 \Pr (S_n \ge \varepsilon)$$

for every $\varepsilon > 0$.

PROOF   For notational simplicity, let $M = \max (S_1, \ldots, S_n)$. Then $S_n \ge \varepsilon$ implies $M \ge \varepsilon$, so that

$$\Pr (M \ge \varepsilon) = \Pr (M \ge \varepsilon, S_n \ge \varepsilon) + \Pr (M \ge \varepsilon, S_n < \varepsilon)$$
$$= \Pr (S_n \ge \varepsilon) + \Pr (M \ge \varepsilon, S_n < \varepsilon) \qquad (6.5)$$

Therefore, it will suffice to show that

$$\Pr (M \ge \varepsilon, S_n < \varepsilon) \le \Pr (S_n \ge \varepsilon) \qquad (6.6)$$

For $k = 1, \ldots, n$, let A_k be the event that $S_k \geq \varepsilon$ and $S_j < \varepsilon$ for $j = 1, \ldots, k - 1$. Then, as in the proof of Theorem 11.6.1, $A_1, \ldots, A_n$ are mutually exclusive, and the union $A = A_1 \cup \cdots \cup A_n$ is the event that $M \geq \varepsilon$. Therefore,

$$\Pr(M \geq \varepsilon, S_n < \varepsilon) = \sum_{k=1}^{n} \Pr(A_k, S_n < \varepsilon) \qquad (6.7)$$

Now A_k is determined by $X_1, \ldots, X_k$ and is therefore independent of $S_n - S_k$. Moreover, A_k and $S_n < \varepsilon$ imply A_k and $S_n - S_k < 0$. It follows that

$$\Pr(A_k, S_n < \varepsilon) \leq \Pr(A_k, S_n - S_k < 0) = \Pr(A_k)\Pr(S_n - S_k < 0)$$
$$= P(A_k)\Pr(S_n - S_k > 0) = \Pr(A_k, S_n - S_k > 0)$$
$$\leq \Pr(A_k, S_n \geq \varepsilon)$$

Here the middle equality follows from the fact that $S_n - S_k$ has a symmetric distribution (Example 11.6.2), and the final inequality follows from the fact that A_k and $S_n - S_k > 0$ imply A_k and $S_n \geq \varepsilon$.

Substitution in (6.7) now yields

$$\Pr(M \geq \varepsilon, S_n < \varepsilon) \leq \sum_{k=1}^{n} \Pr(A_k, S_n \geq \varepsilon)$$
$$= \Pr(M \geq \varepsilon, S_n \geq \varepsilon) = \Pr(S_n \geq \varepsilon)$$

This establishes (6.6), from which the theorem follows. ⫽⫽⫽⫽

Of course, Theorem 11.6.2 may be applied to $-X_1, \ldots, -X_n$ to yield

$$\Pr(\min_{k \leq n} S_k \leq -\varepsilon) \leq 2\Pr(S_n \leq -\varepsilon) \qquad (6.8)$$

for $\varepsilon > 0$. When combined with the conclusion of Theorem 11.6.2, (6.8) yields the following corollary.

Corollary 11.6.1 *Let $X_1, \ldots, X_n$ be as in the statement of Theorem 11.6.1. Then for every $\varepsilon > 0$,*

$$\Pr(\max_{k \leq n} |S_k| \geq \varepsilon) \leq 2\Pr(|S_n| \geq \varepsilon)$$

In the case of a simple symmetric random walk, the arguments used in the proof of Theorem 11.6.2 yield the following equality.

Theorem 11.6.3 *Let $S_0, S_1, S_2, \ldots$ be a simple symmetric random walk, and let $M_n = \max(S_0, S_1, \ldots, S_n)$ for $n = 1, 2, \ldots$. Then*

$$\Pr(M \geq a) = \Pr(S_n \geq a) + \Pr(S_n > a)$$

for positive integers n and a.

PROOF As in the proof of Theorem 11.6.2, we have $\Pr(M_n \geq a) = \Pr(S_n \geq a) + \Pr(M_n \geq a, S_n < a)$, and so it will suffice to show that $\Pr(M_n \geq a, S_n < a) = \Pr(S_n > a)$. Also, as in the proof of Theorem 11.6.2, we may write

$$\Pr(M_n \geq a, S_n < a) = \sum_{k=1}^{n} \Pr(A_k, S_n < a) \qquad (6.9)$$

where A_k denotes the event that $S_k \geq a$ and $S_j < a$ for $j = 1, \ldots, k - 1$. In the case of a simple symmetric random walk A_k implies $S_k = a$, so that A_k and $S_n < a$ will occur simultaneously if and only if A_k occurs and $S_n - S_k < 0$. Therefore,

$$\Pr(A_k, S_n < a) = \Pr(A_k, S_n - S_k < 0) = P(A_k)\Pr(S_n - S_k < 0)$$
$$= P(A_k)\Pr(S_n - S_k > 0) = \Pr(A_k, S_n - S_k > 0)$$
$$= \Pr(A_k, S_n > a)$$

Substitution in (6.9) now yields $\Pr(M_n \geq a, S_n < a) = \Pr(M_n \geq a, S_n > a) = \Pr(S_n > a)$, as asserted. ////

Theorem 11.6.3 has an interesting application. Let $S_0, S_1, S_2, \ldots$ be a simple symmetric random walk, If a is any integer, then the probability that $S_0, S_1, S_2, \ldots$ passes through a is 1 by the results of Section 11.2. In fact, the random walk will visit a infinitely often (Section 11.4). Let N_a be the time at which the random walk first passes through a. That is,

$$N_a = \text{least } n \geq 1 \text{ for which } S_n = a$$

We set $N_a = \infty$ if $S_n \neq a$ for all $n = 1, 2, \ldots$. Since the latter event has probability 0, it need not concern us. We shall call N_a the time of first passage through a.

We shall find the exact distribution of N_a in Section 12.5. Here we shall derive a simple approximation which is valid for large a. For $a > 0$ we have the relation

$$\Pr(N_a \leq n) = \Pr(M_n \geq a) = \Pr(S_n \geq a) + \Pr(S_n > a)$$

by Theorem 11.6.3, since $N_a \leq n$ if and only if $M_n \geq a$. If we let n be the greatest integer which is less than or equal to $a^2 t$, where $t > 0$, and apply the central-limit theorem (Sections 4.5 and 9.4), we find

$$\Pr(S_n \geq a) = \Pr\left(\frac{S_n}{\sqrt{n}} \geq \frac{a}{\sqrt{n}}\right) \to 1 - \Phi\left(\frac{1}{\sqrt{t}}\right)$$

as $a \to \infty$. Moreover, the same limit is obtained for $\Pr(S_n > a)$. Therefore, we have found the limiting distribution of N_a. We summarize our results in the following theorem.

Theorem 11.6.4 *Let $S_0, S_1, S_2, \ldots$ be a simple symmetric random walk, and let N_a be the time of first passage through a. Then as $a \to \infty$,*

$$\lim \Pr (N_a \leq a^2 t) = 2 \left(1 - \Phi \left(\frac{1}{\sqrt{t}} \right) \right)$$

for $t > 0$, where Φ denotes the standard normal distribution function.

11.7 THE STRONG LAW OF LARGE NUMBERS

Let $X_1, X_2, \ldots$ be independent random variables with finite means $\mu_1, \mu_2, \ldots,$ and let $\overline{X}_n = (X_1 + \cdots + X_n)/n$ and $\bar{\mu}_n = (\mu_1 + \cdots + \mu_n)/n$ for $n = 1, 2, \ldots$. In this section we shall give conditions which ensure that

$$\lim_{n \to \infty} (\overline{X}_n - \bar{\mu}_n) = 0 \qquad \text{with probability 1}$$

In particular, we shall show that the convergence in the law of large numbers (Theorem 9.2.2) is convergence with probability 1.

Theorem 11.7.1 *Let $X_1, X_2, \ldots$ be independent random variables with means $\mu_1, \mu_2, \ldots$ and finite variances $\sigma_1{}^2, \sigma_2{}^2, \ldots$. If*

$$\sum_{k=1}^{\infty} \frac{\sigma_k{}^2}{k^2} < \infty \qquad (7.1)$$

then $\lim (\overline{X}_n - \bar{\mu}_n) = 0$ with probability 1 as $n \to \infty$.

PROOF Without loss of generality, we may assume that $\mu_k = 0$ for $k = 1, 2, \ldots$, in which case we must show that

$$\Pr (|\overline{X}_n| \geq \varepsilon, \text{ i.o.}) = 0$$

for every $\varepsilon > 0$. Let $\varepsilon > 0$ be given, and let A_n be the event that $|\overline{X}_n| \geq \varepsilon$. Further, let B_n be the event that $|\overline{X}_k| \geq \varepsilon$ for some k, $2^{n-1} < k \leq 2^n$. Then the occurrence of A_n infinitely often implies the occurrence of B_n infinitely often, and so it will suffice to show that $\Pr (B_n, \text{ i.o.}) = 0$. Now B_n implies that $|S_k| = |X_1 + \cdots + X_k| \geq k\varepsilon$ for some k, $2^{n-1} < k \leq 2^n$, which (in turn) implies that $|S_k| \geq \varepsilon 2^{n-1}$ for some $k \leq 2^n$. It follows from Kolmogorov's inequality (Theorem 11.6.1) that

$$P(B_n) \leq \Pr (\max_{k \leq n} |S_k| \geq \varepsilon 2^{n-1}) \leq 4\varepsilon^{-2} 4^{-n} \sum_{k=1}^{2^n} \sigma_k{}^2$$

Therefore,[1]

$$\sum_{n=1}^{\infty} P(B_n) \le 4\varepsilon^{-2} \sum_{n=1}^{\infty} \left(\sum_{k=1}^{2^n} 4^{-n} \sigma_k^2 \right) = 4\varepsilon^{-2} \sum_{k=1}^{\infty} \left(\sum_{2^n \ge k} 4^{-n} \right) \sigma_k^2$$

Let $j = j_k$ be the smallest integer which is greater than or equal to $\log_2 k$, the logarithm of k to the base 2. Then

$$\sum_{2^n \ge k} 4^{-n} = \sum_{n=j}^{\infty} 4^{-n} = (\tfrac{4}{3}) 4^{-j} \le 2k^{-2}$$

Therefore,

$$\sum_{n=1}^{\infty} P(B_n) \le 8\varepsilon^{-2} \sum_{k=1}^{\infty} \frac{\sigma_k^2}{k^2}$$

which is finite by assumption. Therefore, $\Pr (B_n, \text{i.o.}) = 0$ by the Borel-Cantelli lemmas (Theorem 11.3.1). Theorem 11.7.1 follows. ////

In particular, condition (7.1) is satisfied if $X_1, X_2, \ldots$ have the same variance $\sigma_k^2 = \sigma^2$ for all $k = 1, 2, \ldots$.

Corollary 11.7.1 *Let $X_1, X_2, \ldots$ be independent random variables with common mean μ and common (finite) variance σ^2. Then $\overline{X}_n \to \mu$ as $n \to \infty$ with probability 1.*

Next we show that if $X_1, X_2, \ldots$ are identically distributed, the assumption that they have a finite variance may be dropped.

Lemma 11.7.1 *Let $a_1, a_2, \ldots$ be a sequence of real numbers. If $\lim a_n = a$ as $n \to \infty$, then $\lim \bar{a}_n = a$ as $n \to \infty$, where $\bar{a}_n = (a_1 + \cdots + a_n)/n$ for $n = 1, 2, \ldots$.*

The proof of Lemma 11.7.1 will be left as an exercise. The next theorem is known as the *strong law of large numbers.*

Theorem 11.7.2 *Let $X_1, X_2, \ldots$ be independent and identically distributed random variables, and let F denote their common distribution function. If the mean*

$$\mu = \int_{-\infty}^{\infty} x \, dF(x)$$

is finite, then $\overline{X}_n \to \mu$ with probability 1 as $n \to \infty$.

---

[1] The interchange of orders of summation is justified because the summands are nonnegative. See Apostol (1957), p. 374.

PROOF   We use the method of truncation. Let $Y_k = X_k$ if $-k \leq X_k \leq k$, and let $Y_k = 0$ otherwise. Further, let μ_k and σ_k^2 denote the mean and variance of Y_k, respectively, for $k = 1, 2, \ldots$. Then we may write

$$\overline{X}_n - \mu = (\overline{X}_n - \overline{Y}_n) + (\overline{Y}_n - \bar{\mu}_n) + (\bar{\mu}_n - \mu)$$

and we shall discuss the three terms separately.

We have

$$\mu_k = \int_{-k}^{k} x \, dF(x) \to \mu$$

as $k \to \infty$ by definition of the improper Riemann-Stieltjes integral (Appendix B). Therefore, $\bar{\mu}_n \to \mu$ as $n \to \infty$ by Lemma 11.7.1. Similarly, in Example 11.5.1 we showed that $\Pr(|X_n| \geq n, \text{i.o.}) = 0$ if $X_1, X_2, \ldots$ are independent and identically distributed and have a finite mean. It follows that $X_n - Y_n \to 0$ as $n \to \infty$ with probability 1, so that $\overline{X}_n - \overline{Y}_n \to 0$ with probability 1 as $n \to \infty$, again by Lemma 11.7.1.

To show that $\overline{Y}_n - \bar{\mu}_n \to 0$ with probability 1 as $n \to \infty$, we shall use Theorem 11.7.1. We must verify the condition (7.1). For $k = 1, 2, \ldots$, let B_k be the set of x for which $k - 1 < |x| \leq k$. Then

$$\sigma_k^2 \leq E(Y_k^2) = \int_{-k}^{k} y^2 \, dF(y) = \sum_{j=1}^{k} \int_{B_j} y^2 \, dF(y)$$

so that

$$\sum_{k=1}^{\infty} k^{-2} \sigma_k^2 \leq \sum_{k=1}^{\infty} \sum_{j=1}^{k} k^{-2} \int_{B_j} y^2 \, dF(y)$$

$$= \sum_{j=1}^{\infty} \left(\sum_{k=j}^{\infty} k^{-2} \right) \int_{B_j} y^2 \, dF(y)$$

Moreover,

$$\sum_{k=j}^{\infty} k^{-2} \leq j^{-1} + \sum_{k=j+1}^{\infty} k^{-2} \leq j^{-1} + \int_{j+1}^{\infty} x^{-2} \, dx = 2j^{-1}$$

It follows that

$$\sum_{k=1}^{\infty} k^{-2} \sigma_k^2 \leq \sum_{j=1}^{\infty} \frac{2}{j} \int_{B_j} y^2 \, dF(y)$$

$$\leq \sum_{j=1}^{\infty} 2 \int_{B_j} |y| \, dF(y)$$

$$= \int_{-\infty}^{\infty} 2|y| \, dF(y) = 2E(|X_1|)$$

which is finite by assumption. Therefore, $\overline{Y}_n - \bar{\mu}_n$ converges to 0 with probability 1 by Theorem 11.7.1.   ////

The implications of the strong law of large numbers for gambling and for the frequentistic interpretation of probability theory are similar to those of the weak law of large numbers (Section 9.2). An application of the strong law of large numbers to number theory is sketched in Problems 11.35 to 11.38.

11.8   THE LAW OF THE ITERATED LOGARITHM

In this section we shall investigate the rate of convergence in the strong law of large numbers. Let $X_1, X_2, \ldots$ be independent and identically distributed random variables with finite mean μ. Then we know from the strong law of large numbers (Theorem 11.7.2) that

$$\frac{1}{n}(S_n - n\mu) = \overline{X}_n - \mu \to 0 \qquad (8.1)$$

with probability 1 as $n \to \infty$. We claim that if $X_1, X_2, \ldots$ have a finite positive variance σ^2, then the convergence in (8.1) takes place at the rate

$$a_n = \sqrt{2\sigma^2 n \log (\log n)}$$

for $n \geq 3$. More precisely, we claim that

$$\Pr\left((S_n - n\mu) > (1 + \varepsilon)a_n, \text{ i.o.}\right) = 0 \qquad (8.2a)$$

$$\Pr\left((S_n - n\mu) > (1 - \varepsilon)a_n, \text{ i.o.}\right) = 1 \qquad (8.2b)$$

for every $\varepsilon > 0$. This result is known as *the law of the iterated logarithm*.

Of course, the result (8.2) applies to $-S_n$ as well as to S_n. Combining the results for $\pm S_n$ then yields

$$\Pr\left(|S_n - n\mu| > (1 + \varepsilon)a_n, \text{ i.o.}\right) = 0 \qquad (8.3a)$$

$$\Pr\left(|S_n - n\mu| > (1 - \varepsilon)a_n, \text{ i.o.}\right) = 1 \qquad (8.3b)$$

for every $\varepsilon > 0$.

Theorem 11.8.1   *Let $X_1, X_2, \ldots$ be independent and identically distributed random variables with mean μ and finite positive variance σ^2. Then (8.2) holds.*

PROOF   We shall prove Theorem 11.8.1 only in the case that $X_1, X_2, \ldots$ have a common normal distribution. Moreover, there is no loss of generality in supposing that $\mu = 0$ and $\sigma^2 = 1$. In this case S_n will have the normal distribution with mean 0 and variance n for every n (Example 8.4.7). Let Φ denote the standard normal distribution function. We shall use the relation

$$1 - \Phi(x) \sim \frac{1}{x\sqrt{2\pi}} e^{-\frac{1}{2}x^2} \qquad (8.4)$$

as $x \to \infty$ (Lemma 4.4.2).

We begin with the proof of (8.2a). Let $\varepsilon > 0$ be given, and let $c > 1$ be so close to 1 that $(1 + \varepsilon)^2/c > 1 + 2\varepsilon$. Further for each integer k, let n_k be an integer for which $c^k \leq n_k < c^k + 1$ and observe that $n_k \to \infty$ as $k \to \infty$. For $k = 1, 2, \ldots$, let A_k be the event that

$$S_n > (1 + \varepsilon)a_n \quad \text{for some } n \quad n_{k-1} \leq n < n_k$$

Then since $S_n > (1 + \varepsilon)a_n$ infinitely often implies the occurrence of A_k for infinitely many values of k, it will suffice to show that $\Pr(A_k, \text{i.o.}) = 0$. Now for $n \geq 3$, a_n is an increasing function of n, so that A_k implies

$$\max_{n \leq n_k} S_n \geq (1 + \varepsilon)a_{n_{k-1}}$$

Therefore, by Levy's inequality (Theorem 11.6.2)

$$P(A_k) \leq 2\Pr(S_{n_k} \geq (1 + \varepsilon)a_{n_{k-1}}) = 2[1 - \Phi(d_k)] \tag{8.5}$$

where $d_k = n_k^{-\frac{1}{2}}(1 + \varepsilon)a_{n_{k-1}}$. Now, as $k \to \infty$,

$$d_k^2 = (1 + \varepsilon)^2 n_k^{-1}a_{n_{k-1}}^2 \sim 2(1 + \varepsilon)^2 c^{-1} \log(\log c^{k-1})$$
$$\sim 2(1 + \varepsilon)^2 c^{-1} \log k$$

since $n_k \sim c^k$ and $\log(\log c^{k-1}) = \log[(k-1) + \log c] \sim \log k$ as $k \to \infty$. By choice of c we have $(1 + \varepsilon)^2 c^{-1} > 1 + 2\varepsilon$. Therefore, we have

$$d_k^2 \geq 2(1 + \varepsilon)\log k \tag{8.6}$$

for all sufficiently large values of k. It now follows from (8.4) and (8.6) that

$$1 - \Phi(d_k) \leq \frac{1}{d_k\sqrt{2\pi}}\left(\frac{1}{k}\right)^{1+\varepsilon}$$

for k sufficiently large. Therefore, $\sum_{k=1}^{\infty} P(A_k) < \infty$. Therefore, $\Pr(A_k, \text{i.o.}) = 0$ by Theorem 11.3.1. This establishes (8.2a).

We shall now prove (8.2b). Let $\varepsilon > 0$ be given and choose $\varepsilon' > 0$ so small and $c > 1$ so large that

$$(1 - \varepsilon')^2 \frac{c}{c-1} < 1 \quad \text{and} \quad 1 - \varepsilon' - \frac{2}{c} > 1 - \varepsilon \tag{8.7}$$

For each integer $k = 1, 2, \ldots$ let n_k be an integer for which $c^k \leq n_k < c^k + 1$, and let A_k be the event that

$$S_{n_k} - S_{n_{k-1}+1} > (1 - \varepsilon')a_{n_k}$$

Then $A_1, A_2, \ldots$ are independent events, since different A's are determined by different X's.

As in the proof of (8.2a), we find that $P(A_k) = 1 - \Phi(d_k)$, where

$$d_k^2 = \frac{(1 - \varepsilon')^2 a_{n_k}^2}{n_k - n_{k-1}} \sim \frac{2(1 - \varepsilon')^2 c^k \log (\log c^k)}{c^k - c^{k-1}}$$

$$\sim 2(1 - \varepsilon')^2 \frac{c}{c - 1} \log k$$

as $k \to \infty$. It follows that $d_k \le 2 \log k$ for sufficiently large values of k. Therefore, by (8.4),

$$1 - \Phi(d_k) \ge 1 - \Phi(2 \log k) \sim \frac{1}{d_k \sqrt{2\pi}} \frac{1}{k}$$

for k sufficiently large. It follows that $\sum_{k=1}^{\infty} P(A_k) = \infty$ and, consequently, that $\Pr(A_k, \text{i.o.}) = 1$.

Thus, the probability is 1 that

$$S_{n_k} - S_{n_{k-1}+1} > (1 - \varepsilon')a_{n_k}$$

for infinitely many values of k. Moreover, by (8.2a) applied to $-X_1, -X_2, \ldots$, the probability is 1 that

$$S_{n_{k-1}+1} < -\tfrac{3}{2}a_{n_{k-1}+1}$$

for all but a finite number of k. As $k \to \infty$, $a_{n_k}^{-1} a_{n_{k-1}+1} \to c^{-1}$ by simple algebra, so that

$$(1 - \varepsilon')a_{n_k} - \tfrac{3}{2}a_{n_{k-1}+1} > (1 - \varepsilon' - 2c^{-1})a_{n_k} > (1 - \varepsilon)a_{n_k}$$

for all sufficiently large values of k by the choice of ε' and c. Thus, the probability is 1 that $S_{n_k} > (1 - \varepsilon)a_{n_k}$ for infinitely many values of k. Since this implies that $S_n > (1 - \varepsilon)a_n$ for infinitely many values of n, (8.2b) follows.   ////

While we have proved Theorem 11.8.1 only in the case of normally distributed random variables, its conclusion should be plausible for arbitrary sequences of independent and identically distributed random variables with a finite positive variance. Indeed, by the central-limit theorem (Section 9.4), S_n will have an approximate normal distribution for any such sequence.

REFERENCES

For a more detailed treatment of simple random walks, including a more complete development of the gambler's-ruin problem, see Feller (1968).

PROBLEMS

11.1 Let $X_1, X_2, \ldots$ be any infinite sequence of random variables all of which are defined on the same probability space. Further, let F_n denote the joint distribution function of $X_1, \ldots, X_n$ for $n = 1, 2, \ldots$. Show that

$$F_n(x_1, \ldots, x_n) = F_{n+1}(x_1, \ldots, x_n, \infty) \qquad \text{(P.1)}$$

for all $(x_1, \ldots, x_n) \in R^n$ and all $n = 1, 2, \ldots$.

11.2 A sequence of distribution functions F_n, $n = 1, 2, \ldots$, which satisfy the condition (P.1) is called a *consistent sequence*. Show that if $G_1, G_2, \ldots$ are univariate distribution functions and if $F_n(x_1, \ldots, x_n) = G_1(x_1)G_2(x_2) \cdots G_n(x_n)$ for all $(x_1, \ldots, x_n) \in R_n$ and all $n = 1, 2, \ldots$, then F_n is a consistent sequence.

11.3 A famous theorem, known as *Kolmogorov's consistency theorem*,[1] asserts that if $F_1, F_2, \ldots$ is any consistent sequence of distribution functions, then there is a sequence of random variables $X_1, X_2, \ldots$ such that the joint distribution function of $X_1, \ldots, X_n$ is F_n for every n. Use Kolmogorov's consistency theorem and Problem 11.2 to show the existence of an infinite sequence of independent random variables which have arbitrary preassigned distribution functions.

11.4 For $n = 1, 2, \ldots$, let f_n be an n-variate density function, and let F_n denote the distribution function of f_n. If

$$f_n(x_1, \ldots, x_n) = \int_{-\infty}^{\infty} f_{n+1}(x_1, \ldots, x_n, y) \, dy \qquad \text{(P.2)}$$

for all $(x_1, \ldots, x_n) \in R^n$ and all $n = 1, 2, \ldots$, then $F_1, F_2, \ldots$ is a consistent sequence of distribution functions.

11.5 For each $y \in R$, let g_y be a univariate density function. Suppose also that $g_y(x)$ is bounded and jointly continuous in (x, y). Let H be any univariate distribution function, and define $f_n(x_1, \ldots, x_n) = \int_{-\infty}^{\infty} g_y(x_1)g_y(x_2) \cdots g_y(x_n) \, dH(y)$ for $(x_1, \ldots, x_n) \in R^n$ and $n = 1, 2, \ldots$. Show that $f_1, f_2, \ldots$ satisfy condition (P.2).

11.6 Use the result of Problem 11.5 to deduce the existence of random variables $Y, X_1, X_2, \ldots$ with the following properties. The distribution of Y is normal and conditionally given $Y = y$, the distribution of $X_1, \ldots, X_n$ is that of independent normal random variables with common mean y and common variance 1.

11.7 Find a joint density for $X_1, \ldots, X_n$ in Problem 11.6.

11.8 Show the existence of random variables $X_1, X_2, \ldots$ with the following property. For every $n = 1, 2, \ldots$, $X_1, \ldots, X_n$ have joint density f_n, where $f_n(x_1, \ldots, x_n) = n!/(1 + x_1 + \cdots + x_n)^{n+1}$ if $x_i > 0$ for $i = 1, \ldots, n$ and $f_n(x_1, \ldots, x_n) = 0$ for other values of $(x_1, \ldots, x_n)$.

NOTE: Problems 11.9 to 11.14 refer to the gambler's-ruin problem, described in Section 11.2.

[1] For a proof, see Neveu (1965), chap. 3.

11.9 Let N be the time at which the game ends. That is, let N = least $n \geq 0$ for which either $S_n = -a$ or $S_n = b$ or ∞ if no such n exists. Observe that $\Pr(N < \infty) = 1$ by Lemma 11.2.2. Show that $E(N) < \infty$.

11.10 Let $\eta_a = E(N)$. Show that $\eta_a = 1 + p\eta_{a+1} + q\eta_{a-1}$ for $0 < a < c$.

11.11 Show that $E(N) = ab$ if $p = \frac{1}{2}$.

11.12 Show that

$$E(N) = \frac{a}{q-p} - \frac{c}{q-p}\frac{1-\beta^a}{1-\beta^c}$$

if $p \neq \frac{1}{2}$.

11.13 Show that $\Pr(B_n^a \mid X_1 = 1) = P(B_{n-1}^{a+1})$ and that $\Pr(B_n^a \mid X_1 = -1) = P(B_{n-1}^{a-1})$ for $0 < a < c$ and $n = 1, 2, \ldots$.

11.14 Use Problem 11.13 and Equation (2.2) to show that $P(B_a \mid X_1 = 1) = P(B_{a+1})$ and $\Pr(B_a \mid X_1 = -1) = P(B_{a-1})$ for $0 < a < c$.

11.15 Let $X_1, X_2, \ldots$ be any sequence of independent and identically distributed random variables for which $\Pr(X_1 = 0) < 1$. Show that $\Pr(-a < S_n < b$ for all $n = 1, 2, \ldots) = 0$ for any $a > 0$ and $b > 0$.

> NOTE: Problems 11.16 to 11.22 study the probability that a simple random walk will ever return to the origin.

11.16 Let $S_0, S_1, S_2, \ldots$ be a simple random walk, and let B be the event that $S_n = 0$ for some $n = 1, 2, \ldots$. Show that $P(B) = 1 - |p - q|$. *Hint:* Write $P(B) = \Pr(B \mid X_1 = 1) \Pr(X_1 = 1) + \Pr(B \mid X_1 = -1) \Pr(X_1 = -1)$.

11.17 Let $u_n = \Pr(S_{2n} = 0)$, and let $v_n = \Pr(S_k \neq 0$ for $k = 1, \ldots, 2n - 1$ and $S_{2n} = 0)$ for $n = 1, 2, \ldots$. Further, let $u_0 = 1$ and $v_0 = 0$. Show that

$$u_n = \sum_{k=1}^{n} v_k u_{n-k} \qquad \text{(P.3)}$$

for $n = 1, 2, \ldots$. Equation (P.3) is known as the *renewal equation*.

11.18 Let U and V denote the generating functions of $u_0, u_1, u_2, \ldots$ and $v_1, v_2, \ldots$, respectively. Use (P.3) to show that $U(s) - 1 = U(s)V(s)$ for $0 < s < 1$.

11.19 Show that $U(s) = (1 - 4pqs)^{-1/2}$ for $0 < s < 1$. *Hint:* Use Problem 1.60.

11.20 Show that $V(s) = 1 - \sqrt{1 - 4pqs}$ for $0 < s < 1$.

11.21 Use Problem 11.20 to rederive the result of Problem 11.16.

11.22 Show that $v_n = \binom{\frac{1}{2}}{n}(-1)^{n-1}(4pq)^n$ for $n = 1, 2, \ldots$.

11.23 Let $A_1, A_2, \ldots$ be an infinite sequence of events. We define $\liminf A_n = \bigcup_{n=1}^{\infty} \bigcap_{k=n}^{\infty} A_k$. Show that $(\limsup A_n)' = \liminf A_n'$.

11.24 Let $S = (0,1)$ be the open unit interval. Let $A_{2n} = (0, 1 - 1/n)$ and $A_{2n-1} = (0, 1/n)$ for $n = 1, 2, \ldots$. Find $\limsup A_n$ and $\liminf A_n$.

11.25 Let $X_1, X_2, \ldots$ be independent random variables, let $J_1, J_2, \ldots$ be intervals, and let A_n be the event that $X_n \in J_n$ for $n = 1, 2, \ldots$. Show that $A_1, A_2, \ldots$ are independent events.

11.26 Let $X_1, X_2, \ldots$ be independent random variables which are uniformly distributed over (0,1). What is the probability that $X_n < 1/n$, i.o.; what is the probability that $X_n < 1/n^2$, i.o.?

11.27 Let $X_1, X_2, \ldots$ be independent random variables all of which have the standard exponential distribution. Let B_a be the event that $X_n > \log n + a \log (\log n)$, i.o. for $a > 0$. For what values of a does $P(B_a) = 1$?

11.28 Let $S_0, S_1, S_2, \ldots$ be a simple symmetric random walk, and let N_k be the time of the kth return to the origin. Show that N_1 and $N_2 - N_1$ are independent random variables.

11.29 Find the generating function of N_2. Find $\Pr(N_2 = n)$ for $n = 1, 2, \ldots$.

11.30 Show that $X_n \to X$ with probability 1 as $n \to \infty$ if and only if $\sup_{k \geq n} |X_k - X| \to 0$ in probability as $n \to \infty$.

11.31 Show that $X_n \to X$ with probability 1 as $n \to \infty$ if and only if the following condition is satisfied. For every $\varepsilon > 0$ and $\delta > 0$, there is an integer $n_0 = n_0(\varepsilon, \delta)$ for which $\Pr(|X_k - X| \geq \varepsilon$ for some $k = n_0, \ldots, n) \leq \delta$ for all $n \geq n_0$.

11.32 Let U be a random variable which is uniformly distributed over (0,1). Let A_{nk} be the event that $k - 1 < nU < k$ for $k = 1, \ldots, n$ and $n = 1, 2, \ldots$. Further, let $X_1, X_2, \ldots$ be $I_{A_{11}}, I_{A_{21}}, I_{A_{22}}, \ldots$. Show that $X_n \to 0$ in probability but X_n does not converge to 0 with probability 1 as $n \to \infty$.

11.33 Prove Lemma 11.7.1.

11.34 Let $X_1, X_2, \ldots$ be independent random variables, and let $X_k = \pm k^\alpha$ each with probability $\frac{1}{2}$. Show that $\bar{X}_n \to 0$ with probability 1 as $n \to \infty$ if and only if $\alpha < \frac{1}{2}$. *Hint:* Use the Lindeberg-Feller theorem to show that if $\alpha \geq \frac{1}{2}$, then $\bar{X}_n$ does not converge to 0 in probability as $n \to \infty$.

NOTE: Problems 11.35 to 11.38 sketch an application of the strong law of large numbers to number theory. For $x > 0$, let $w(x)$ be the greatest integer which is less than or equal to x.

11.35 For $0 < x < 1$, let $w_1(x) = w(10x)$ and for $k \geq 2$, let

$$w_k(x) = w \left(10^k \left[x - \sum_{j=1}^{k-1} 10^{-j} w_j(x) \right] \right)$$

Show that $x = \sum_{k=1}^{\infty} 10^{-k} w_k(x)$ for $0 < x < 1$. $w_k(x)$ is the kth decimal in the decimal expansion of x.

11.36 Let $S = (0,1)$, let $\mathscr{S}$ be the class of Borel subsets of S, and let $P(A)$ be the length of A for every subinterval $A \subseteq S$. Further, let $W_k = w_k(s)$ for $s \in S$. Show that W_1 and W_2 have the discrete uniform distribution on the integers $0, 1, \ldots, 9$. Show also that W_1 and W_2 are independent. (*Hint:* See Example 3.3.3c.)

11.37 Show that $W_1, W_2, \ldots$ are independent and identically distributed.

11.38 For fixed j, $0 < j < 9$, let $X_k = 1$ if $W_k = j$ and let $X_k = 0$ otherwise. Further, let $S_n = X_1 + \cdots + X_n$ for $n = 1, 2, \ldots$. Thus, S_n is the number of j's among the first n decimals of a randomly selected number. Show that $S_n/n \to 0.1$ with probability 1 as $n \to \infty$. That is, if a number is selected at random from the interval (0,1), the proportion of j's among the first n decimals converges to 0.1 with probability 1 as $n \to \infty$ for every $j = 0, \ldots, 9$.

12

MARTINGALES[1]

12.1 GAMBLING SYSTEMS

In this section we shall consider gambling strategies, or *gambling systems* as we shall call them. Given a particular system, we shall define random variables to represent the gambler's fortune as it evolves in time, and we shall prove that unless the gambler has an unlimited amount of time, *no gambling system will convert a sequence of fair games into a favorable one.*

Consider a gambler who may play a sequence of games each of which he wins with probability $\frac{1}{2}$ and loses with probability $\frac{1}{2}$. Let $X_1, X_2, \ldots$ be independent random variables with common distribution

$$\Pr(X_i = 1) = \tfrac{1}{2} = \Pr(X_i = -1) \tag{1.1}$$

and interpret the event that $X_i = 1$ ($X_i = -1$) as the event that the gambler wins (loses) the ith game for $i = 1, 2, \ldots$. Observe that each game is fair in the sense that if the gambler wagers any amount w on the ith game, his expected winnings on the ith game are $w \Pr(X_i = 1) - w \Pr(X_i = -1) = 0$.

[1] This chapter treats a special topic and may be omitted.

Let us allow the gambler to employ a system by which we mean a rule for varying his bets according to his fortune. The only restriction to which we shall subject the gambler is that he not be allowed to look into the future. That is, the amount he bets on the ith game may depend on the outcomes of the first $i - 1$ games, but it may not depend on the outcome of the ith or any later games. We define a *gambling system* to be a sequence of nonnegative functions $w_1, w_2, \ldots$, where w_1 is a constant, and, for $k \geq 2$, w_k is a function whose domain is R^{k-1}. We shall call the random variable

$$W_k = w_k(X_1, \ldots, X_{k-1}) \qquad (1.2)$$

the *gambler's bet* on the kth game.

Let Y_0 be a constant which represents the gambler's initial fortune. Then we can represent the gambler's fortune after n plays of the game by the random variable

$$Y_n = Y_0 + \sum_{k=1}^{n} W_k X_k \qquad (1.3)$$

since the gambler wins the amount $W_k X_k$ on the kth game for $k = 1, 2, \ldots$. We shall call Y_n *the gambler's fortune at time n*.

EXAMPLE 12.1.1 The following system has fascinated gamblers for years: *Double your bets until you win a game; then quit.* Formally, let $W_1 = w$, a constant, and let

$$W_k = \begin{cases} w2^{k-1} & \text{if } X_i = -1 \text{ for } i = 1, \ldots, k - 1 \\ 0 & \text{otherwise} \end{cases} \qquad (1.4)$$

for $k = 2, 3, \ldots$. *Using this strategy, a gambler is certain to win.* Indeed, the gambler is certain to win at least one game (Section 4.2); and if the first game the gambler wins is the nth, then he will have lost

$$w + 2w + \cdots + 2^{n-2}w = (2^{n-1} - 1)w$$

on the first $n - 1$ games, but he will win $2^{n-1}w$ on the nth game. Therefore, the probability is 1 that the gambler will win w.

Of course, there is a catch. A gambler must have both unlimited time and unlimited capital to employ the strategy of this example, for he will lose the first n games with probability $2^{-n} > 0$ for every $n = 1, 2, \ldots$. For example, if a gambler starts with an initial reserve of $Y_0 = 2^m - 1$ dollars and no credit, and if he bets 1 dollar ($w = 1$) on the first game, then (1.4) must be modified to

$$W_k' = \begin{cases} 2^{k-1} & \text{if } X_i = -1 \text{ for } i < k \text{ for } k \leq m \\ 0 & \text{otherwise} \end{cases} \qquad (1.4a)$$

In this case the gambler will lose all his money if he loses the first m games, which happens with probability 2^{-m}; and, as above, he will win 1 dollar if he wins at least 1 of the first m games. Therefore, his expected winnings are

$$1 \Pr(\text{win}) - (2^m - 1) \Pr(\text{lose}) = 1(1 - 2^{-m}) - (2^m - 1)2^{-m} = 0$$

Therefore, the expected winnings using the system (1.4a) are 0. The probability of winning, $1 - 2^{-m}$, may be quite high, however.   ////

We shall now prove that in the absence of unlimited time no gambling system will convert a sequence of fair games into a favorable one.

Theorem 12.1.1   *Let* $X_1, X_2, \ldots$ *be independent random variables with common distribution given by* (1.1), *and let* Y_n *be defined by* (1.2) *and* (1.3) *for* $n = 1, 2, \ldots$. *Further, let* $\mathbf{X}_n = (X_1, \ldots, X_n)$ *for* $n = 1, 2, \ldots$. *Then*

$$E(Y_{n+1} \mid \mathbf{X}_n) = Y_n \qquad (1.5)$$

$$E(Y_n) = Y_0 \qquad (1.6)$$

for $n = 1, 2, \ldots$.

PROOF   Let us first prove (1.5). We have $Y_{n+1} = Y_n + W_{n+1}X_{n+1}$ by (1.3), so that

$$E(Y_{n+1} \mid \mathbf{X}_n) = E(Y_n \mid \mathbf{X}_n) + E(W_{n+1}X_{n+1} \mid \mathbf{X}_n)$$

by Theorem 10.3.1. Now Y_n and W_{n+1} are determined by $\mathbf{X}_n$, and X_{n+1} is independent of $\mathbf{X}_n$. Therefore,

$$E(Y_n \mid \mathbf{X}_n) = Y_n$$

and $E(W_{n+1}X_{n+1} \mid \mathbf{X}_n) = W_{n+1}E(X_{n+1} \mid \mathbf{X}_n) = W_{n+1}E(X_{n+1}) = 0$

by Theorems 10.3.2 and 10.3.3. We also used the fact that $E(X_{n+1}) = 0$. This establishes (1.5), and a similar argument with unconditional expectations replacing conditional expectations will show that $E(Y_1) = Y_0$. Equation (1.6) now follows from (1.5) and Theorem 10.3.4 since

$$E(Y_{n+1}) = E[E(Y_{n+1} \mid \mathbf{X}_n)] = E(Y_n)$$

for $n = 1, 2, \ldots$. In fact, $E(Y_n) = E(Y_{n-1}) = \cdots = E(Y_1) = Y_0$.   ////

The interpretations of (1.5) and (1.6) are the following. Equation (1.5) asserts that given the outcomes of the first n games, one's expected winnings on the $(n + 1)$st game, are zero, while (1.6) asserts that one's expected fortune after any n plays is the same as one's initial fortune. That is, one cannot increase one's expected fortune by playing a finite number of fair games.

Theorem 12.1.1 leaves open the possibility of converting a sequence of fair games into a favorable game by playing an (unbounded) random number of games, as in Example 12.1.1. We shall return to this question in Section 12.4.

12.2 MARTINGALES

In the previous section, we defined a sequence of random variables $Y_0, Y_1, \ldots$ to represent the fortunes of a gambler who plays a sequence of fair games, and we found that they had the property

$$E(Y_{n+1} \mid X_1, \ldots, X_n) = Y_n \qquad (2.1)$$

for every $n = 1, 2, \ldots$. This property is worthy of abstraction.

Let $X_1, X_2, \ldots$ be a finite or infinite sequence of random variables or random vectors. The X_i need not be independent or identically distributed; they need not even be of the same dimension. For $n = 1, 2, \ldots$, let $\mathbf{X}_n = (X_1, \ldots, X_n)$, and let D_n be a subset of the range of $\mathbf{X}_n$ for which $\Pr(\mathbf{X}_n \in D_n) = 1$. Further, let $w_1, w_2, \ldots$ be a sequence of real-valued functions with domains $D_1, D_2, \ldots$, and let $Y_1, Y_2, \ldots$ be a sequence of random variables defined by

$$Y_n = w_n(X_1, \ldots, X_n)$$

for $n = 1, 2, \ldots$. We shall say that the sequence $Y_1, Y_2, \ldots$ is a *submartingale with respect to* $X_1, X_2, \ldots$ *if and only if*

$$E(|Y_n|) < \infty \qquad (2.2)$$

$$E(Y_{n+1} \mid \mathbf{X}_n = \mathbf{x}_n) \geq w_n(\mathbf{x}_n) \qquad (2.3)$$

for all $\mathbf{x}_n \in D_n$ for every $n = 1, 2, \ldots$. We suppose that the conditional expectations in (2.3) may be defined by one of the recipes of Section 10.3. In the sequel we shall write (2.3) in the equivalent form

$$E(Y_{n+1} \mid \mathbf{X}_n) \geq Y_n \qquad (2.3a)$$

(see Section 10.3). Further, we shall say that $Y_1, Y_2, \ldots$ is a *martingale with respect to* $X_1, X_2, \ldots$ if and only if there is equality in (2.3) and (2.3a). Thus $Y_1, Y_2, \ldots$ is a martingale with respect to $X_1, X_2, \ldots$ if and only if (2.1) and (2.2) hold for every $n = 1, 2, \ldots$, and $Y_1, Y_2, \ldots$ is a submartingale with respect to $X_1, X_2, \ldots$ if and only if (2.2) and (2.3a) hold for every $n = 1, 2, \ldots$. Observe that $Y_1, Y_2, \ldots$ is a martingale with respect to $X_1, X_2, \ldots$ if and only if $Y_1, Y_2, \ldots$ and $-Y_1, -Y_2, \ldots$ are both submartingales with respect to $X_1, X_2, \ldots$. When there is no danger of confusion, we shall omit the qualifying phrase "with respect to $X_1, X_2, \ldots$."

The sequence $Y_1, Y_2, \ldots$ of the previous section is a martingale. Several additional examples will now be given.

EXAMPLE 12.2.1 Many interesting martingales may be built from independent random variables.

a Let $X_1, X_2, \ldots$ be independent random variables with common expectation $E(X_k) = 0$ for $k = 1, 2, \ldots$. Then the sequence of partial sums

$$S_n = X_1 + \cdots + X_n \qquad n = 1, 2, \ldots$$

is a martingale. Indeed, $E(|S_n|) \leq E(|X_1|) + \cdots + E(|X_n|) < \infty$ for $n = 1, 2, \ldots$. Moreover,

$$E(S_{n+1} \mid \mathbf{X}_n) = E(S_n \mid \mathbf{X}_n) + E(X_{n+1} \mid \mathbf{X}_n)$$

and, as in the proof of Theorem 12.1.1, $E(S_n \mid \mathbf{X}_n) = S_n$ because S_n is determined by $\mathbf{X}_n$, and $E(X_{n+1} \mid \mathbf{X}_n) = E(X_{n+1}) = 0$ since X_{n+1} is independent of $\mathbf{X}_n$.

b Let $X_1, X_2, \ldots$ be independent with means $E(X_k) = 0$ and finite variances $\sigma_k^2 = E(X_k^2)$ for $k = 1, 2, \ldots$. Further, let $s_n^2 = \sigma_1^2 + \cdots + \sigma_n^2$ be the variance of S_n for $n = 1, 2, \ldots$. Then

$$Y_n = S_n^2 - s_n^2 \qquad n = 1, 2, \ldots$$

defines a martingale. Observe first that $E(|Y_n|) \leq E(S_n^2) + s_n^2 \leq 2s_n^2 < \infty$ for $n = 1, 2, \ldots$. Moreover, since $S_{n+1}^2 = S_n^2 + 2S_nX_{n+1} + X_{n+1}^2$ and

$$E(S_nX_{n+1} \mid \mathbf{X}_n) = S_nE(X_{n+1} \mid \mathbf{X}_n) = S_nE(X_{n+1}) = 0$$

by Theorems 10.3.2 and 10.3.3, we have $E(S_{n+1}^2 \mid \mathbf{X}_n) = E(S_n^2 \mid \mathbf{X}_n) + E(X_{n+1}^2 \mid \mathbf{X}_n) = S_n^2 + \sigma_{n+1}^2$, where the final equality also follows from Theorems 10.3.2 and 10.3.3. It follows that $E(Y_{n+1} \mid \mathbf{X}_n) = E(S_{n+1}^2 \mid \mathbf{X}_n) - s_{n+1}^2 = S_n^2 + \sigma_{n+1}^2 - s_{n+1}^2 = S_n^2 - s_n^2 = Y_n$ for $n = 1, 2, \ldots$, as required.

c Now let $X_1, X_2, \ldots$ be independent nonnegative random variables with common expectation $E(X_k) = 1$ for $k = 1, 2, \ldots$. Then

$$Y_n = \prod_{k=1}^{n} X_k \qquad n = 1, 2, \ldots$$

defines a martingale. In fact, $E(Y_n) = \prod_{k=1}^{n} E(X_k) = 1 < \infty$ for $n = 1, 2, \ldots$, and

$$E(Y_{n+1} \mid \mathbf{X}_n) = Y_nE(X_{n+1} \mid \mathbf{X}_n) = Y_nE(X_{n+1}) = Y_n$$

by Theorems 10.3.2 and 10.3.3.

d As a special case of part *c*, let $X_1, X_2, \ldots$ be independent and identically distributed with common moment-generating function M. Then for any t for which $M(t)$ is finite,

$$Y_n = \frac{e^{tS_n}}{M(t)^n} \qquad n = 1, 2, \ldots$$

defines a martingale. ////

Lest the reader think that all martingales are sums or products of independent random variables in disguise, we shall consider some examples of a different nature.

EXAMPLE 12.2.2 Polya's urn scheme Suppose that repeated drawings are made from an urn which contains red and black balls. Suppose that after each drawing, the ball drawn is replaced, along with c balls of the same color, where c is a positive integer. Let Y_n denote the proportion of red balls in the urn after the nth draw. We shall show that $Y_1, Y_2, \ldots$ is a martingale with respect to a sequence $X_1, X_2, \ldots$ which will be defined below.

Suppose that there are r red balls and b black balls in the urn at the time of the first draw, where r and b are positive integers. Let $X_n = 1$ if the nth ball drawn is red, and let $X_n = 0$ if the nth ball drawn is black. Further, let r_n and b_n denote the number of red balls and the number of black balls in the urn after the nth draw. Then

$$Y_n = \frac{r_n}{r_n + b_n}$$

for $n = 1, 2, \ldots$; and r_n, b_n, and X_n evolve according to the equations

$$r_{n+1} = \begin{cases} r_n + c & \text{if } X_{n+1} = 1 \\ r_n & \text{if } X_{n+1} = 0 \end{cases} \qquad b_{n+1} = \begin{cases} b_n & \text{if } X_{n+1} = 1 \\ b_n + c & \text{if } X_{n+1} = 0 \end{cases}$$

Here we set $r_0 = r$ and $b_0 = b$. Now $\Pr(X_{n+1} = 1 \mid \mathbf{X}_n) = Y_n$ for $n = 1, 2, \ldots$, so that

$$E(Y_{n+1} \mid \mathbf{X}_n) = \frac{r_n + c}{r_n + b_n + c} \frac{r_n}{r_n + b_n} + \frac{r_n}{r_n + b_n + c} \frac{b_n}{r_n + b_n}$$

$$= \frac{r_n}{r_n + b_n} = Y_n$$

for $n = 1, 2, \ldots$, as required. $/\!/\!/\!/$

EXAMPLE 12.2.3 Likelihood ratios Let $X_1, X_2, \ldots$ be any sequence of random variables with absolutely continuous joint distributions. Let f_n denote a joint density for $X_1, \ldots, X_n$, and let us suppose that $f_1, f_2, \ldots$ satisfy the consistency condition

$$f_n(x_1, \ldots, x_n) = \int_{-\infty}^{\infty} f_{n+1}(x_1, \ldots, x_n, y) \, dy \qquad (2.4)$$

for *all* $(x_1, \ldots, x_n) \in R^n$ for all $n = 1, 2, \ldots$ (see Section 6.3 and Problem 11.4). Let $g_1, g_2, \ldots$ be any other sequence of density functions which satisfies the consistency condition (2.4), and suppose, for simplicity, that $f_n(x_1, \ldots, x_n)$ is positive for all $x = (x_1, \ldots, x_n) \in R^n$.

$$Y_n = \frac{g_n(X_1, \ldots, X_n)}{f_n(X_1, \ldots, X_n)} \qquad n = 1, 2, \ldots$$

Then $Y_1, Y_2, \ldots$ is a martingale. To see this observe that a conditional density for X_{n+1} given $\mathbf{X}_n = \mathbf{x}_n$ is

$$h(y \mid \mathbf{x}_n) = \frac{f_{n+1}(\mathbf{x}_n, y)}{f_n(\mathbf{x}_n)}$$

for $-\infty < y < \infty$. Therefore,

$$E(Y_{n+1} \mid \mathbf{X}_n = \mathbf{x}_n) = \int_{-\infty}^{\infty} \frac{g_{n+1}(\mathbf{x}_n, y)}{f_{n+1}(\mathbf{x}_n, y)} h(y \mid \mathbf{x}_n) \, dy$$

$$= \int_{-\infty}^{\infty} \frac{g_{n+1}(\mathbf{x}_n, y)}{f_n(\mathbf{x}_n)} \, dy = \frac{g_n(\mathbf{x}_n)}{f_n(\mathbf{x}_n)}$$

where the final step follows from the consistency of the sequence $g_1, g_2, \ldots$. The martingale equality (2.1) now follows by replacing $\mathbf{x}_n$ with $\mathbf{X}_n$. ////

EXAMPLE 12.2.4 In this example we present a general method for constructing martingales. Let $Z, X_1, X_2, \ldots$ be random variables which are defined on the same probability space and suppose that Z has a finite expectation. Then the sequence

$$Y_n = E(Z \mid X_1, \ldots, X_n) \qquad n = 1, 2, \ldots$$

is a martingale. In fact, $|Y_n| \le E(|Z| \mid \mathbf{X}_n)$, so that $E(|Y_n|) \le E[E(|Z| \mid \mathbf{X}_n)] = E(|Z|)$, which is finite by assumption. Moreover,

$$E(Y_{n+1} \mid \mathbf{X}_n) = E[E(Z \mid \mathbf{X}_{n+1}) \mid \mathbf{X}_n] = E(Z \mid \mathbf{X}_n) = Y_n$$

for $n = 1, 2, \ldots$ by Theorem 10.4.1. ////

12.3 ELEMENTARY PROPERTIES OF MARTINGALES

We shall now develop some elementary properties of martingales and submartingales.

Lemma 12.3.1 *If $Y_1, Y_2, \ldots$ is a submartingale with respect to $X_1, X_2, \ldots$, then*

$$E(Y_n) \le E(Y_{n+1}) \qquad (3.1)$$

for every $n = 1, 2, \ldots$. If $Y_1, Y_2, \ldots$ is a martingale, then there is equality in (3.1).

PROOF The inequality (3.1) follows from the submartingale inequality (2.3a) and Theorem 10.3.4. In fact, $E(Y_{n+1}) = E[E(Y_{n+1} \mid X_n)] \geq E(Y_n)$ for $n = 1, 2, \ldots$. Moreover, there is equality if $Y_1, Y_2, \ldots$ is a martingale by (2.1).

/////

EXAMPLE 12.3.1 Polya's urn scheme revisited In the notation of Example 12.2.2, the probability that a red ball is drawn on the nth draw is $\Pr(X_n = 1)$. Moreover,

$$\Pr(X_n = 1) = E[\Pr(X_n = 1 \mid X_{n-1})] = E(Y_{n-1}) = E(Y_1) = \frac{r}{r+b}$$

where the penultimate equality follows from Lemma 12.3.1. Thus, the unconditional probability of drawing a red ball is the same for every draw.

/////

Lemma 12.3.2 Let $Y_1, Y_2, \ldots$ be a submartingale with respect to $X_1, X_2, \ldots$. If n and k are positive integers for which $n < k$, then

$$E(Y_k \mid X_n) \geq Y_n$$

with equality if $Y_1, Y_2, \ldots$ is a martingale.

PROOF We shall prove Lemma 12.3.2 by induction. By definition, the lemma is true if $k - n = 1$. Suppose that the lemma is true when $k - n < m$, and consider the case that $k - n = m$. Then, by Theorem 10.4.1,

$$E(Y_k \mid X_n) = E[E(Y_k \mid X_{n+1}) \mid X_n]$$

By induction, we have $E(Y_k \mid X_{n+1}) \geq Y_{n+1}$, and by (2.3a) we have $E(Y_{n+1} \mid X_n) \geq Y_n$. Consequently, $E(Y_k \mid X_n) \geq E(Y_{n+1} \mid X_n) \geq Y_n$. This completes the induction from which the first assertion of the lemma follows. The second can be established by a similar argument which uses (2.1) in place of (2.3a). Alternatively, the second assertion of the lemma can be established by applying the first to the submartingales $Y_1, Y_2, \ldots$ and $-Y_1, -Y_2, \ldots$.

/////

We shall now give a method for constructing submartingales from martingales.

Lemma 12.3.3 Let $Y_1, Y_2, \ldots$ be a martingale with respect to a sequence $X_1, X_2, \ldots$. Then $|Y_1|, |Y_2|, \ldots$ is a submartingale with respect to $X_1, X_2, \ldots$. If, in addition, $E(Y_n^2) < \infty$ for all $n = 1, 2, \ldots$, then $Y_1^2, Y_2^2, \ldots$ is a submartingale with respect to $X_1, X_2, \ldots$.

PROOF The lemma uses (2.1) and Theorem 10.3.1. In fact, we have $E(|Y_{n+1}| \mid \mathbf{X}_n) \geq |E(Y_{n+1} \mid \mathbf{X}_n)| = |Y_n|$ for $n = 1, 2, \ldots$. This establishes the first assertion of the lemma, and the second follows from a similar argument.

////

12.4 THE OPTIONAL-STOPPING THEOREM

Let $Y_1, Y_2, \ldots$ be a martingale with respect to a sequence $X_1, X_2, \ldots$, and let us regard $Y_1, Y_2, \ldots$ as the fortunes of a gambler. Then, by Lemma 12.3.1, $E(Y_n) = E(Y_1)$ for every $n = 1, 2, \ldots$, so that the gambler's expected fortune does not increase with time. However, we saw in Example 12.1.1 how a gambler might actually guarantee himself a net gain by playing a random number of games. This is an exciting possibility. Unfortunately, it can be realized only by gamblers who have unlimited credit, as we shall show in this section.

Let $X_1, X_2, \ldots$ be any sequence of random variables or random vectors, and let N be a random variable which is determined as a function of the sequence X_1, $X_2, \ldots$. We shall say that N is a *stopping time with respect to the sequence* $X_1, X_2, \ldots$ if and only if the following conditions are satisfied:

1 N assumes only positive integer values or the value ∞.

2
$$\Pr(N < \infty) = 1 \qquad (4.1)$$

3 For every $n = 1, 2, \ldots$, the event that $N = n$ is determined by $\mathbf{X}_n = (X_1, \ldots, X_n)$ in the sense that there is a subset B_n of the range of $\mathbf{X}_n$ for which

$$N = n \qquad \text{if and only if} \qquad \mathbf{X}_n \in B_n \qquad (4.2)$$

If we think of $X_1, X_2, \ldots$ as the outcomes of a sequence of games, as in Section 12.1, then we may regard N as a rule which tells a gambler how long to continue playing the games, that is, stop after the Nth play. Condition (4.1) then requires that the gambler stop at some finite time, and condition (4.2) requires that the decision to stop after the nth play depend only on the outcomes of the first n plays and not on the outcomes of any later plays. That is, (4.2) requires that a gambler not be allowed to look into the future.

EXAMPLE 12.4.1 Let $S_0, S_1, S_2, \ldots$ be a simple random walk. That is, let $X_1, X_2, \ldots$ be independent and identically distributed random variables which assume the values 1 and -1 with probabilities p and $q = 1 - p$, respectively, and let $S_n = X_1 + \cdots + X_n$ for $n = 1, 2, \ldots$.

a   If $p \geq \frac{1}{2}$ and a is a positive integer, then the random variable N_a defined by $N_a = $ least $n \geq 1$ for which $S_n = a$ or ∞ if no such n exists is a stopping time (with respect to $X_1, X_2, \ldots$). Indeed, N_a assumes only positive integer values or the value ∞, and $\Pr(N_a < \infty) = 1$ by Theorem 11.2.3. Moreover, the event that $N_a = n$ occurs if and only if $S_k < a$ for $k = 1, \ldots, n - 1$ and $S_n = a$, and the latter event is determined by $X_1, \ldots, X_n$ in the sense of (4.2). We call N_a the time of first passage through a.

b   If $p = \frac{1}{2}$, then the random variable N_0 defined by $N_0 = $ least $n \geq 1$ for which $S_n = 0$ or ∞ if no such n exists is a stopping time. Again N_0 assumes only positive integer values or the value ∞, and $\Pr(N_0 < \infty) = 1$ by Theorem 11.2.3. Moreover, the event that $N_0 = n$ occurs if and only if $S_k \neq 0$ for $k = 1, \ldots, n - 1$ and $S_n = 0$, and the latter event depends only on $X_1, \ldots, X_n$ in the sense of (4.2). We call N_0 the time of first return to the origin.

c   If $p \neq \frac{1}{2}$, then N_0 is not a stopping time, since $\Pr(N_0 < \infty) = 1 - |p - q|$ (Problem 11.16).

d   If $N = $ least $n \geq 1$ for which $X_{n+1} = 1$ or ∞ if no such n exists, then N is not a stopping time with respect to $X_1, X_2, \ldots$, because condition (4.2) is violated.   ////

Lemma 12.4.1   *Let N be a stopping time with respect to a sequence $X_1, X_2, \ldots$. Then for $n = 1, 2, \ldots$, the events that $N \leq n$ and $N > n$ are determined by $\mathbf{X}_n = (X_1, \ldots, X_n)$. Moreover, if k is any positive integer, then $M = \min(N, k)$ is a stopping time with respect to $X_1, X_2, \ldots$.*

PROOF   For simplicity, we shall assume that $X_1, X_2, \ldots$ are random variables. Let A_j denote the event that $N = j$. Then, by assumption, there are subsets $B_j \subset R^j$ for which

$$A_j = \mathbf{X}_j^{-1}(B_j) = \mathbf{X}_n^{-1}(B_j \times R^{n-j})$$

for $j \leq n$. Therefore, the event that $N \leq n$ is

$$\bigcup_{j=1}^{n} A_j = \bigcup_{j=1}^{n} \mathbf{X}_n^{-1}(B_j \times R^{n-j}) = \mathbf{X}_n^{-1}\left(\bigcup_{j=1}^{n} B_j \times R^{n-j}\right)$$

Thus, the event that $N \leq n$ is determined by $X_1, \ldots, X_n$ in the sense of (4.2). The event that $N > n$ is the complement of the event that $N \leq n$, and so it is determined by $X_1, \ldots, X_n$ too.

Now let k be a positive integer, and let $M = \min(N, k)$. Then the event that $M = j$ is the same as the event that $N = j$ if $j < k$; it is the event that

$N > k - 1$ if $j = k$; and it is impossible if $j > k$. In any case, the event that $M = j$ is determined by $X_1, \ldots, X_j$, as required. ////

Now suppose that $Y_1, Y_2, \ldots$ is a martingale and that N is a stopping time with respect to the same sequence $X_1, X_2, \ldots$. Then we can define a random variable Y_N by letting $Y_N = Y_n$ if $N = n$, where $n = 1, 2, \ldots$, and letting $Y_N = 0$ if $N = \infty$. Equivalently, we can define Y_N by the formula

$$Y_N = \sum_{n=1}^{\infty} Y_n I_{A_n} \qquad (4.3)$$

where A_n denotes the event that $N = n$ and I_A denotes the indicator of the event A. If we regard $Y_1, Y_2, \ldots$ as the fortunes of a gambler who plays a sequence of games and quits after playing N games, we may regard Y_N as the fortune of the gambler at the time he terminates his play. The results of this section give conditions under which

$$E(Y_N) = E(Y_1) \qquad (4.4)$$

In the gambling terminology, (4.4) asserts that the gambler does not increase his expected fortune by using a stopping time.

Theorem 12.4.1 *Let $Y_1, Y_2, \ldots$ be a submartingale, and let N be a stopping time with respect to the same sequence $X_1, X_2, \ldots$. If there is an integer k for which $N \leq k$ for all possible realizations of $X_1, X_2, \ldots$, then*

$$E(Y_N) \leq E(Y_k) \qquad (4.5a)$$

If, in addition, $Y_1, Y_2, \ldots$ is a martingale, then

$$E(Y_N) = E(Y_1) \qquad (4.5b)$$

PROOF   As in (4.3), let A_n denote the event that $N = n$, and let I_A denote the indicator of the event A. Then since $N \leq k$, we must have $I_{A_n} = 0$ for $n > k$. Therefore, by (4.3),

$$E(Y_N) = \sum_{n=1}^{k} E(Y_n I_{A_n})$$

Suppose first that $Y_1, Y_2, \ldots$ is a submartingale. Then $Y_n \leq E(Y_k \mid \mathbf{X}_n)$ for $n = 1, \ldots, k$ by Lemma 12.3.2. Moreover, since I_{A_n} is determined by $\mathbf{X}_n$, we also have $E(Y_k I_{A_n} \mid \mathbf{X}_n) = I_{A_n} E(Y_k \mid \mathbf{X}_n)$ by Theorem 10.3.3. Therefore,

$$Y_n I_{A_n} \leq I_{A_n} E(Y_k \mid \mathbf{X}_n) = E(Y_k I_{A_n} \mid \mathbf{X}_n)$$

for $n = 1, \ldots, k$. It follows that

$$E(Y_N) \leq \sum_{n=1}^{k} E(Y_k I_{A_n}) = E[Y_k(I_{A_1} + \cdots + I_{A_k})]$$

Finally, since $N \leq k$, we must have $I_{A_1} + \cdots + I_{A_k} = 1$. Inequality (4.5a) follows.

If $Y_1, Y_2, \ldots$ is a martingale, we can apply (4.5a) to Y_N and to $-Y_N$ to deduce that $E(Y_N) = E(Y_k)$. Moreover, by Lemma 12.3.1, we have $E(Y_k) = E(Y_1)$, so that (4.5b) follows.   ////

We shall now relax the condition that N be bounded. Let $Y_1, Y_2, \ldots$ be a martingale, and let N be a stopping time with respect to the same sequence $X_1, X_2, \ldots$. Then for each integer $k = 1, 2, \ldots$, we define the random variable

$$N_k = \min (N, k)$$

N_k is a stopping time by Lemma 12.4.1, and $N_k \leq k$. Therefore,

$$E(Y_{N_k}) = E(Y_1)$$

for every $k = 1, 2, \ldots$. Now as $k \to \infty$, $N_k \to N$, and so it seems reasonable to hope that $E(Y_{N_k}) \to E(Y_N)$, leaving $E(Y_N) = E(Y_1)$.

To implement this program, we shall have to impose some additional conditions on the martingale $Y_1, Y_2, \ldots$ and the stopping time N. We shall require that

$$E(|Y_N|) < \infty \qquad (4.6a)$$

$$\lim_{k \to \infty} E(Y_k I_{B_k}) = 0 \qquad (4.6b)$$

where B_k denotes the event that $N > k$.

Theorem 12.4.2   *Let $Y_1, Y_2, \ldots$ be a martingale, and let N be a stopping time with respect to the same sequence $X_1, X_2, \ldots$. If the conditions (4.6) are satisfied, then $E(Y_N) = E(Y_1)$.*

   PROOF   Since $Y_{N_k} = Y_N$ if $N \leq k$ and $Y_{N_k} = Y_k$ if $N > k$, we may write $Y_{N_k} = Y_N(1 - I_{B_k}) + Y_k I_{B_k}$, where B_k is the event that $N > k$. Therefore,

$$E(Y_1) = E[Y_N(1 - I_{B_k})] + E(Y_k I_{B_k}) \qquad (4.7)$$

by Theorem 12.4.1. Since the second term on the right side of (4.7) tends to 0 as $k \to \infty$ by assumption (4.6b), it will suffice to show that the first approaches $E(Y_N)$. Let F denote the joint distribution function of N and Y_N. Then

$$E[Y_N(1 - I_{B_k})] = \int_0^k \int_{-\infty}^\infty y \, dF(x, y)$$

which converges to

$$\int_0^\infty \int_{-\infty}^\infty y \, dF(x, y) = E(Y_N)$$

as $k \to \infty$ by definition of the improper Riemann-Stieltjes integral. The theorem follows.   ////

We shall now list some simple conditions which are sufficient to ensure the validity of conditions (4.6a) and (4.6b).

EXAMPLE 12.4.2 If there is a constant c for which $|Y_k| \leq c$ for all $k = 1, 2, \ldots$, then conditions (4.6) are satisfied. In this case $E(|Y_N|) \leq c < \infty$, and $|E(Y_k I_{B_k})| \leq c \Pr(N > k)$, which tends to 0 as $k \to \infty$ by (4.1).

If $Y_1, Y_2, \ldots$ are regarded as the fortunes of a gambler, the condition of this example may be interpreted as requiring that the gambler and his opponent have only a finite amount of capital and limited credit. ////

EXAMPLE 12.4.3 Some other easily checked conditions which imply the validity of (4.6b) are the following.

a If there is a constant c for which $N > k$ implies $|Y_k| \leq c$, then $|E(Y_k I_{B_k})| \leq c \Pr(N > k) \to 0$ as $k \to \infty$ by (4.1).

b If there is a constant c for which $E(Y_n^2) \leq cn$ for $n = 1, 2, \ldots$, and if $E(N) < \infty$, then (4.6b) is satisfied. Indeed, by Schwarz's inequality (Problem 8.18), we have

$$E(Y_k I_{B_k})^2 \leq E(Y_k^2) \Pr(N > k) \leq ck \Pr(N > k)$$

which tends to 0 as $k \to \infty$ if $E(N) < \infty$. ////

We shall now specialize by considering sums of independent random variables. The following result is known as *Wald's lemma*.

Theorem 12.4.3 *Let $X_1, X_2, \ldots$ be independent random variables with common expectation $E(X_i) = \mu$, $i = 1, 2, \ldots$. Let N be a stopping time with respect to $X_1, X_2, \ldots$, and let $S_n = X_1 + \cdots + X_n$ for $n = 1, 2, \ldots$. If $E(N) < \infty$, then*

$$E(S_N) = \mu E(N) \qquad (4.8)$$

PROOF It is possible to deduce Theorem 12.4.3 from Theorem 12.4.2, but a direct proof is just as simple, and we shall give a direct proof. Suppose first that $X_1, X_2, \ldots$ all nonnegative random variables. Let A_n denote the event that $N = n$, and let B_n denote the event that $N > n$ for $n = 1, 2, \ldots$. Then

$$E(S_N) = \sum_{n=1}^{\infty} E(S_n I_{A_n}) = \sum_{n=1}^{\infty} \sum_{k=1}^{n} E(X_k I_{A_n}) = \sum_{k=1}^{\infty} \sum_{n=k}^{\infty} E(X_k I_{A_n})$$

$$= \sum_{k=1}^{\infty} E\left[X_k \left(\sum_{n=k}^{\infty} I_{A_n}\right)\right] = \sum_{k=1}^{\infty} E(X_k I_{B_{k-1}}) \qquad (4.9)$$

Here the interchange of orders of summation is justified because all the summands are nonnegative.[1] The justification of the interchange of expectation and summation is more difficult, but it can also be justified for nonnegative random variables.[2] Now B_{k-1} is determined by $X_1, \ldots, X_{k-1}$ by Lemma 12.4.1, and therefore B_{k-1} is independent of X_k. It follows that

$$E(X_k I_{B_{k-1}}) = E(X_k)P(B_{k-1}) = \mu \Pr (N \geq k)$$

Therefore,

$$E(S_N) = \sum_{k=1}^{\infty} \mu \Pr (N \geq k) = \mu E(N)$$

by Corollary 8.1.2. This completes the proof of (4.8) in the special case that $X_1, X_2, \ldots$ are nonnegative.

In the general case, when $X_1, X_2, \ldots$ are no longer assumed to be nonnegative, we first apply the special case to $|X_1|, |X_2|, \ldots$ and deduce that the series in (4.9) converges absolutely. The absolute convergence of the series is also sufficient to justify the interchange of orders of summation and the interchange of expectation and summation. The proof of Theorem 12.4.3 in the general case then proceeds as in the special case.   ////

12.5   APPLICATIONS OF THE OPTIONAL-STOPPING THEOREM

The implications of Theorem 12.4.2 for gambling may be summarized as follows: subject to the conditions (4.6) one cannot convert a sequence of fair games into a favorable game by using a stopping time. In this section we shall see how this general principle can be used to simplify many probability calculations. We begin by rederiving some of the results of Section 11.2.

EXAMPLE 12.5.1   The gambler's-ruin problem   Let $S_0, S_1, S_2, \ldots$ be a simple random walk. That is, let $X_1, X_2, \ldots$ be independent and identically distributed random variables which assume the values 1 and -1 with probabilities p and $q = 1 - p$, respectively, and let $S_n = X_1 + \cdots + X_n$ for $n = 1, 2, \ldots$. We shall regard X_i as our winnings on the ith of a sequence of independent games, in which case S_n is our cumulative winnings after n games. Let a and b be positive integers. We regard a as our initial capital and b as our opponent's initial capital. We agree to

---

[1] See, for example, Apostol (1957), p. 374.
[2] See, for example, Neveu (1965), pp. 37–42.

continue playing until we have either won all our opponent's money or lost all our money, and we require the probability of winning.

Let N be the duration of the game. That is, let $N =$ least $n \geq 1$ for which either $S_n = -a$ or $S_n = b$ or let $N = \infty$ if no such n exists. We showed in Lemma 11.2.2 that $\Pr(N < \infty) = 1$. Moreover, N satisfies condition (4.2), since the event that $N = n$ occurs if and only if $-a < S_k < b$ for $k = 1, \ldots, n-1$ and $S_n = -a$ or b and the latter event depends on $X_1, \ldots, X_n$ in the sense of (4.2). Therefore, N is a stopping time with respect to $X_1, X_2, \ldots$. Let $c = a + b$. For later reference, we observe that $|S_N| \leq \max(a,b) < c$ and

$$N > k \qquad \text{implies} \qquad |S_k| \leq c \qquad (5.1)$$

This remark will be useful in checking condition (4.6).

Let us first consider the case that $p = \frac{1}{2}$. In this case $E(X_i) = 0$, so that S_1, $S_2, \ldots$ is a martingale. In view of (5.1) and Example 12.4.3a it follows from Theorem 12.4.2 that $E(S_N) = E(S_1) = 0$. Let π denote the probability that we win all our opponent's money. Then $\pi = \Pr(S_N = b)$, so that $E(S_N) = \pi b - (1 - \pi)a$ by direct computation. Therefore, the probability of winning is

$$\pi = \frac{a}{a + b}$$

A similar argument will give $E(N)$. Since the common variance of $X_1, X_2, \ldots$ is $\sigma^2 = E(X_i^2) = 1$, the sequence $Y_n = S_n^2 - n, n = 1, 2, \ldots$ is a martingale (Example 12.2.1b). Again, it follows from (5.1), Problem 11.9, and Theorem 12.4.2 that $E(Y_N) = E(Y_1) = 0$, so that $E(S_N^2) = E(N)$. By direct computation, $E(S_N^2) = \pi b^2 + (1 - \pi)a^2 = ab$. Therefore, $E(N) = ab$.

When $p \neq \frac{1}{2}$, the games are no longer fair, so that the technique used above might not appear to be applicable. It is, however. We only have to be a little clever. Let $\beta = q/p$. Then

$$E(\beta^{X_i}) = p\beta + q\beta^{-1} = 1$$

so that

$$Y_n = \beta^{S_n} \qquad n = 1, 2, \ldots$$

defines a martingale by Example 12.2.1c. Since $|Y_N| \leq \max(\beta^c, \beta^{-c})$ and $N > k$ implies $|Y_k| \leq \max(\beta^b, \beta^{-a})$ by (5.1), it follows from Example 12.4.3a and Theorem 12.4.2 that $E(Y_N) = E(Y_1) = 1$. Also, $E(Y_N) = \pi\beta^b + (1 - \pi)\beta^{-a}$ by direct computation. Thus,

$$\pi = \frac{1 - \beta^a}{1 - \beta^{a+b}} \qquad (5.2)$$

Letting $b \to \infty$ in (5.2), we find that the probability of losing to an infinitely rich opponent is

$$\lim_{b \to \infty} (1 - \pi) = \begin{cases} \beta^a & \text{if } p > \frac{1}{2} \\ 1 & \text{if } p \leq \frac{1}{2} \end{cases}$$

Moreover, since we lose to an infinitely rich opponent if and only if $S_n = -a$ for some $n = 1, 2, \ldots$, we have

$$\Pr(S_n = -a \text{ for some } n = 1, 2, \ldots) = 1 \tag{5.3}$$

if $p \leq \frac{1}{2}$. That is, in the language of Section 11.2, a passage through $-a$ is certain if $p \leq \frac{1}{2}$ and $a > 0$. By symmetry, a passage through a is certain if $p \geq \frac{1}{2}$ and $a > 0$.

/////

EXAMPLE 12.5.2   We can use Theorem 12.4.2 to compute the distribution of the time of first passage through a (Example 12.4.1a). Suppose $p \geq \frac{1}{2}$, and let a be a positive integer. Then the time of first passage through a, $N_a =$ least $n \geq 1$ for which $S_n = a$ or ∞ if no such n exists, is a stopping time by Example 12.4.1a. For later reference, we observe that $S_{N_a} = a$ and

$$N_a > k \qquad \text{implies} \qquad S_k < a \tag{5.4}$$

We shall compute the generating function of N_a. For $t > 1$, we have the identity

$$E(t^{X_i}) = pt + qt^{-1}$$

Let $\gamma = \gamma(t) = pt + qt^{-1}$, and observe that $\gamma(t) > 1$ for $t > 1$. It follows from Example 12.2.1c that the sequence

$$Y_n = \frac{t^{S_n}}{\gamma^n} \qquad n = 1, 2, \ldots$$

is a martingale. By (5.4), $|Y_{N_a}| \leq t^a/\gamma^{N_a} \leq t^a$ for $t > 1$, and $N_a > k$ implies $|Y_k| < t^a/\gamma^k \leq t^a$ for $t > 1$. Therefore, conditions (4.6) are satisfied, and it follows from Theorem 12.4.2 that $E(Y_{N_a}) = E(Y_1) = 1$. Since $S_{N_a} = a$ with probability 1, we have

$$E(\gamma^{-N_a}) = t^{-a} \tag{5.5}$$

for $t > 1$. Now for $0 < s < 1$, let

$$t^{-1} = \frac{1}{2qs}(1 - \sqrt{1 - 4pqs^2})$$

Then t is a solution of the equation $pt + qt^{-1} = s$, and it is easily verified that $t > 1$. Substitution in (5.5) now yields

$$E(s^{N_a}) = \left(\frac{1 - \sqrt{1 - 4pqs^2}}{2qs}\right)^a \tag{5.6}$$

for $0 < s < 1$. Thus we have found the generating function $P(s) = E(s^{N_a})$, $0 < s < 1$, of the random variable N_a.

The generating function P uniquely determines the distribution of N_a (Section 8.4.1). For example, the expectation of N_a is $E(N_a) = P'(1-)$. After some calculations, we find that $E(N_a) = a/(p - q)$ if $p > \frac{1}{2}$ and $E(N_a) = \infty$ if $p = \frac{1}{2}$.

If $a = 1$, the generating function P can be expanded by the generalized binomial theorem (Section 1.7) as

$$P(s) = \sum_{n=1}^{\infty} \frac{-1}{2q} \binom{\frac{1}{2}}{n} (-4pq)^n s^{2n-1}$$

for $0 < s < 1$, and it follows that

$$\Pr(N_1 = 2n - 1) = \frac{1}{2q} \binom{\frac{1}{2}}{n} (-1)^{n-1} (4pq)^n$$

for $n = 1, 2, \ldots$. For general a, the result is

$$\Pr(N_a = 2n - a) = \frac{a}{2n - a} \binom{2n - a}{n - a} p^n q^{n-a}$$

for $n = 1, 2, \ldots$. We omit the details.　　　　////

12.6   THE SUBMARTINGALE INEQUALITY

The techniques used in Examples 12.5.1 and 12.5.2 extend to arbitrary martingales and submartingales, but in general, they only yield inequalities. Inequalities (6.1) and (6.2a) below are known as the *submartingale and martingale inequalities*.

Theorem 12.6.1   *Let $Y_1, Y_2, \ldots$ be a nonnegative submartingale with respect to a sequence $X_1, X_2, \ldots$. Then*

$$\Pr(\max_{n \le k} Y_n \ge a) \le \frac{1}{a} E(Y_k) \qquad (6.1)$$

for every $a > 0$ and every $k = 1, 2, \ldots$.

PROOF   Given a and k, let N be the least integer n, $1 \le n \le k$, for which $Y_n \ge a$ if there is such an n, and let $N = k$ if no such n exists (that is, if $Y_n < a$ for $n = 1, \ldots, k$). Then N is a stopping time. Indeed, $N \le k < \infty$; and for $n = 1, \ldots, k - 1$, the event that $N = n$ occurs if and only if $Y_n \ge a$ and $Y_j < a$ for $j = 1, \ldots, n - 1$. The latter event is determined by $Y_1, \ldots, Y_n$, which in turn are determined by $X_1, \ldots, X_n$. The event that $N = k$ is the complement of the event that $N \le k - 1$ and is therefore determined by $X_1, \ldots, X_{k-1}$; and the event that $N = n$ is impossible for $n > k$.

Let A be the event that $\max(Y_1, \ldots, Y_k) \ge a$. Then, by definition of N, A occurs if and only if $Y_N \ge a$. Therefore, $P(A) \le a^{-1} E(Y_N)$ by Markov's inequality (Section 9.1). Moreover, since N is a stopping time and $N \le k$, $E(Y_N) \le E(Y_k)$ by Theorem 12.4.1. Inequality (6.1) follows.　　　　////

Corollary 12.6.1   *Let $Y_1, Y_2, \ldots$ be a martingale with respect to a sequence $X_1, X_2, \ldots$. Then*

$$\Pr\left(\max_{n \leq k} |Y_n| \geq a\right) \leq a^{-1}E(|Y_k|) \qquad (6.2a)$$

for $a > 0$ and $k = 1, 2, \ldots$. If, in addition, $E(Y_k^2) < \infty$, then

$$\Pr\left(\max_{n \leq k} |Y_n| \geq a\right) \leq a^{-2}E(Y_k^2) \qquad (6.2b)$$

for $a > 0$.

PROOF   Inequality ($6.2a$) is a consequence of Theorem 12.6.1 and Lemma 12.3.3, which asserts that if $Y_1, Y_2, \ldots$ is a martingale, then $|Y_1|, |Y_2|, \ldots$ is a submartingale. The proof of ($6.2b$) is similar.   ////

EXAMPLE 12.6.1   Inequality ($6.2b$) contains Kolmogorov's inequality (Section 11.6) as a special case. Indeed, if $X_1, X_2, \ldots$ are independent random variables with expectations $E(X_k) = 0$ and finite variances $E(X_k^2)$ for $k = 1, 2, \ldots$, then $S_n = X_1 + \cdots + X_n$, $n = 1, 2, \ldots$, defines a martingale, so that

$$\Pr\left(\max_{n \leq k} |S_n| \geq a\right) \leq a^{-2}E(S_k^2) \qquad (6.3)$$

by ($6.2b$). Of course, (6.3) is Kolmogorov's inequality.   ////

PROBLEMS

NOTE: The results of Section 12.1 do require the independence of the outcomes $X_1, X_2, \ldots$. Problems 12.1 to 12.5 show how to construct gambling systems which work when the outcomes exhibit a particular type of dependence.[1]

Let an ordered random sample of size n be drawn without replacement from an urn which contains n red balls and n white balls. Suppose also that you win the ith game if the ith ball drawn is red and that you lose otherwise. Suppose also that you are allowed to wager any amount between 1 and 10 dollars on each draw.

12.1   Let $X_i = 1$ if the ith ball drawn is red, and let $X_i = -1$ otherwise. Show that $\Pr(X_i = 1) = \frac{1}{2} = \Pr(X_i = -1)$ for $i = 1, \ldots, n$. That is, each game is fair.

12.2   Suppose $n = 2$ and consider the following strategy. You wager $w_1 = 1$ dollar on the first draw. If the first ball drawn is red, then you also wager $w_2 = 1$ dollar on the second draw; but if the first ball drawn is black, then you wager $w_2 = 10$ dollars on the second draw. Show that your expected winnings are 1.5 dollars.

---

[1] A more practical application of these ideas may be found in H. O. Thorp, "Beat the Dealer," Blaisdell, New York, 1962.

12.3   For any n, one may employ the following strategy. Wager $w = 10$ dollars on those draws which are drawn when the urn contains more red balls than white balls, and wager $w = 1$ dollar on all other draws. Compute the expected winnings which result from this strategy for:

      (a) $n = 3$ (b) $n = 4$ (c) $n = 5$

12.4   The strategy of Problem 12.3 is, in fact, optimal. That is, it produces the largest expected gain. Verify this in the case $n = 2$.

12.5   Let $X_1, X_2, \ldots$ be independent random variables with common mean $\mu = 0$. Which of the following sequences are martingales with respect to $X_1, X_2, \ldots$?

      (a) $Y_n = S_n/n$, $n = 1, 2, \ldots$

      (b) $Y_n = S_n/\sqrt{n}$, $n = 1, 2, \ldots$

      (c) $Y_n = X_1 X_2 \cdots X_n$, $n = 1, 2, \ldots$

      (d) $Y_n = \exp S_n$, $n = 1, 2, \ldots$

12.6   Let $Y, X_1, X_2, \ldots$ be random variables with the following properties. Y has the standard exponential distribution, and for any $n = 1, 2, \ldots$ the conditional distribution of $X_1, \ldots, X_n$ given $Y = y > 0$ is that of independent, exponentially distributed random variables with common parameter y. Show that $Y_n = (n + 1)/(1 + X_1 + \cdots + X_n)$ is a martingale with respect to $X_1, X_2, \ldots$. *Hint:* Compute $E(Y \mid X_1, \ldots, X_n)$.

12.7   Let $X, X_1, X_2, \ldots$ be random variables. Suppose that X has the normal distribution with mean μ and variance σ^2 and that given $X = x$, $X_1, X_2, \ldots$ are independent, normally distributed random variables with mean x and variance 1. Show that $Y_n = (\mu\sigma^{-2} + S_n)/(n + \sigma^{-2})$, $n = 1, 2, \ldots$ is a martingale with respect to $X_1, X_2, \ldots$.

12.8   Let $X_1, X_2, \ldots$ be independent, standard normal random variables, and let $S_n = X_1 + \cdots + X_n$. Show that for any $\alpha \in R$, $Y_n(\alpha) = \exp(\alpha S_n - \tfrac{1}{2}n\alpha^2)$ is a martingale with respect to $X_1, X_2, \ldots$.

12.9   Let $Y_n(\alpha)$ be as in Problem 12.8, and let H be any distribution function on R. Show that

$$Y_n = \int_{-\infty}^{\infty} Y_n(\alpha)\, dH(\alpha)$$

is a martingale with respect to $X_1, X_2, \ldots$.

12.10   Let $X_1, X_2, \ldots$ be independent, standard, normal random variables, and let $S_n = X_1 + \cdots + X_n$ for $n \geq 1$. Show that

$$Y_n = \frac{\exp \dfrac{S_n^2}{2n + 2}}{\sqrt{n + 1}}$$

is a martingale with respect to $X_1, X_2, \ldots$. *Hint:* Apply Problem 12.9 with H equal to the standard normal distribution function.

12.11   Let N be a positive integer or infinite-valued random variable which is determined as a function of a sequence $X_1, X_2, \ldots$. Suppose also that $\Pr(N < \infty) = 1$. Show that N is a stopping time with respect to $X_1, X_2, \ldots$ if and only if the event that $N \leq n$ is determined by $X_1, \ldots, X_n$ for every $n = 1, 2, \ldots$.

12.12 Let M and N be stopping times with respect to the same sequence $X_1, X_2, \ldots$. Show that min (N, M) and max (N, M) are also stopping times with respect to $X_1, X_2, \ldots$.

12.13 Let $X_1, X_2, \ldots$ be independent and identically distributed random variables which assume the values 1 and -1 with probabilities p and $q = 1 - p$, respectively: which of the following random variables are stopping times?

(a) N = least $n \geq 1$ for which $X_n = 1$ or ∞ if no such n exists.

(b) N = least $n \geq 1$ for which $|S_n| \geq 10$ or ∞ if no such n exists.

(c) N = least $n \geq 1$ for which $S_{n+1} = 0$ or ∞ if no such n exists?

Justify your assertions.

12.14 Compute $E(S_N)$ for the random variable N of part (a) in Problem 12.13.

12.15 Let $Y_1, Y_2, \ldots$ be a martingale, and let N be a stopping time with respect to a sequence $X_1, X_2, \ldots$. Suppose also that there is a constant b for which Pr $(Y_n \geq b) = 1$ for all $n = 1, 2, \ldots$. Show that $E(Y_N) \leq E(Y_1)$. Interpret your result in terms of gambling.

> NOTE: Problems 12.16 to 12.22 sketch an application of the optional stopping theorem to *renewal theory*. We suppose that events occur in time and that the times between successive events are independent and identically distributed, nonnegative random variables $X_1, X_2, \ldots$. We also suppose that X_1 has a positive, finite mean μ. The time at which the nth event occurs is then $S_n = X_1 + \cdots + X_n$. A particular application is to the theory of queues, where S_n is regarded as the time at which the nth customer enters a place of business to be serviced.

12.16 Let $t > 0$ and N_t = least positive integer n for which $S_n > t$ or ∞ if no such n exists. Thus, $N_t - 1$ events have occurred by time t. Show that N is a stopping time. *Hint:* Pr $(N > n)$ = Pr $(S_n \leq t)$.

12.17 Show that $E(N_t)$ is finite for every $t > 0$. *Hint:* Given t, there is an integer r for which Pr $(S_r > t) > 0$; then Pr $(N > nr) \leq$ Pr $(S_r \leq t)^n$.

12.18 Show that $E(N_t) \geq t/\mu$. *Hint:* Use Wald's lemma.

12.19 Suppose that there is a constant c for which Pr $(X_1 \leq c) = 1$. Show that $E(N_t) \leq (t + c)/\mu$.

12.20 The function V defined by $V(t) = E(N_t)$ for $t > 0$ is known as the *renewal function*. Calculate $V(t)$ in the special case that X_1 has the exponential distribution with parameter $\beta > 0$.

12.21 Suppose that there is a constant c for which Pr $(X_1 \leq c) = 1$. Show that $V(t) \sim t\mu^{-1}$ as $t \to \infty$.

12.22 Show that $V(t)$ may also be written in the form $V(t) = \sum_{n=0}^{\infty}$ Pr $(S_n \leq t)$.

12.23 Let $Y_1, Y_2, \ldots$ be a nonnegative martingale with common expectation $E(Y_n) = 1$. Show that Pr $(Y_k > a$ for some $k \geq 1) \leq 1/a$ for $a > 1$.

12.24 Let $X_1, X_2, \ldots$ be independent, standard normal random variables, and let $S_n = X_1 + \cdots + X_n$ for $n \geq 1$. Further, let $c_n(a)^2 = (n + 1)[a^2 + \log (n + 1)]$ for $n \geq 1$. Show that Pr $(|S_n| \geq c_n(a)$, for some $n \geq 1) \leq e^{-a^2/2}$. *Hint:* Use Problems 12.23 and 12.10.

APPENDIX A

SET THEORY

A set A is a collection of objects or elements a. The notation $a \in A$ means that a is one of the elements which constitute A. Two sets are equal if and only if they contain the same elements. That is, $A = B$ if and only if $a \in A$ if and only if $a \in B$.

We define a set by specifying which objects are elements of A and which are not. For example, we may define a set Z by specifying that Z consists of all nonnegative integers. Similarly, we may define another set R by specifying that R consists of all real numbers. The notation Z and R will be reserved for these two sets throughout this appendix.

Some useful notational devices for specifying sets are the following. If $a_1, \ldots, a_n$ are objects, then $\{a_1, \ldots, a_n\}$ will denote the set whose elements are $a_1, \ldots, a_n$. Also, if A is a set, and if $\pi(a)$ is a proposition which is either true or false for every $a \in A$, then $\{a \in A : \pi(a)\}$ will denote the set of those elements $a \in A$ for which $\pi(a)$ is true. For example, $\{0,1,2,3,4\}$ denotes the set whose elements are the integers 0, 1, 2, 3, and 4. This set can also be written $\{a \in Z : a \leq 4\}$. We shall use the following notation for *intervals*. For real numbers a and b with $a < b$, let

$$(a,b) = \{x \in R : a < x < b\}$$
$$(a,b] = \{x \in R : a < x \leq b\}$$
$$[a,b) = \{x \in R : a \leq x < b\}$$
$$[a,b] = \{x \in R : a \leq x \leq b\}$$

(a,b) will be called the *open* interval from a to b; $(a,b]$ will be called the *left-open* and *right-closed* interval from a to b; $[a,b)$ will be called the *left-closed* and *right-open* interval from a to b; and $[a,b]$ will be called the *closed* interval from a to b. We use (a,b) to denote both the open interval from a to b and the ordered pair whose first component is a and whose second component is b. The meaning of the notation (a,b) will always be clear from the context in which it is used.

If A and B are sets, we shall say that A is a *subset* of B and write $A \subset B$ if and only if $a \in A$ implies $a \in B$. For example, $Z \subset R$. Observe that $A = B$ if and only if $A \subset B$ and $B \subset A$.

There is a distinguished set $\varnothing$, called the *empty set*. This set contains no elements and is a subset of every other set.

If A and B are sets, then the set $B - A = \{a \in B: a \notin A\}$ is called the *difference*. It is also known as the *complement of A with respect to B*. When there is no danger of confusion, the qualifying phrase "with respect to B" will be omitted and we shall write A' for $B - A$.

If A and B are sets, we define their *union* and *intersection* $A \cup B$ and $A \cap B$ as follows: $A \cup B$ consists of all objects a for which either $a \in A$ or $a \in B$, or both; and $A \cap B$ consists of all objects a for which both $a \in A$ and $a \in B$. The notation AB will also be used for $A \cap B$. For example, if $A = \{0,1,2,3,4\}$ and $B = \{3,4,5,6,7\}$, then $A \cup B = \{0,\ldots,7\}$ and $A \cap B = \{3,4\}$.

More generally, if I is a set and if A_i is a set for each $i \in I$, then we define the union and intersection of the collection A_i, $i \in I$, as follows. The union $\bigcup_I A_i$ consists of all objects a for which $a \in A_i$ for some $i \in I$; and the intersection $\bigcap_I A_i$ consists of all objects a for which $a \in A_i$ for all $i \in I$. If $I = 1, 2, \ldots, n$ is the set of the first n integers, then we shall write

$$\bigcup_{i=1}^{n} A_i \quad \text{and} \quad \bigcap_{i=1}^{n} A_i$$

for $\bigcup_I A_i$ and $\bigcap_I A_i$, respectively. Similarly, if $I = \{1,2,\ldots\}$ consists of all positive integers, then we shall write $\bigcup_{i=1}^{\infty} A_i$ and $\bigcap_{i=1}^{\infty} A_i$ for $\bigcup_I A_i$ and $\bigcap_I A_i$. If there is a larger set S for which $A_i \subset S$ for all $i \in I$, and if we denote complement with respect to S by a prime, then we have De Morgan's laws:

$$\left(\bigcup_I A_i\right)' = \bigcap_I A_i' \quad \text{and} \quad \left(\bigcap_I A_i\right)' = \bigcup_I A'$$

If A and B are sets, we define the *cartesian product of A and B* to be the set of all ordered pairs (a,b) with $a \in A$ and $b \in B$. The cartesian product will be denoted by $A \times B$. More generally, if $A_1, A_2, \ldots, A_n$ are sets, then we define the *cartesian product* of $A_1, \ldots, A_n$ to be the set of all ordered n-tuples $(a_1, \ldots, a_n)$ with $a_i \in A_i$ for $i = 1, \ldots, n$. The cartesian product of $A_1, \ldots, A_n$ will be denoted by $A_1 \times \cdots \times A_n$. In the special case that all the sets $A_1, \ldots, A_n$ are the same, say $A_i = A$, $i = 1, \ldots, n$, we shall denote $A_1 \times \cdots \times A_n$ by A^n. An important special case occurs when $A_i = R$, the set of all real numbers, for $i = 1, \ldots, n$. In this case R^n consists of all ordered n-tuples $(x_1, \ldots, x_n)$ of real numbers. We shall refer to R^n as *n-dimensional euclidean space*.

APPENDIX B

INTEGRATION

In this appendix we present the Riemann-Stieltjes integral. Since the ordinary Riemann integral is a special case of the Riemann-Stieltjes integral, this appendix may also serve as a review of the elements of integration. Proofs of assertions not proved in this appendix can be found in Apostol (1957), chap. 9.

Let a and b be real numbers with $a < b$. By a *partition* of the interval $[a,b]$ we mean a finite set $P = \{x_0, x_1, \ldots, x_n\}$, where $a = x_0 < x_1 < \cdots < x_n = b$. If P and Q are two partitions of $[a,b]$, we shall say that P is a *refinement* of Q if and only if $Q \subset P$. If P_1 and P_2 are any two partitions, then the union $P = P_1 \cup P_2$ is a refinement of both P_1 and P_2. We define the *norm* of a partition P to be $|P| = \max \{x_i - x_{i-1} : i = 1, \ldots, n\}$.

If for every partition P of $[a,b]$, $s(P)$ is a real number which is determined by P, then we shall say that $s(P)$ converges to s_0 as P becomes *infinitely fine* if and only if for every $\varepsilon > 0$ there is a partition P_ε for which $|s(P) - s_0| \le \varepsilon$ whenever P is a refinement of P_ε. In this case we shall write $s_0 = \lim s(P)$.

Now let F and g be bounded functions which are defined on $[a,b]$. If

$$P = \{x_0, x_1, \ldots, x_n\}$$

is a partition of $[a,b]$, and if $x_{i-1} \le t_i \le x_i$ for $i = 1, \ldots, n$, then we can form the sum

$$s_F(P,g) = \sum_{i=1}^{n} g(t_i)[F(x_i) - F(x_{i-1})]$$

Of course, $s_F(P,g)$ depends on $t_1, \ldots, t_n$ as well as F, P, and g. If $\lim s_F(P,g)$ exists as P becomes infinitely fine, and if the limit is independent of the choice of $t_1, \ldots, t_n$, then we shall say that g *is* (*Riemann-Stieltjes*) *integrable with respect to F over* $[a,b]$, and we define the (*Riemann-Stieltjes*) *integral of g with respect to F over* $[a,b]$ *to be*

$$\int_a^b g \, dF = \lim s_F(P,g) \qquad (B.1)$$

The notation $\int_a^b g(x) \, dF(x)$ will also be used for the integral.

In the special case that $F(x) = x$ for $a \le x \le b$, we shall say that g is integrable over $[a,b]$ and write

$$\int_a^b g(x) \, dx = \int_a^b g \, dF$$

It can be shown that if g is continuous and F is nondecreasing on $[a,b]$, then g is integrable with respect to F over $[a,b]$. In particular, if g is continuous on $[a,b]$, then g is integrable over $[a,b]$.

The following results are to be anticipated.

Theorem B.1 *Let g_1 and g_2 be integrable with respect to F over $[a,b]$, and let α_1 and α_2 be constants. Then $g = \alpha_1 g_1 + \alpha_2 g_2$ is integrable with respect to F over $[a,b]$ and*

$$\int_a^b g \, dF = \alpha_1 \int_a^b g_1 \, dF + \alpha_2 \int_a^b g_2 \, dF$$

Theorem B.2 *Let g be integrable with respect to F_1 and F_2 over $[a,b]$, and let α_1 and α_2 be constants. Then g is integrable with respect to $F = \alpha_1 F_1 + \alpha_2 F_2$ over $[a,b]$ and*

$$\int_a^b g \, dF = \alpha_1 \int_a^b g \, dF_1 + \alpha_2 \int_a^b g \, dF_2$$

Theorem B.3 *Let g be integrable with respect to F over $[a,b]$. If $a < c < b$, then g is integrable with respect to F over $[a,c]$ and $[c,b]$. Moreover,*

$$\int_a^b g \, dF = \int_a^c g \, dF + \int_c^b g \, dF$$

Theorem B.4 *Let F be nondecreasing on $[a,b]$, and let g and h be integrable with respect to F over $[a,b]$. If $g(x) \le h(x)$ for $a \le x \le b$, then*

$$\int_a^b g \, dF \le \int_a^b h \, dF$$

EXAMPLE B.1

a Let $F_0(x) = 0$ for $x < 0$, and let $F_0(x) = 1$ for $x \geq 0$. If $a < 0 < b$, and if g is any continuous function on $[a,b]$, then

$$\int_a^b g \, dF = g(0) \qquad (B.2)$$

In fact, given $\varepsilon > 0$, there is a $\delta > 0$ for which $|g(x) - g(y)| \leq \varepsilon$ whenever $a \leq x \leq b$, $a \leq y \leq b$, and $|x - y| \leq \delta$. Let P be any partition of norm $|P_\varepsilon| \leq \delta$, and let P be any refinement of P_ε. Write $P = \{x_0, x_1, \ldots, x_n\}$, and choose i such that $x_{i-1} < 0 \leq x_i$; further, let $x_{i-1} \leq t_i \leq x_i$ for $i = 1, \ldots, n$. Then since $F(x_j) - F(x_{j-1}) = 0$ for $j \neq i$, we have $s_{F_0}(P,g) = g(t_i)$; moreover, since $|t_i| \leq \delta$, we also have $|g(t_i) - g(0)| \leq \varepsilon$. Equation (B.2) follows.

b Let $\alpha_1, \ldots, \alpha_n$ and $t_1, \ldots, t_n$ be any constants with $a < t_1 < t_2 < \cdots < t_n < b$. Then the function F defined by

$$F(x) = \sum_{i=1}^n \alpha_i F_0(x - t_i)$$

for $a \leq x \leq b$ is a step function with jumps of height α_i at the points t_i for $i = 1, \ldots, n$. By Theorem B.2, Example B.1*a*, and translation we have

$$\int_a^b g \, dF = \sum_{i=1}^n \alpha_i g(t_i)$$

for every continuous function g on $[a,b]$. ////

We shall say that F is continuously differentiable on $[a,b]$ if F' exists and is continuous on (a,b) and, in addition, $F'(x)$ approaches a finite limit as either $x \to a$ or $x \to b$.

EXAMPLE B.2   Suppose that F is continuously differentiable on $[a,b]$. Let $f = F'$.

If g is any function on $[a,b]$ for which fg is integrable over $[a,b]$, then g is integrable with respect to F over $[a,b]$ and

$$\int_a^b g \, dF = \int_a^b fg \, dx \qquad (B.3)$$

In particular, (B.3) holds if g is continuous on $[a,b]$. In fact, if $P = \{x_0, x_1, \ldots, x_n\}$ is any partition of $[a,b]$, we may write $F(x_i) - F(x_{i-1}) = f(s_i)(x_i - x_{i-1})$ with $x_{i-1} < s_i < x_i$ for $i = 1, \ldots, n$ by the mean-value theorem. Thus if $x_{i-1} \leq t_i \leq x_i$, $i = 1, \ldots, n$, then

$$s_F(P,g) = \sum_{i=1}^n g(t_i)f(s_i)(x_i - x_{i-1})$$

$$= \sum_{i=1}^n f(t_i)g(t_i)(x_i - x_{i-1}) + \sum_{i=1}^n g(t_i)[f(s_i) - f(t_i)](x_i - x_{i-1}) \qquad (B.4)$$

Now as P becomes infinitely fine, the second summation in (B.4) approaches $\int_a^b fg \, dx$, by assumption, and the last summation converges to 0 by the assumed continuity of f. Therefore,

$$\lim s_F(P,g) = \int_a^b fg \, dx$$

as asserted. ////

Two other theorems of interest give the formulas for integration by parts and change of variables.

Theorem B.5 *Let g be integrable with respect to F over $[a,b]$. Then F is integrable with respect to g over $[a,b]$, and*

$$\int_a^b g \, dF + \int_a^b F \, dg = F(b)g(b) - F(a)g(a)$$

Theorem B.6 *Let g be integrable with respect to F over $[a,b]$. Further, let h be an increasing function on an interval $[\alpha,\beta]$ with $h(\alpha) = a$ and $h(\beta) = b$. Finally, let*

$$f(x) = g(h(x)) \qquad \text{and} \qquad G(x) = F(h(x))$$

for $\alpha \leq x \leq \beta$. Then f is integrable with respect to G and

$$\int_\alpha^\beta f \, dG = \int_a^b g \, dF$$

Corollary B.1 *Let the hypotheses of Theorem B.6 be satisfied with $F(x) = x$, $a \leq x \leq b$, and let h be continuously differentiable on $[\alpha,\beta]$. Then*

$$\int_a^b g \, dx = \int_\alpha^\beta g(h(x))h'(x) \, dx$$

PROOF The corollary follows from Theorem B.6 and Example B.2 by taking $F(x) = x$, $a \leq x \leq b$. ////

Suppose that g has an infinite discontinuity at a point c and that g is integrable with respect to F over $[a,b]$ for every $b < c$. If $\int_a^b g \, dF$ approaches a finite limit as $b \to c$, and if F is continuous at c, we say that g is *improperly integrable with respect to F over $[a,c]$* and we define the *improper integral of g with respect to F over $[a,c]$* to be

$$\int_a^c g \, dF = \lim_{b \to c} \int_a^b g \, dF \qquad (B.5)$$

The integral $\int_c^b g \, dF$ is defined similarly when g has an infinite discontinuity at c, and g is integrable with respect to F over $[a,b]$ for every $a > c$. If g has an infinite discontinuity at c, and if $\int_a^c g \, dF$ and $\int_c^b g \, dF$ are both defined, where $a < c < b$, then we say that g is *improperly integrable over* $[a,b]$ and we define the *improper integral of g with respect to F* over $[a,b]$ to be

$$\int_a^b g \; dF = \int_a^c g \; dF + \int_c^b g \; dF$$

We now extend the definition of the integral to infinite intervals of integration. Let g be a real-valued function which is defined on the interval $[a,\infty)$, and suppose that g is integrable with respect to F over $[a,b]$ for every $b > a$. If $\int_a^b g \; dF$ approaches a finite limit as $b \to \infty$, we say that g is *improperly integrable with respect to F over* $[a,\infty)$, and we define the *improper integral of g with respect to F over* $[a,\infty)$ to be

$$\int_a^\infty g \; dF = \lim_{b \to \infty} \int_a^b g \; dF \qquad (B.6)$$

Integrals of the form $\int_{-\infty}^a g \; dF$ are defined in a similar manner. If both $\int_{-\infty}^a g \; dF$ and $\int_a^\infty g \; dF$ are defined, then we say that g is *improperly integrable with respect to F over* $(-\infty,\infty)$ and we define

$$\int_{-\infty}^\infty g \; dF = \int_{-\infty}^a g \; dF + \int_a^\infty g \; dF \qquad (B.7)$$

If $|g|$ is improperly integrable with respect to F over $(-\infty,\infty)$, then we say that the integral on the left side of (B.7) *converges absolutely*.

Theorems B.1 to B.4 remain valid if the term "integrable" is replaced by "improperly integrable" throughout. This can be seen by taking limits.

Table C.1 THE BINOMIAL PROBABILITIES $b(k;n,p)$

				p		
n	k	0.10	0.20	0.30	0.40	0.50
2	0	0.810	0.640	0.490	0.360	0.250
	1	0.180	0.320	0.420	0.480	0.500
	2	0.010	0.040	0.090	0.160	0.250
3	0	0.729	0.512	0.343	0.216	0.125
	1	0.243	0.384	0.441	0.432	0.375
	2	0.027	0.096	0.189	0.288	0.375
	3	0.001	0.008	0.027	0.064	0.125
4	0	0.656	0.401	0.240	0.130	0.063
	1	0.292	0.401	0.412	0.346	0.250
	2	0.049	0.154	0.265	0.346	0.375
	3	0.004	0.026	0.076	0.154	0.250
	4	0.000	0.002	0.008	0.026	0.063
5	0	0.590	0.328	0.168	0.078	0.031
	1	0.328	0.410	0.360	0.259	0.156
	2	0.073	0.205	0.309	0.346	0.313
	3	0.008	0.051	0.132	0.230	0.313
	4	0.000	0.006	0.028	0.077	0.156
	5	0.000	0.000	0.002	0.010	0.031

Table C.1   THE BINOMIAL PROBABILITIES $b(k;n,p)$
(*Continued*)

n	k	p				
		0.10	0.20	0.30	0.40	0.50
6	0	0.531	0.262	0.118	0.047	0.016
	1	0.354	0.393	0.303	0.187	0.094
	2	0.098	0.246	0.324	0.311	0.234
	3	0.015	0.082	0.185	0.276	0.313
	4	0.001	0.015	0.060	0.138	0.234
	5	0.000	0.002	0.010	0.037	0.094
	6	0.000	0.000	0.001	0.004	0.016
7	0	0.478	0.210	0.082	0.028	0.008
	1	0.372	0.367	0.247	0.131	0.055
	2	0.124	0.275	0.318	0.261	0.164
	3	0.023	0.115	0.227	0.290	0.273
	4	0.003	0.029	0.097	0.194	0.273
	5	0.000	0.004	0.025	0.077	0.164
	6	0.000	0.000	0.004	0.017	0.055
	7	0.000	0.000	0.000	0.002	0.008
8	0	0.430	0.168	0.058	0.017	0.004
	1	0.383	0.336	0.198	0.090	0.031
	2	0.149	0.294	0.296	0.209	0.109
	3	0.033	0.147	0.254	0.279	0.219
	4	0.005	0.046	0.136	0.232	0.273
	5	0.000	0.009	0.047	0.124	0.219
	6	0.000	0.001	0.010	0.041	0.109
	7	0.000	0.000	0.001	0.008	0.031
	8	0.000	0.000	0.000	0.001	0.004
9	0	0.387	0.134	0.040	0.010	0.002
	1	0.387	0.302	0.156	0.060	0.018
	2	0.172	0.302	0.267	0.161	0.070
	3	0.045	0.176	0.267	0.251	0.164
	4	0.007	0.066	0.172	0.251	0.246
	5	0.001	0.017	0.074	0.167	0.246
	6	0.000	0.003	0.021	0.074	0.164
	7	0.000	0.000	0.004	0.021	0.070
	8	0.000	0.000	0.000	0.004	0.018
	9	0.000	0.000	0.000	0.000	0.002
10	0	0.349	0.107	0.028	0.006	0.001
	1	0.387	0.268	0.121	0.040	0.010
	2	0.194	0.302	0.233	0.121	0.044
	3	0.057	0.201	0.267	0.215	0.117
	4	0.011	0.088	0.200	0.251	0.205
	5	0.001	0.026	0.103	0.201	0.246
	6	0.000	0.006	0.037	0.111	0.205
	7	0.000	0.001	0.009	0.042	0.117
	8	0.000	0.000	0.001	0.011	0.044
	9	0.000	0.000	0.000	0.002	0.010
	10	0.000	0.000	0.000	0.000	0.001

Table C.2 THE POISSON PROBABILITIES $p(k;\beta)$

k	β					
	0.5	1	2	3	4	5
0	0.607	0.368	0.135	0.050	0.018	0.007
1	0.303	0.368	0.271	0.149	0.073	0.034
2	0.076	0.184	0.271	0.224	0.147	0.084
3	0.013	0.061	0.180	0.224	0.195	0.140
4	0.002	0.015	0 090	0.168	0.195	0.175
5		0.003	0.036	0.101	0.156	0.175
6		0.001	0.012	0.050	0.104	0.146
7			0.003	0.022	0.060	0.104
8			0.001	0.008	0.030	0.065
9				0.003	0.013	0.036
10				0.001	0.005	0.018
					0.002	0.008
					0.001	0.003
						0.001

Table C.3 THE STANDARD NORMAL DISTRIBUTION FUNCTION

x	$\Phi(x)$	x	$\Phi(x)$	x	$\Phi(x)$
0.00	0.500	1.05	0.853	2.05	0.980
0.05	0.520	1.10	0.864	2.10	0.982
0.10	0.540	1.15	0.875	2.15	0.984
0.15	0.560	1.20	0.885	2.20	0.986
0.20	0.579	1.25	0.894	2.25	0.988
0.25	0.599	1.30	0.903	2.30	0.989
0.30	0.618	1.35	0.911	2.35	0.991
0.35	0.637	1.40	0.919	2.40	0.992
0.40	0.655	1.45	0.926	2.45	0.993
0.45	0.674	1.50	0.933	2.50	0.994
0.50	0.691	1.55	0.939	2.55	0.995
0.55	0.709	1.60	0.945	2.60	0.995
0.60	0.726	1.645	0.950	2.65	0.996
0.65	0.742	1.70	0.955	2.70	0.997
0.70	0.758	1.75	0.960	2.75	0.997
0.75	0.773	1.80	0.964	2.80	0.997
0.80	0.788	1.85	0.968	2.85	0.998
0.85	0.802	1.90	0.971	2.90	0.998
0.90	0.816	1.95	0.974	2.95	0.998
0.95	0.829	1.96	0.975	3.00	0.999
1.00	0.841	2.00	0.977		

APPENDIX D

REFERENCES

APOSTOL, T.: "Mathematical Analysis," Addison-Wesley, Reading, Mass., 1957.

ARROW, K., S. KARLIN, and H. SCARF (eds.): "Studies in the Mathematical Theory of Inventory and Production," Stanford University Press, Stanford, Calif., 1958.

———, ———, and ——— (eds.): "Studies in Applied Probability and Management Science," Stanford University Press, Stanford, Calif., 1962.

BEYER, W.: "CRC Handbook of Tables for Probability and Statistics," Chemical Rubber, Cleveland, 1966.

BLACKWELL, D., and M. GIRSHICK: "Theory of Games and Statistical Decisions," Wiley, New York, 1954.

BLACKWOOD, O., T. OSGOOD, and A. RUARK: "An Outline of Atomic Physics," Wiley, New York, 1957.

COCHRAN, W. G.: "Sampling Techniques," Wiley, New York, 1963.

CONSTANT, F. W.: "Theoretical Physics," Addison-Wesley, Reading, Mass., 1958.

DAVID, F. N.: "Games, Gods, and Gambling: The Origins and History of Probability and Statistical Ideas from the Earliest Times to the Newtonian Era," Hafner, New York, 1962.

DE FINETTI, B.: Probability: Interpretation in "International Encyclopedia of the Social Sciences," vol. 12, pp. 496–504, Free Press, New York, 1968.

DEGROOT, M. H.: "Optimal Statistical Decisions," McGraw-Hill, New York, 1970.

ESTES, W. K.: The Statistical Approach to Learning Theory, in S. Koch (ed.), "Psychology: A Study of Science," vol. 2, McGraw-Hill, New York, 1959.

FELLER, W.: "An Introduction to the Theory of Probability and Its Applications," vol. 2, Wiley, New York, 1966.

————: "An Introduction to the Theory of Probability and Its Applications," 3d ed., vol. 1, Wiley, New York, 1968.

HOGG, R., and A. CRAIG: "Introduction to Mathematical Statistics," Macmillan, New York, 1970.

KARLIN, S.: "A First Course in Stochastic Processes," Academic, New York, 1966.

NEVEU, J.: "Mathematical Foundations of the Calculus of Probability," trans. A. Feinstein, Holden-Day, San Francisco, 1965.

PARZEN, E.: "Modern Probability Theory and Its Applications," Wiley, New York, 1960.

RIORDEN, J.: "An Introduction to Combinatorial Analysis," Wiley, New York, 1958.

RUDIN, W.: "Principles of Mathematical Analysis," 2d ed., McGraw-Hill, New York, 1964.

SELBY, S.: "Standard Mathematical Tables," 14th ed., Chemical Rubber, Cleveland, 1965.

SMOKLER, H. E., and H. E. KYBURG, JR. (eds.), "Studies in Subjective Probability," Wiley, New York, 1964.

THOMAS, G. B., JR.: "Calculus and Analytical Geometry," alt. ed., Addison-Wesley, Reading, Mass., 1972.

TODHUNTER, I.: "A History of the Mathematical Theory of Probability from the Time of Pascal to That of Laplace," Macmillan, London, 1865.

TUCKER, H.: "A Graduate Course in Probability," Academic, New York, 1967.

ANSWERS TO SELECTED PROBLEMS

CHAPTER 1

1.1 (a) $S = \{(H,H),(H,T),(T,H),(T,T)\}$;
  $S = \{(H,H,H),(H,H,T),(H,T,H),(T,H,H),(T,T,H),(T,H,T),(H,T,T),(T,T,T)\}$,
  (c) $S = \{(x,y): x$ and y are integers, $1 \le x \le 52, 1 \le y \le 52,$ and $x \ne y\}$

1.2 (a) 4; 8, (c) $52 \times 51 = 2652$ *1.4* $\frac{3}{8}; \frac{1}{2}$

1.6 26^4 *1.8* (a) 10^7, (b) $(10)_7$ *1.10* $\binom{10}{2}$

1.14 (a) $\binom{10}{4}$, (b) $10\binom{9}{3}$ *1.20* (a) $\dfrac{(4)_2}{(52)_2}$, (b) $\dfrac{(13)_2}{(52)_2}$, (c) $\dfrac{13(4)_2}{(52)_2}$, (d) $\dfrac{4(13)_2}{(52)_2}$

1.24 No; the probability that at least one student will be selected more than once is $1 - (10)_5 \times 10^{-5} = 0.6976.$

1.26 (a) $\dfrac{\binom{4}{2}\binom{48}{11}}{\binom{52}{13}}$, (b) $\dfrac{\binom{4}{2}\binom{48}{11} + \binom{4}{3}\binom{48}{10} + \binom{4}{4}\binom{48}{9}}{\binom{52}{13}}$

1.28 $\dfrac{4^{13}}{\dbinom{52}{13}}$

1.32 (a) $\dfrac{\dbinom{6}{2}\dbinom{4}{1}}{\dbinom{10}{3}}$, (b) $\dfrac{\dbinom{6}{2}\dbinom{4}{1}+\dbinom{6}{3}\dbinom{4}{0}}{\dbinom{10}{3}}$

1.34 $\frac{17}{70}$

1.40 $\dbinom{5}{2}\left(\dfrac{1}{3}\right)^2\left(\dfrac{2}{3}\right)^3$

1.42 (a) $\dfrac{\dbinom{5}{2}13^2 39^3}{52^5}$, (b) $\dfrac{13\times39^4}{52^5}$

1.44 (a) $\dfrac{(4)_2\times2}{(6)_3}$, (b) $\dfrac{1-(4)_3}{(6)_3}$, (c) $\dfrac{(4)_3}{(6)_3}$

1.46 At least 1 six in 6 tosses

1.48 (a) $\frac{1}{16}$, (b) $\frac{5}{16}$

1.50 (a) $\dfrac{\dbinom{5}{3}}{\dbinom{9}{6}}$, (b) $\dfrac{4\dbinom{5}{2}}{\dbinom{9}{6}}+\dfrac{\dbinom{5}{3}}{\dbinom{9}{6}}$

1.52 $\dfrac{\dbinom{m+r-1}{r}\dbinom{n-m+k-r-1}{k-r}}{\dbinom{n+k-1}{k}}$

CHAPTER 2

2.2 (a) $S=[0,\infty)=\{x\in R:0\le x<\infty\}$, (b) $S=(0,\infty)$

2.7 (a) $AB-C$, (b) $A'\cap(B\cup C)$, (c) $(A'\cup B)-A'B$, (d) $A\cap[(B\cup C)-BC]$

2.10 $e^{-1};(2\cdot5)e^{-1}$ 2.12 $\frac{4}{5};\frac{1}{10}$ 2.14 (a) $\frac{1}{10}$, (b) $\frac{3}{5}$

2.16 $\dfrac{e^{\frac{1}{4}}-e^{-\frac{1}{4}}}{e-1}$

2.20 1

2.24 (a) $1-\dfrac{\dbinom{90}{10}}{\dbinom{100}{10}}$, (b) $1-\dfrac{\dbinom{90}{10}+\dbinom{10}{1}\dbinom{90}{9}}{\dbinom{100}{10}}$

2.26 $(39)_2(52)_2{}^{-1}-(39)_8(52)_8{}^{-1}$

2.28 $\dfrac{2j-1}{36}$

2.30 $1-\dbinom{13}{5}4^5\dbinom{52}{5}^{-1}$

2.32 54%

2.34 $\dfrac{4\left[\displaystyle\sum_{k=7}^{13}\dbinom{13}{k}\dbinom{39}{13-k}\right]}{\dbinom{52}{13}}$

2.40 $\dfrac{4\dbinom{39}{13}}{\dbinom{52}{13}}$

2.42 $\dfrac{4\dbinom{13}{6}\dbinom{39}{7}-\dbinom{4}{2}\dbinom{13}{6}^2\dbinom{26}{1}}{\dbinom{52}{13}}$

CHAPTER 3

3.2 (a) $\dfrac{3^2(5)_2}{(8)_3}$, (b) $\dfrac{3^2(5)_2}{(8)_3}$

3.4 (a) $\dfrac{\binom{4}{2}4^2}{5^4}$, (b) $\dfrac{\sum_{k=2}^{4}\binom{6}{k,\,2,\,4-k}4^{4-k}}{\sum_{k=2}^{6}\binom{6}{k}5^{6-k}}$

3.6 $\dfrac{\binom{13}{5}}{\binom{26}{5}}$

3.8 $\dfrac{\binom{10}{2}}{\binom{47}{2}}$

3.12 $\frac{20}{45}$

3.16 $\dfrac{\binom{4}{1}}{\sum_{k=1}^{4}\binom{4}{k}}$

3.18 (a) 0.75, (b) 0.25

3.20 $\dfrac{4\binom{13}{5}+4\binom{13}{4}\binom{39}{1}\frac{9}{47}}{\binom{52}{5}}$

3.24 (a) $\dfrac{4}{3^{k+1}}$, (b) $\dfrac{\binom{n}{k}3^{k+1}}{4^{n+1}}$

for $k \geq 1$ and $n \geq 1$

3.36 (a) 0.38, (b) 0.38

3.38 $\dfrac{\sum_{k=1}^{6}k^2}{441}$

3.40 $\frac{2}{7}$

3.42 $P(A) = 0$ or 1

3.44 A and B are independent in (a)

3.48 $\frac{5}{9}$

CHAPTER 4

4.2 $\sum_{k=2}^{4}\binom{4}{2}\left(\dfrac{5}{18}\right)^k\left(\dfrac{13}{18}\right)^{4-k}$

4.6 $\sum_{k=6}^{8}\binom{8}{k}\left(\dfrac{3}{4}\right)^k\left(\dfrac{1}{4}\right)^{8-k} = 0.6785$

4.8 (a) $\binom{10}{8}(0.85)^8(0.15)^2$, (b) $\sum_{k=7}^{10}\binom{10}{k}(0.85)^k(0.15)^{10-k} = 0.95$

4.10 $\dfrac{4!}{4^4}$

4.12 $\binom{n+k}{n}2^{-n-k+1}$ for $k < n$

4.13 $\frac{1}{4};\frac{1}{2};\frac{1}{4}$

4.14 $\binom{6}{3}2^{-6}$

4.15 $\binom{6}{2,2,2}\left(\dfrac{1}{4}\right)^4\left(\dfrac{1}{2}\right)^2$

4.22 $(r-1)/(k-1)$

4.24 $1 - \sum_{k=0}^{10}p(k;\,2\cdot5)$

4.30 $n \geq 96$

4.34 Yes; if the die were balanced, the probability of obtaining at least 2500 aces would be approximately $1 - \Phi(12\cdot25) < 0.001$.

CHAPTER 5

5.4 $\dfrac{1}{2} + \dfrac{1}{\pi} \arcsin 0.5$

5.5   (a) $X(s_1,s_2) = \sqrt{s_1^2 + s_2^2}$, (b) $\Pr(X \le r) = r^2$

5.10 $f(x) = \begin{cases} \dfrac{13(39)_{x-1}}{(52)_x} & x = 1, 2, \ldots, 40 \\ 0 & \text{otherwise} \end{cases}$

5.14 $f(x) = \begin{cases} \dfrac{1}{x(x+1)} & x = 1, 2, \ldots \\ 0 & \text{otherwise} \end{cases}$

5.16 0.328; at least $\dfrac{\log 100}{-\log 0.8}$ 5.18 $2.5e^{-1}$ 5.24 (a) $c = \frac{3}{8}$, (b) $c = \frac{1}{2}$

5.30 $\dfrac{1}{\pi}(\arctan 3 - \arctan 1)$ 5.32 (a) 0.3, (b) 0.3

5.40 $F(x) = \begin{cases} 0 & x < 0 \\ x^\alpha & 0 \le x < 1 \\ 1 & x \ge 1 \end{cases}$ 5.42 $f(x) = \begin{cases} \dfrac{1}{2\sqrt{x}} & 0 < x < 1 \\ 0 & \text{otherwise} \end{cases}$

5.44 $P(X \le 1) = 0.08$; $P(X \le 2) = 0.323$

5.48 $a = 480.4$; $b = 519.6$ 5.54 $m = \dfrac{\alpha - 1}{\beta + 1}$

CHAPTER 6

6.6   (b) $g(x) = xe^{-x}$ for $x > 0$ and $g(x) = 0$ for $x \le 0$, $h(y) = e^{-|y|}$, $-\infty < y < \infty$, (c) no, X and Y are not independent.

6.9   No   6.14 $\Pr(X = a, Y = b)$

6.18 X has the univariate Cauchy distribution, and (X,Y) has the bivariate Cauchy distribution.

6.22 (a) $g(w,x) = \begin{cases} 12(1-x)^2 & 0 < w < x < 1 \\ 0 & \text{otherwise} \end{cases}$

$h(y,z) = \begin{cases} 12y^2 & 0 < y < z < 1 \\ 0 & \text{otherwise} \end{cases}$

(b) no

6.24 (a) Multinomial with parameters n and $p_1, \ldots, p_6$, (b) binomial with parameters n and $p_1 + p_2 + p_3$, (c) multinomial with parameters n and $p_1 + p_2 + p_3, p_4, p_5, p_6$.

6.32 $\frac{1}{2}$

CHAPTER 7

7.4 Both Y and Z have density $f(x) = \begin{cases} \dfrac{1}{\pi\sqrt{1 - x^2}} & -1 < x < 1 \\ 0 & \text{otherwise} \end{cases}$

7.6 $f(x) = \begin{cases} \dfrac{1}{\alpha} x^{1/\alpha - 1} & 0 < x < 1 \\ 0 & \text{otherwise} \end{cases}$

7.8 $\Phi(X)$ has the uniform distribution on $(0,1)$

7.10 $g(y) = \dfrac{1}{\sigma} \sqrt{\dfrac{1}{2\pi y}} \exp\left(-\dfrac{1}{2}\dfrac{y + \mu^2}{\sigma^2}\right) \cosh \dfrac{\mu\sqrt{y}}{\sigma^2} \quad y > 0$

7.14 $g(y) = (e - 1)e^{-y}, y = 0, 1, 2, \ldots$

7.20 $f_1(y) = ne^{-ny}, y > 0; f_2(y) = ne^{-y}(1 - e^{-y})^{n-1}, y > 0;$
$f(y_1, y_2) = n(n - 1)(e^{-y_1} - e^{-y_2})^{n-2} e^{-y_1 - y_2}, 0 < y_1 < y_2 < \infty$

7.24 $g(y) = 1 - |y|, -1 < y < 1$

CHAPTER 8

8.2 0 8.4 $\frac{161}{36}$ 8.6 1

8.8 $k/(n + 1)$ 8.12 $0; \frac{1}{2}$

8.16 (a) k/n, (b) $n(1 - 1/n)^k$ 8.20 $\frac{1}{4}$

8.22 $\alpha^{-1} \sum_{j=1}^{k} (N - j + 1)^{-1}$ 8.24 $\mu = 0; \sigma^2 = 2\beta^{-2}$

8.26 (a) $\mu = \begin{cases} \dfrac{\alpha}{\beta - 1} & \beta > 1 \\ \infty & 0 < \beta \le 1 \end{cases}$ (b) $\sigma^2 = \dfrac{\alpha(\alpha + \beta - 1)}{(\beta - 1)^2(\beta - 2)}$ for $\beta > 2$

8.28 $\mu = \dfrac{n - 1}{n + 1}; \sigma^2 = \dfrac{2(n - 1)}{(n + 2)(n + 1)^2}$ 8.30 σ^2

8.32 $\alpha_i = \dfrac{\sigma_i^{-2}}{\sum_{j=1}^{n} \sigma_j^{-2}}$ 8.34 $1 - e^{-1}$ 8.36 Brand A

8.42 (a) $s = \dfrac{q - p}{\sqrt{nqp}}; k = \dfrac{1 - 6pq}{npq}$, (b) $s = \dfrac{1}{\sqrt{\beta}}; k = \dfrac{1}{\beta}$, (c) $s = 0; k = -1.2$,

(e) $s = 2; k = 6$

8.44 (a) $M(t) = \dfrac{\beta^2}{\beta^2 - t^2}, -\beta < t < \beta$ 8.46 $\Pr(X = 1) = \frac{1}{2} = \Pr(X = -1)$

8.50 $f(x) = \dfrac{\beta_1\beta_2}{\beta_1 - \beta_2}(e^{-\beta_2 x} - e^{-\beta_1 x})$ for $x > 0$

8.54   (a) $f_k = (n)_k p^k$, (b) $f_k = k!\, q^{k-1} p^{-k}$, (c) $f_k = \beta^k$

8.56 $\dfrac{(n + 1)_{j+1}}{j + 1}$ 8.58 $-\frac{1}{2}$ 8.62 $\dfrac{\sigma^2 - \tau^2}{\sigma^2 + \tau^2}$

CHAPTER 9

9.2   (a) $\Pr\left(|\bar{X} - p| \ge 0.1\right) \le \dfrac{100pq}{n}$, (b) $n \ge 500$

9.16   0.996

9.18   (a) $\Pr\left(|\bar{X} - p| \ge 0.1\right) \approx 2\left[1 - \Phi\left(0.1\sqrt{\dfrac{n}{pq}}\right)\right]$, (b) $n \ge 96$

9.20   0.774   9.22 $\lim\limits_{n \to \infty} \Pr\left(Z_n \le z\right) = 1 - e^{-z}$ for $z > 0$

CHAPTER 10

10.4 $\Pr\left(X = x \mid Y = y\right) = \dfrac{1}{y}$, $x = y + 1, \ldots, 2y$

10.6 $g(x \mid y) = \dfrac{1}{2}\dfrac{1 + y^2}{(\sqrt{1 + x^2 + y^2})^3}$, $-\infty < x < \infty$

10.8 $g(x \mid y) = \dfrac{\Gamma(\alpha + \gamma)}{\Gamma(\alpha)\Gamma(\gamma)}\dfrac{x^{\alpha-1}(1 - x - y)^{\gamma-1}}{(1 - y)^{\alpha+\gamma-1}}$ for $0 < x < 1 - y < 1$

10.10 (a) normal with mean $\frac{1}{2}z$ and variance $\frac{1}{2}$, (b) normal with mean x and variance 1.

10.12 $h(y) = \dfrac{\alpha\Gamma(\alpha + \beta)\Gamma(\beta + y - 1)}{\Gamma(\beta)\Gamma(\alpha + \beta + y)}$ for $y = 1, 2, \ldots$

10.14 $h(y) = \dfrac{\beta^\alpha\Gamma(\alpha + y)}{y!\,\Gamma(\alpha)(1 + \beta)^{\alpha+y}}$ for $y = 0, 1, 2, \ldots$

10.16 $h(y) = \dfrac{1}{2 \max(1,y)^2}$ for $0 < y < \infty$

10.18 $g(x \mid y) = 2 \max(1,y)^2/x^3$ for $x > \max(1,y)$

10.20 Let $\alpha = \dfrac{\arcsin 0.5}{4\pi}$; the conditional distribution of X is uniform on the finite set

$\{\alpha, \frac{1}{4} - \alpha, \frac{1}{2} + \alpha, \frac{3}{4} - \alpha\}$.

10.22 $\sqrt{1+x^2}$

10.26 $(y+1)2^{-y} - y2^{-y-1}$

10.28 $\dfrac{\beta(1-x)}{\beta+\gamma}$; $\dfrac{\beta\gamma(1-x)^2}{(\beta+\gamma)^2(\beta+\gamma+1)}$

10.30 $\dfrac{mz}{m+n}$

10.32 $\dfrac{\beta}{\alpha-1}$; $\dfrac{\alpha\beta^2}{(\alpha-1)^2\,(\alpha-2)}$

10.36 $a = r\sqrt{\dfrac{D(Y)}{D(X)}}$; $b = E(Y) - aE(X)$

10.38 $(N - x_1 - \cdots - x_m)s$ and $(N - x_1 - \cdots - x_m)s(1-s)$,

where $s = \dfrac{q_1}{q_1 + \cdots + q_n}$

10.40 $k'p$ and $k'pq\,\dfrac{r-k'}{r-1}$, where $r = r_1 + \cdots + r_n$, $p = \dfrac{r_1+r_2}{r}$ and $q = 1-p$

10.42 $h(y_1, y_3 \mid y_2) = \dfrac{1}{y_2(1-y_2)}$ for $0 < y_1 < y_2 < y_3 < 1$

10.46 $\delta(y) = \dfrac{\alpha+y}{\beta+t}$; $\dfrac{\alpha}{\beta(\beta+t)}$

10.48 $\delta(y) = \dfrac{2}{y+2}$

10.50 $\delta(y) = \dfrac{\mu\sigma^{-2} + n\bar{y}}{\sigma^{-2} + n}$

INDEX